Construction Methods and Planning

Second edition

J.R. Illingworth
Consultant in Construction Methods and Technology

E & FN SPON
ALERE FLAMMAM
Taylor & Francis Group

London and New York

First published 1993 by E & FN Spon, an imprint of Chapman & Hall
Reprinted 1994, 1996

Reprinted 1998 by E & FN Spon, an imprint of Routledge

Second edition published 2000 by E & FN Spon
11 New Fetter Lane, London EC4P 4EE

Simultaneously published in the USA and Canada
by E & FN Spon
29 West 35th Street, New York, NY 10001

E & FN Spon is an imprint of the Taylor & Francis Group

© 2000 J.R. Illingworth

Typeset in 10/12½ pt Sabon by J&L Composition Ltd, Filey, North Yorkshire
Printed and bound in Great Britain by St Edmundsbury Press, Bury St Edmunds, Suffolk

British Library Cataloguing in Publication Data
A catalogue record for this book is available from the British Library

Library of Congress Cataloging in Publication Data
Illingworth, J. R. (John R.)
Construction methods and planning / J.R. Illingworth. – 2nd ed.
     p. cm.
     Includes bibliographical references and index.
ISBN 0–419–24980–X (pbk.: alk. paper)
1. Building – Planning. 2. Construction industry – Management.
I. Title.
TH153.I43 2000
690′.028–dc21

99–42019
CIP

ISBN 0–419–24980–X

Contents

Contents

Contents

Contents

Preface to second edition

The success of the first edition – reprinted three times – and the fact that time has elapsed since it was published in 1993 have made a second edition necessary. Legislation relating to construction has changed considerably, while the references to each chapter have needed updating. At the same time, it has become necessary to update parts of the text or to introduce new case studies where they enhance the overall picture of the section in question. In considering the extent of such alterations, note has been taken of comments made by reviewers at the time of the publication of the first edition.

A key object in this new edition has been to deal with the considerable changes in legislation since the book was first published. Many regulations of that time have now been revoked and replaced. The key new elements of new legislation are as follows: The Management of Health and Safety at Work Regulations, 1992, which became operative from 1 January 1993; the Construction (Design and Management) Regulations 1994, effective from 31 March 1995; the Construction (Health, Safety and Welfare) Regulations 1996. More recently, the Lifting Operations and Lifting Equipment Regulations 1998 became operative from 5 December 1998.

There has also been a new Act of Parliament, The Housing Grants, Construction and Regeneration Act, which came into force in 1996. This, however, is mainly related to financial matters which involve the Quantity Surveyor and Site Management as distinct from planning.

All the above legislation is described in more detail in the revised Appendix A. Mention must be made, at this point, of the importance of the Construction (Design and Management) Regulations. For the first time, health, safety and welfare for a contract becomes the responsibility of all parties involved from the Client down to the smallest subcontractor. Emphasis is placed on communication and co-ordination. This is a long overdue piece of legislation. Appendix A explains the impact in detail.

Several chapters have considerable modification in the light of new developments. All references and text have been updated to cover the new items of legislation. At the same time two new chapters have been added. One deals with the refurbishment of domestic accommodation. The second deals with facade retention, much in the news these days. Both chapters are based on the author's experience in these fields.

A final matter concerns the inclusion or not of the use of computerization in a book of this nature. After careful consideration, it has been decided only to make passing reference to it where appropriate, but not to go into any detail. As was made clear in the first edition, the purpose of this work is to teach students and young engineers and builders the basis of knowledge in relation to the practice of construction, together with an early appreciation of the importance of safety in all aspects of the construction process. Fundamental principles are the building blocks of getting the method approach right in terms of time, cost and quality before applying the computer to management control.

J.R. Illingworth
London 1999

Preface to the first edition

In the construction industry, it has long been a common belief that the planning of construction methods is a subject that cannot be taught – only learned by experience. Such a situation was undoubtably true when the importance of detailed planning of construction methods was recognized in the 1950s. At that time I was entrusted with the setting up of a construction planning organization for an international contractor. No guidance as to how one set about such things was in existence and one had to start from scratch. Knowledge came from bitter experience.

While there is no doubt that knowledge gained from experience is an essential part of continuous professional development, and construction method planning is no exception, I have felt for many years that there ought to be available an authoritative introduction to the planning process and the factors that affect it. In other words, something that set the scene and provided the basis on which future experience could be built. Ideally, such an introduction would need to be readily understood at undergraduate and postgraduate levels or for entrants to professional qualifications. For construction as a whole to grow in efficiency, the need for careful planning of the construction process needs early introduction into the student's mind, no matter what discipline is being followed. This need is especially true if he/she aspires to a managerial role as the ultimate objective. At the same time, construction planning needs to be recognized as a worthwhile career in its own right.

This book has been designed to fulfil the needs stated above, and to provide the starting-off point for those beginning a career in the construction field. It has been based on lectures given over some years as part of the first degree course in building construction management offered by the University of Reading. As experience has shown that these lectures have been well received by the students, it was decided to produce this book for the benefit of a wider audience and take the opportunity to extend the overall scope.

While the level is pitched at students for first degrees in building construction management, the subjects covered will also provide a practical introduction to construction for those beginning careers in structural design, architecture, quantity surveying, building surveying and building services. In today's situation of depression in the construction industry, this book should also be of value to the many small firms specializing in specific trade skills. Many such firms lack a background appreciation of how their activities could be made more profitable by improved pre-planning of their work.

The more experienced reader will detect that some of the illustrations used look rather elderly. The reason for their use is quite straightforward. Having been in the practical side of construction for more than forty years, many as the manager of a construction method planning organization, I have continually collected photos and performance data which illustrated especially well particular aspects of the planning process. If, in a particular situation, an older photo illustrates the point more clearly than a more modern one, I have not hesitated to use it.

A similar situation arises with case study examples involving cost figures. However up-to-date costs are when a book is written, it is a safe bet that they will be out of date by the time the book is published. In consequence, the figures used are as they were at the time of recording. If attempts are made to update such figures, it is sometimes the case that the results of a comparison would would work out differently between the then and now situations. It is the principle of the method that matters, not the figures used. Equally, invented case studies are never satisfactory.

Where the studies used are of some age, there is a benefit in the sense that the student can follow the method of calculation and the components necessary for an accurate comparison, yet follow up with updating all items to appreciate the influence of inflation, changing relationships between plant and labour, together with the impact of changing method developments.

Many organizations have willingly contributed material for this book and their contributions are acknowledged elsewhere. My thanks to them all.

Finally, my thanks are due to the academic staff of the Department of Construction Management and Engineering at the University of Reading, who convinced me that writing this book would provide a much needed addition to available teaching material. I hope that this proves to be true, as I have always enjoyed giving something back to an industry that gave me such a worthwhile career.

J.R. Illingworth
London 1992

Acknowledgements

The Author wishes to express his gratitude to the following organizations and individuals for their help in providing illustrations and permission to use material from their literature and publications: British Cement Association; British Concrete Council; British Pre-cast Concrete Federation; British Steel (Sections, Plates and Commercial Steels); Chartered Institute of Building: Construction News; Cordek Ltd; Jim Elliot & Co; Lilley Construction Ltd; National Access and Scaffolding Confederation; D. Neal; Precision Metal Forming Ltd; Stent Foundations Ltd; Stuarts Industrial Flooring Ltd; Tarmac Construction Ltd; Westpile Ltd.

The Author also wishes to express his thanks to the following organizations and individuals for permission to publish photographs, diagrams and tables as individually acknowledged in the text: ACE Machinery Ltd; Concrete Society; Cordek Ltd; Construction Confederation; Davis Landon & Everest; A.J. Goldsmith; ICI Agricultural Division; Donald I. Innes; Kvaerner Cementation Foundations; Kwikform (UK) Ltd (with special thanks for the Millennium Dome information); Lilley Construction Ltd; Mabey Ltd; Rapid Metal Developments Ltd; J. Rusling Ltd; Scaffolding (Great Britain) Ltd; Sanderson (Fork Lifts) Ltd; Tarmac Topmix Ltd; Trent Structural Concrete Ltd.

PART ONE

Factors Affecting Construction Method Assessment

Introduction

Before commencing any study of the principles of the planning of construction methods, it is essential that the meaning of a number of words and phrases are defined and understood as they will be used in this book. In the so-called world of information technology, it is remarkable that specific words seem to be capable of endless meanings, and as a result can readily be misinterpreted by different people. It is hoped that the definitions and meanings given here will make some contribution to stabilizing the meanings of words and phrases associated with the planning of construction activities.

Probably the most abused word today is 'Management'. It is all things to all men. In the construction industry, the pundits stress the need for better management. Everything involves management – indeed, everyone is a manager or potential manager, or should aspire to be one. The impression is given that no other discipline is needed any more for construction to be efficient.

To manage, management

The Chambers Dictionary defines the verb 'to manage' as: to conduct; to control; to administer; to be at the head of. It follows that any manager is at the head of a group of individuals who perform a function. Thus a site manager, for example, conducts, controls, administers and is at the head of a construction team on site. It is obvious that if he is to do his job effectively and efficiently, he will require a suitably qualified team to provide back-up services in both the administrative and technical fields. On the technical side, such services conveniently divide into two distinct parts. These may be termed Aids to management and Adjuncts to management.

Aids to management

While not necessarily a complete list, the following matters are the main components of Aids to management: Method Statements on which the job was priced; cost breakdown of the Tender sum; specialist staff for the supervision of subcontractors; the availability of work study facilities to evaluate performance and provide recommendations for improvements in efficiency; and the use of programme systems appropriate to the contract in question. Such systems will need to allow rapid assimilation of progress situations and provide weekly or short term methods of giving easily understood communication to first line supervision. In addition, visual cost control graphics for all subcontractors are vital where the majority of work and trade elements is sub-let. (See Chapter 18 for operational details.)

These components cannot materialize unless specialized staff with the necessary experience are available.

Adjuncts to management

All the management in the world has no significance if the methods of construction adopted are incorrect for the situation in question and do not take account of all the factors arising on a particular site. It is in this context that Adjuncts to management fall.

In this field we have pre-qualification submissions and pre-tender planning, encompassing plant and method selection, labour requirements and an assessment of the type and scale of temporary works. The temporary works assessment would include the requirements for access, storage, temporary offices and the provision of welfare facilities, to name some. In overseas locations, one may need to assess the need for a construction camp and all the costs attached to it.

Unless the above matters have been adequately priced at the tender stage, no level of management can assess whether the contract is capable of making a profit at the end of the day.

In order to provide adequately for effective Aids and Adjuncts to management, a number of specialized staff have to become involved. Indeed, as construction becomes more specialized, the demand for specialist staff becomes more critical. This view has even more force when one considers the extent of sub-letting of work today, both in relation to trade activity and specialist services. Yet again, the move towards Quality Assurance clauses in contract documents requires the main contractor to have adequately qualified technical staff to see that sub-contractors perform to the standard required.

In conclusion, then, there are other worthwhile occupations apart from the managing role, which have given many people worthwhile and rewarding careers.

To plan, planning

Planning is probably one of the most overworked words in the English language. Even when the word is related solely to the construction industry, difficulties continue to arise. It is often taken to mean the preparation of designs and other information for a project. It can be taken to mean the approval of a project by the local authority. Even if planning can be restricted to activities related to the actual construction, differences of interpretation still occur.

In a recent article in the *Chartered Builder*[1] the planner is seen as the individual who prepares the Programme, sells it to the site manager and ensures that it will act as an early warning system to trends that may not be desirable. In an article in *Building Technology and Management* in 1987[2] entitled 'Planning techniques for construction projects' it rapidly becomes clear that what is being discussed are methods of programming construction works, both by manual and computerized approaches. There is no doubt that this view of the meaning of the word 'planning' is a commonly accepted one in certain parts of the construction industry.

The over-emphasis on the programme is, in this author's view, the wrong way to look at things if the greatest efficiency in contract organization and control is to be achieved. All the programme and management techniques in the world are of little value unless the construction methods to be used are the most cost effective for the situation and safely carried out. Once the methods of construction are settled, the programme can be prepared showing the results of planning the construction method for the benefit of site management.

It is pleasing to see, therefore, that a new publication from the Chartered Institute of Building[3] has gone some way to helping to rectify this situation. Superseding a previous publication, *Programmes in construction – a guide to good practice*, the title is now *Planing and programming in construction – a guide to good practice*. In the first chapter, 'Terminology', recognition is at last given to the existence of a person variously known as construction planner or planning engineer or construction programmer. It is also stated that the first name is the preferred term. It is just rather a pity that the last name was not deleted.

In this same document, 'planning' is defined as 'the process of analysing, formulating and organizing the intended actions for the carrying out of a project.' Even better, in this author's view, is the definition of 'programming': 'the process of producing a time related schedule of the planning decisions'. Altogether this is a much more realistic appreciation of how the construction industry, as far as contractors are concerned, see the terminology. The following terms and methods for planning reflect this author's views on why construction method planning needs persons of skill and experience if the first step in construction efficiency is to be properly achieved.

The construction process

In any form of construction, there are only two fundamental activities: (a) the handling of materials and equipment and (b) by the skill of the work force in the positioning of the materials and equipment (assembly) to produce the desired completed whole. It follows that the overwhelming majority of plant on site will be concerned with the handling of materials and equipment (Figure 1). When performance is related to plant and labour activities, in specified areas, a time scale for these elements emerges.

While non-handling plant comes into the picture – compressors, welding equipment, power operated hand tools, and such things as lighting equipment, to name a few – such items have to be seen as aids to the handling and assembly process.

Efficient construction is much dependent on safe places of work and the safe use of construction plant. Indeed, such dependence is required by law. The main requirements, in the legal sense, are given in Appendix A. To achieve such safe working practices, temporary works are a major requirement. While not all temporary works relate to site safety, taken as a whole they

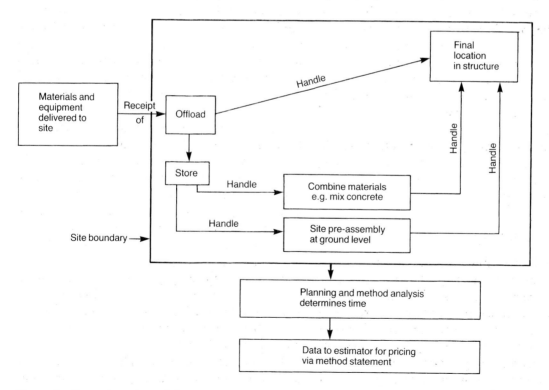

Figure 1   Representation of the construction process.

represent a significant part of the contract cost. This aspect is dealt with in more detail in Chapters 3 and 4.

What can be said at this point is that:

$$\text{Plant and labour} + \text{Production performance (Handling method)} = \text{Element time}$$

Construction planning

The planning of construction methods may properly be defined as 'understanding what has to be built, then establishing the right method, the right plant and the right labour force to carry out the works, safely and to the quality required, in the most economical way to meet the clients requirements'. Within this definition, it should be noticed that no reference is made to programming or programming methods. Within the meaning of planning as set out in this book programmes are not part of the planning **process** – they represent the results of planning, in visual form, for the benefit of site management.

Understanding what has to be built

Given the definition above, it is now desirable to examine the components of it in more detail.

Drawings

At whatever stage planning takes place: pre-qualification, tender, pre-contract or contract, the first essential requirement is to have a thorough understanding of what has to be built. This may seem obvious, but it is surprising how many people tend to skim through drawings and believe they know what is required when, in fact, the fundamental issues have been missed. Examples are: missing the fact that certain areas of work may have a critical influence on the planning as a whole, or details shown which may have an effect out of all proportion to their individual value.

The planning of the construction methods to be used must start with a detailed assessment of the work to be carried out. In civil engineering contracts, it is usual for a complete set of working drawings to be available at the tender stage. Such information allows a detailed consideration of the work to be done together with the temporary works needed. Building contracts, by comparison, rarely provide at the tender stage much more than the outline drawings required for final planning approval. To become better informed, the tenderer is required to visit the architect's office to see what other drawings are available. Copies are rarely allowed to be taken, even if an offer to pay is made. The reason, of course, is that if such drawings were altered later – after the tender was submitted – the contractor would have ready made evidence for a claim. Full assessment of what has to be built is much more difficult and more reliance has to be placed on interpreting the information contained in the Specification and the Bills of Quantities. If these happen to be in outline form, at the tender stage, the experience of the planning engineer becomes crucial, if as accurate a result as possible is to be achieved.

Specification and Bills of Quantities

Assuming that the Bills are reasonably complete, they will provide a good idea of the physical scale of the works. Item descriptions will clarify the form and type of construction. By contrast, Specifications need careful study from a number of points of view.

1. If particular items are excessively long they usually mean a fussy architect or engineer.
2. Many architects and engineers use 'standard' specifications from job to job. As a result, problems tend to arise in relation to their interpretation properly fitting the contract in question. Experience with a particular architect or engineer will give a lead on the degree to which clauses will be enforced or not. With others, it is essential that the right questions are asked at the tender stage, in order to clarify the degree to which the specification holds good.

3. The implications of individual clauses are important. For example, the striking times of formwork. They may drastically affect the quantities required and hence the cost of the formwork.
4. Descriptions in bills of quantities can be misleading and require checking against drawings – if available.
5. Specifications may have a major effect on the way in which the work is carried out. For example, factory floors required to be poured 'checker board' fashion instead of continuous bay or large area methods which are more economical to the client. See Part Two, Chapter 8.

Particular specification and special conditions

In addition to what might loosely be called the standard specification, a particular contract may also have a particular specification or special conditions related to the specific contract. Where these apply, they need careful study as they will have to be followed.

Examples of special conditions are:

1. Limitations on access if the works are within an existing installation: or if the surrounding streets are particularly busy at rush hours.
2. Specified sequence in which the client wants handover. (This may not be that giving the most efficient construction sequence).
3. Limitations may be imposed on the plant that can be used.

Depending on their nature, so will the planning of the construction method approach be affected.

The remaining items in the definition of construction planning and how to evaluate them form the main body of this book and are dealt with in the following chapters. Before proceeding, however, there are a number of matters that need attention as they are common to all aspects of planning construction methods. They encompass the requirements of the law, codes and standards and the prime need to achieve safe working practices.

Influence of the law, codes and standards

Legislation

Construction operations are affected by a great deal of legislation. While much of it relates to periodic inspections and testing of plant and equipment, those involved in any way in the planning of construction methods must be aware of the implications of any legislation which is likely to affect decision making. Appendix A lists the main areas of legislation likely to affect construction method selection and planning, together with summaries of their contents.

Codes and standards

Appendix B contains a list of Codes and Standards having direct relevance to construction method assessment.

Construction safety

Much of the legislation listed in Appendix A is related to ensuring the safety of those working on construction sites and the general public who may be at risk from such activities.

Safety legislation has, or should have, a significant influence on the method planning of construction work. If ignored, those who do so may well find substantial fines coming to the company, and, in certain cases, the individual in charge. In addition, very large sums of money may be awarded against the contractor by the Civil Courts in the event of serious accidents or fatalities.

As far as the construction method planner is concerned, a particular philosophy needs to be followed. This can be stated as 'Safety requirements, both in the legal and moral sense, must be an integral part of the method planning and execution of the construction work'. It is quite inadequate to see safety as a secondary matter superimposed after the planning and methods have been decided.

Safety costs money, in most cases, and for its

adequate implementation the planner and the estimator need to ensure that adequate monies have been included in the tender sum. If this is not done, site managers may well be tempted to take short cuts to avoid extra expense.

For example, when working over water it may be considered adequate to provide barriers and insist on all operatives wearing life jackets. The Construction (Health, Safety and Welfare) Regulations 1996 [4], however, require that, in addition, rescue facilities must be available. (Basically a boat with competent boatman ready at all times that work is going on above the water). The point, of course, is that anyone falling could be rendered unconscious by striking their head while falling and, on hitting the water, be swept away down a river or out to sea.

Safe working is, to a very high degree, a matter of human attitudes. In choosing plant and methods, the planner needs to be very aware of the vagaries of human nature and use methods which can be made as safe as possible.

Health and Safety at Work etc. Act 1974[5]

This is the head legislation in relation to health and safety at the workplace. Appendix A sets out the contents of this Act in more detail, but certain sections have particular relevance in relation to construction method planning. Section 2 covers the general duties of employers to employees. In the direct sense of the planning and execution of construction, section 2(2)C is of particular importance:

> the provision of such information, instruction, training and supervision as is necessary to ensure, so far as is reasonably practicable, the health and safety at work of his employees.

In the field of construction method assessment, particular attention needs to be given to 'provision of such information, instruction ...' This phrase affects all levels and grades of operatives and adds up, really to one word – communication!

Adequate information given to individuals about what they are required to do is an essential component of safety – especially in circumstances where an erection procedure or the carrying out of a particular method may contain hazards which are not readily apparent to the work force. In such circumstances, pictorial presentation of the safe sequence of events – with the minimum of words – issued to all involved is desirable if compliance with section 2(2)C of the Act is to be achieved. As a spin off, past experience of such action shows that other benefits arise. In giving prior information in association with a verbal briefing, not only can an accident free result be achieved, but the learning time for the operation is reduced and productivity increased.

Section 3 of the Act deals with the contractor's liability of care to the public, both on and adjacent to the site. Site managers and planning staff need to be fully conversant with the requirements of this section. Construction methods adopted which may impinge on the public have to be organized in such a way as to give adequate protection to the public. In so doing special temporary works needs will usually arise. For example, where tall structures are involved, the planning must adequately consider, and allow for, effective measures to stop objects falling from a height and landing onto public areas outside the site boundary.

Another example relates to the unloading of materials from vehicles in the street. If the public can pass under the loads, measures need to be taken to prevent public access while the loads pass over the pavement. Failure in such situations could lead to a visiting Inspector putting a prohibition notice on the site until the situation was rectified. This could effectively stop all work on site.

Less obviously, every care needs to be exercised to avoid accidents to children. They are apt to believe that a construction site is an adventure playground. Useful advice in this respect is given in Health and Safety Executive guidance note

GS 7 *Accidents to children on construction sites*[6].

Section 4 of the Act requires that the person having control of a construction site – e.g. the main contractor or managing contractor – shall see that the site is safe for others to work on when not in his (the main contractor's) employ. In other words, subcontractors must be provided with safe access and any plant or substances on the premises must be safe and without risk to health.

The effect of this section is also to put a duty of care on the main contractor not only to see that what he provides for his subcontractors is safe, but also that the subcontractors are themselves complying with section 2 of the Act.

Construction Regulations

Construction Regulations, made under the Factories Act 1961[7] are listed in detail in Appendix A. At this point it is important to recognize that all have an impact on the construction methods adopted – mainly in relation to safety, laying down specific provisions that have to be complied with, in respect of temporary works. As such they affect scaffolding specifications, scales of accommodation on site, requirements regarding lifting operations, safety of excavations and the safe use of dangerous substances on site. Those planning construction methods must be fully aware of all such Regulations and make adequate allowance in the planning for their implementation.

Safe operation of construction plant

Section 6 of the Health and Safety at Work etc. Act requires that any person who designs, manufactures, imports or supplies any article for use at work must ensure, as far as is reasonably practical, that the article is safe and without health risks when properly used. To ensure that this is so, adequate testing and inspection must take place. Additionally, the supplier must have available adequate information about the use for which it was designed and has been designed and tested. Such information is to include any conditions necessary to ensure that, when put to use, it will be safe and without risk to health.

These conditions will cover matters which the contractor has to provide if the plant or equipment is to be used safely. Examples are design and construction of concrete block foundations for tower cranes and adequate support to outriggers of mobile cranes. Both items are of course temporary works. In planning the construction method, such requirements must be designed by persons competent to do so and the installation and removal costs allowed for in the tender figure. This is dealt with in more detail in Chapter 3.

The construction planner

Effective construction planning demands that it is carried out by competent and experienced personnel. While the estimator must remain in charge of the pricing, most contracts, today, are too complex for the estimator to handle the technical content of pre-tender appraisal. It is here that the construction planner has a major function.

When the contract is won, the site planning is equally important. It is usually at this point that the site manager will have his first chance to examine the methods on which the job was priced. What is important to recognize, at this point, is that planning must be seen as an adjunct to management, as already stated. Thus the site manager must have the right to question the planning and methods on which the contract was priced. To do so, site managers need the technical ability to assess what has been put forward. In so doing, the manager concerned must, for his part, accept that if he wants to change the method from that priced in the tender, he must demonstrate that his alternative is demonstrably cheaper, or at least as cheap, as that priced in the tender submission.

When final agreement has been reached, the planning function can be established on site and develop the final method to be adopted. This, in turn, will lead to the contract master programme and such subsidiary programmes as may be necessary (Chapter 17).

Once the methods to be adopted have been agreed by site management, the construction planner continues to have an important role on site. First, in developing control and short term programmes for all organizations contributing to the contract under the main or managing contractor. Second, the construction planner's intimate knowledge of how the work was originally planned makes him uniquely placed to examine the potential consequences of alterations to what has to be built. In this area, he is a valuable source of information to the site quantity surveyor when assessing where possible contractual claims may arise.

Today, with the ever increasing use of trade subcontractors as well as specialist service con-

tractors, the experienced construction planner has the option of either seeing the role as satisfying in its own right, or as an essential ingredient of experience for becoming a successful site manager.

The role of planning in claims

Tendering for complex commercial and industrial work in competition, often with inadequate information at the tender stage, may and often does leave the contractor vulnerable to loss if proper site management control of variations and the like is not carried out.

Those responsible for contract planning have a particular role to play in this context. Initially, they need to become fully conversant with the extent of information available at the tender planning stage. With this knowledge, they will be in a position to suggest where claims are jus-

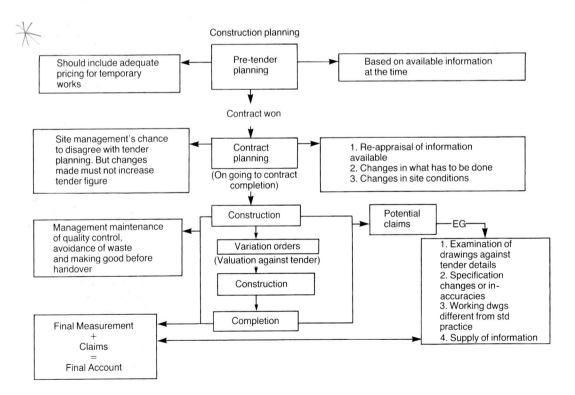

Figure 2   The impact of construction planning – from tender to contract completion.

tifiable, within the terms of the contract, because of variations between the tender information and the work as actually detailed and executed. The way in which the planning engineer can function in this area is set out in some detail in Chapter 19.

The construction method planning process

At this point it is possible to examine in more detail the way in which construction method planning needs to be carried out. To be successful, it has to be recognized that those who do the planning need to be well versed in construction technology, be knowledgeable in relation to plant availabilities and performance, the interrelation with labour and with a wide experience of factors which will affect the requirements and performance needed. This knowledge can be described as 'factors affecting construction method assessment'. This basic data and knowledge has then to be used in specific real life situations to determine methods, sequence of construction and time

scale for the benefit of the estimator at the tender stage and the site manager once the contract has been won. The application of the tools etc. of planning to actual situations can be entitled 'Establishing methods and their planning control' and accordingly, this book is divided into two parts. Finally, Figure 2 provides a useful summary of the impact of construction method planning from tender to contract completion.

References

1. Mace, D. (1990) Problems of programming in the building industry. *Chartered Builder*. March/April pp. 4–6.
2. Eshete, S. and Langford, D. (1987) Planning techniques for construction projects. *Building technology & Management* February/March
3. *Planning and Programming in Construction – a guide to good practice.* (1991) Chartered Institute of Building. Ascot.
4. *Construction (Health, Safety and Welfare) Regulations 1996.* HMSO, London.
5. *Health & Safety at Work etc Act 1974.* HMSO London.
6. Health & Safety Executive. (1977) *Guidance Note GS 7. Accidents to children on construction sites.* HMSO London.
7. *Factories Act 1961.* HMSO London.

Chapter one

Site inspections

Regardless of the information provided in the contract documents, bills of quantities, specifications and soil reports, a visit to the site for a detailed study is an essential preliminary to any planning or pricing action. The reality of a site is often quite different from the impression given by the drawings supplied. This can be particularly true in relation to scale and the three-dimensional nature of the site. Boundaries become more tangible in their effect on the siting and choice of plant, and a totally different concept of access may result.

Site inspection forms

To be effective, all site inspections need to follow a structured form to make sure that nothing important has been missed. To achieve this, a standard form which can be used for any situation is the best approach. A typical example is given in Figure 1.1. It should be noted, however, that this example has been specifically prepared for planning purposes. The information recorded is not necessarily complete as far as the estimator is concerned. If the estimator and the planner do the site inspection together, it will be necessary either for the estimator to prepare a list of additional items that he will need to investigate or, better still, have his own checklist format. Any inspection carried out should be augmented by site photography; carefully chosen to give as good an indication of the main site features as possible.

Completing the form

The following listings correspond to the numbering of each section of the form. They amplify the sort of answers one should be looking for when visiting the site.

1. Access (off site)
Width or weight restrictions, which may affect plant choice, on existing roads. (In extreme cases might a new length of road pay for itself in terms of site cost efficiency?)

Are any bridges nearby subject to height or load restrictions? If road construction is inadequate for the traffic envisaged, will the local authority allow use if the contractor pays for reinstatement at the end of the contract? Will additional access over private land be needed as well as that specified? If so, what is the reaction of the landowner to wayleave provision?

2. Access (on site)
In providing temporary roads or allowing for part use of permanent construction sub-bases on site for the siting and movement of plant and equipment, what obstruction exist?

(a) Overhead: Power lines, telephone cables, trees etc.
(b) Ground level: Surface level drainage pipes, on or near surface distribution mains in industrial plants and areas of natural features to be protected.

13

SITE INSPECTION FORM (QUESTIONNAIRE)

PRE TENDER PLANNING

REGION _____

SITE NAME _____

JOB CODE NO. _____

SITE LOCATION _____

NEAREST MAIN LINE STATION/AIRPORT
_____

INSPECTION BY _____

DATE _____

1. ACCESS (Off Site)

   (a) Restrictions on existing roads and bridges to site for heavy plant.

   (b) Restrictions on use of local roads (Rein-statement costs for example if damaged).

   (c) Are Wayleaves required.

2. ACCESS (On Site)

   (a) Overhead obstructions.

   (b) Ground level obstructions.

   (c) Underground obstructions.

   (d) Buildings and trees.

   (e) Watercourses.

   (f) General.

3. BOUNDARY CONDITIONS

   (a) Adjacent buildings and trees (Heights and obstruction value).

   (b) Proximity of Railway lines and public roads (Effect on Tower Cranes).

   (c) Safety of the public (Need for covered ways, etc. Cranes over-swing with loads).

(d) School playgrounds, Parks etc.

(e) Adjacent watercourses.

(f) Adjacent or nearby Airfields.

(g) Rights of Access for others and ditto maintenance of services.

(h) Adjacent Mains and Sewers.

4. NOISE
   (a) Nearby Hospitals, Schools, Offices, etc.

   (b) Restrictions on nightwork or weekends.

5. PUBLIC UTILITY SUPPLIES FOR CONSTRUCTION
   (a) Water – Availability and Location.

   (b) Electricity (Lt. and Ht.) – Availability and location (Check if cable costs involved).

   (c) Telephone – Availability and Location.

   (d) Foul Sewers – Availability and Location.

6. VISUAL SURFACE GROUND CONDITIONS AND LOCAL GEOLOGICAL FEATURES

7. LOCAL WORKING WEEK
   (Degree of operation of 5 day week).

8. LOCAL HOLIDAYS

9. LOCAL WEATHER CONDITIONS

10. REMARKS

Figure 1.1 Site Inspection Form (Questionnaire).

(c) Underground: Drainage, sewers, cables and water and gas mains. In addition, old basements and foundations.

(d) Buildings and trees: Do any buildings and trees have to remain during the contract? Is access to them to be maintained? How do tall trees on site affect the possible choice of plant and its siting? Estimates of their heights will be needed.

(e) Water courses: Extent and impact on new works. Is bridging needed? Are they liable to sudden changes in level during the year? Is diversion possible if the method planning would be improved by so doing?

(f) Generally: What may be the extent of temporary works needed?

3. Boundary conditions

(a) What adjacent buildings and trees are there? Estimates of heights needed for comparison with heights needed by tower cranes, if required to swing outside the site boundary.

(b) Relationship of new works to boundary. How much room will there be between new construction and the site boundary? This information is essential in determining room for plant movement and scaffolding. If railway lines exist adjacent to the site, special conditions apply. Location of running lines and railway property boundaries should be established. (More detail about this is given in Chapter 2.)

(c) Safety of the public. In connection with any contractor's obligations in relation to public safety[1] [2], consider: Are rights of way to be maintained across the site during the contract or adjacent to it? Will pavement gantries be required to allow safe pedestrian access under any external scaffolding?

(d) Adjacent public areas. Are there schools, playgrounds or parks adjoining the site? Will they be within the range of possible falling objects?

(e) Adjacent watercourses. Are adjacent rivers or streams likely to have an impact on excavation on site? That is, on possible water level in the existing strata.

For example, a site in London was within half a mile of the river Thames. All the underlying strata was sand and gravel mixed. When excavating for a basement conditions were dry, as was a deeper excavation for the lift pits. Later in the day, the lift excavation was found to contain 300 mm of water. Approximately six hours later the excavations were dry again. Studies showed that the flooding was cyclic. It turned out that the Thames tidal flow was responsible. At high tide water was flowing in the sand/gravel and raising the water table to match the tide but one hour later due to the flow time through the sand/gravel bed. As a result, it was necessary to cofferdam the lift pits with short lengths of sheet piling.

(f) Airfields. Any adjacent airfields may affect plant location and selection. Do runway approach paths pass over the site? Consult airfield controller about height restrictions on cranes, their location and the need for red warning lights on possible obstructions. Are there any other restrictions?

(g) Rights of access for others. During the course of the contract do others have rights of access? Do services have to be maintained for others at all times or specified periods?

(h) Adjacent mains and sewers etc. What risk may arise from deep excavations or other activities within the site to mains etc. outside the site?

4. Noise

Noise, today, is a pollutant and is covered by legislation[3]. Maximum permitted noise levels are laid down by the local authority. If these are not already recorded in the contract documents, the local authority should be asked for the set noise level in the particular instance. The limits given can seriously affect plant choice for certain activities. Local schools, hospitals and nearby offices will play a part in the level fixed. Any restrictions on night work should also be checked. If allowed, it may only be within certain times and at much reduced noise levels.

5. Public utility supplies for construction

Each public utility company will need to be contacted concerning availability of supply for construction purposes. What may be adequate for the existing situation may not be adequate when supplies for construction are added on. Availability and location is essential knowledge for all services. This is especially so where operations involving large quantities of water or electricity are concerned. For example, a large demand for water arises in the installation of ground anchors.

6. Surface and below ground conditions

A visual inspection of the site can often provide useful information to the construction planner. If the site inspection is done in dry weather, a false impression can arise. What seems a dry site may become a wet boggy one in the winter. Inspection of the types of grasses growing can give a clue to periods of soggy surface conditions. Wild flowers can also give useful clues. Nature books on these topics are readily available[4][5]. Local knowledge can also be very informative in this respect.

An examination of the lie of the land and the flow of any watercourses will allow consideration, at a later stage, of the effect of new construction on the natural drainage of the site. The need for diversion channels will become apparent if the new work is seen to block existing drainage channels.

Above all, a careful examination of the boreholes taken for the consulting engineer is essential for knowledge of any difficulties that may exist in below ground conditions when excavating for foundations and trenches for drainage and services. If, for example, some boreholes show running sand, conventional methods will be useless (Figure 1.2). Sheet piling may be needed if there is a suitable cut-off below the

sand. (See Sheet Piling in Chapter 4). The other main alternative would be to use well-pointing to drain the waterlogged sand or gravel. See Chapter 15. In some types of gravel – large size – the only solution to stop the flow of water would be ground freezing, which is very expensive.

7. Local working week

Such knowledge is essential to the construction method planner, in the assessment of performance and time. What is needed here are not just the days per week worked, but also the the hours worked in the week, including any agreed overtime.

8. Local holidays

While the total holiday entitlement is fixed nationally, local convention often fixes summer holidays, for example, at different times from other areas. When the planning has reached the stage of setting it down in programme form, such knowledge is essential.

9. Local weather conditions

Adequate knowledge of local weather conditions is important to the construction method planner in making allowance in performance for delays due to the weather. Today, the meteorological office can provide this information on demand. None the less, it is no bad thing to also talk to local people, especially in country areas.

Examples of failures in site inspections

Failure to check adequately the availability of public supplies for site use is a common source of problems at a later stage when the contract is under way. The following two examples make the point very clearly.

1. Power source for tower cranes

When tendering for three blocks of multistorey flats, the construction planner envisaged a tower crane on each block to meet the client's contract

Figure 1.2 An attempt to use proprietary trench support system in running sand.

period. As the site was almost in the centre of the town, it was not thought necessary to confirm with the local electricity board whether the existing cables could supply the load needed by the site. Allowance was made for the usual provision of transformers and a short connection to the mains.

In the event, when application for a supply was made, the contractor was told that no three-phase supply existed in the near vicinity. The site was a slum clearance area and no power source had existed in the old buildings. While the new development would be provided with three-phase cabling, it was not going to be installed until the new blocks neared completion. The contractor was invited to pay for the installation of the cable early – over a quarter of a mile in length! The installation of a suitable generator was seen as a more economic solution – although not allowed for in the tender.

2. Water supply for site mixers

During a site inspection for a new housing estate, it was confirmed that a water main existed at the site entrance. When brickwork started, two mortar mixers were installed, with a joint pipe connected to the main in the road. Unfortunately, no one had checked the mains delivery capacity, it turning out that the feed pressure would only allow one mixer to receive water at a time. Hardly good planning.

References

1. *Health & Safety at Work etc Act 1974*, Section 3. HMSO, London.
2. Health and Safety Executive. Protecting the public – your next move. HS(a)151, HMSO, London.
3. *Control of Pollution Act 1974*, Sections 60 & 61 HMSO, London.
4. Aichele, D. (1975) *Wild Flowers – a field guide in colour.* Octopus Books Ltd London.
5. Fitter, A., Fitter, R. and Blamey, M. (1974) *Wild Flowers of Britain and Northern Europe* Collins, London.

Chapter two

The influence of the site and its boundaries on plant and method selection

The site and its boundaries play an important part in the selection of plant and methods, the siting of facilities and storage areas, the sequence of operations and the problems of access. At the same time, it will be apparent from a careful study of the site inspection form that the site and boundary conditions are likely to be a major factor in temporary works needs.

As every site is unique, it may be thought that arriving at any basic rules for a site would be impossible. In fact, all sites can be classified into one of three types. Each has special characteristics of its own which affect the planning process and decisions made. Once these are understood, the characteristics particular to the site in question can be taken into account, together with the boundary situation determined from the site inspection report.

Basic types of site

The three basic site types can be defined as shown in Table 2.1. Each is now examined in more detail.

Open field sites

A typical example of a housing estate is shown in Figure 2.1. It is clear that what has to be built occupies a relatively small portion of the site area. Large road and parking areas, together with what eventually will become gardens exist. In the case of industrial sites, there are normally perimeter roads and extensive allowance for the parking of trade vehicles and cars.

In such circumstances, the siting of temporary hutting, storage areas and general construction access is rarely a problem. Method planning can usually be geared to the most efficient and cost effective method. However, a number of key items will need analysis.

Table 2.1 Types of site

Type	Description
1. Open field	What has to be built occupies only a portion of the site. Ample access is available and plenty of space for materials storage and accommodation.
Examples	Housing estates, factory developments, and many civil engineering projects.
2. Long and thin	Very restricted in width – considerable length. Access at few locations.
Examples	Railway work, motorways and gas and fuel pipelines.
3. Restricted	New construction occupies the whole or a very high percentage of the site. Access restricted.
Examples	Central redevelopment, industrial improvements or additions and some railway or motorway work.

Each of these types has an influence on the construction planning for whatever has to be built.

1. How does access to site relate to the most effective work sequence? Are there problems? For example, on a private development site, the sales negotiator will want the show house as near to the main site access as possible. High volumes of construction traffic passing into the site, making dirty conditions, conflicts with such a requirement. How can the problem be overcome, on the site in question?

2. Where are the main sewer and surface water outfalls in relation to site entry and site roads? Will temporary access be needed for their construction, or can the permanent road bases be utilized? This latter may prove a cheaper solution, even though some remedial cost will be needed to the permanent road sub-base.

3. Where are the main service runs? On housing estates, water, gas, telephones and electricity, together with, today, cables for television and radio. On industrial work, the above items plus additional specialized services – such as compressed air, industrial gases, provision for trade effluents and dust or fume extraction. These latter items will normally be within the site boundary.

   How will such items affect the main construction sequence? Bear in mind that public utility companies are not notable for coming when they are wanted. The forward planning must take such matters into account.

4. Is any sequence of construction or phasing specified? How does this relate to site access, drainage and service runs and economical construction?

5. Where can storage areas, site hutting and welfare facilities be located? Bear in mind that they not only have to be convenient to the work as a whole, but they also need siting to minimize temporary service runs and save further cost by using the formations of car parking facilities or hard standings required in the permanent works. Can they remain in one location for the contract duration? If not, how many moves will be needed.

While the above is not necessarily complete, it will provide a lead as to the matters which, in the method planning sense, need careful attention in this category of site.

Long and thin sites

Long and thin sites present special problems of their own (Figure 2.2). By their very nature,

OPEN FIELD SITE

Show area

Site entrance

N

Figure 2.1 Open field site.

access is the key problem which has to be solved in relation to the work sequence.

As defined in Table 2.1, long and thin sites arise in relation to motorway construction, railways and gas and fuel grid pipelines. An even more awkward example is when motorway and railway widening works have to be undertaken. With the exception of widening works, all situations involve construction across virgin land. While motorways and railways will be a permanent feature of the landscape, gas and fuel grid lines only require temporary occupation of the land they cross. After restoration, fuel pipelines become invisible, with the land handed back to the farmer or landowner for resumption of normal use. In the case of gas pipelines, the line normally will be marked and sections may need fencing.

Motorways and new railways
Site width is normally determined by the tops of cuttings, the bottoms of embankments, or the minimum width to accommodate road or rail in level conditions. The first activity will, therefore, be the installation of boundary fencing on each side of the works. Animals must be prevented from straying and the public stopped from entering construction areas. Any extra

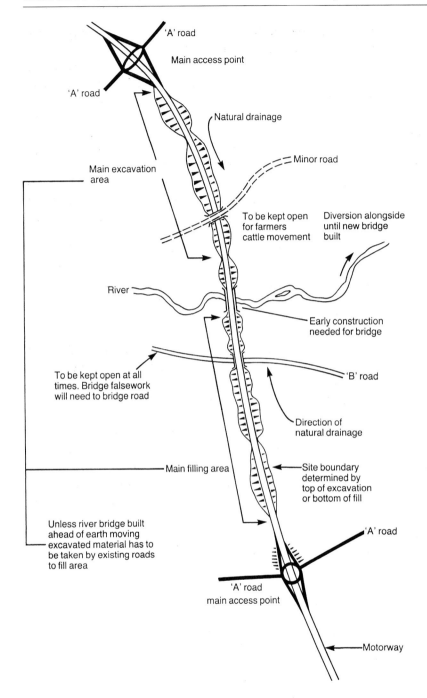

Figure 2.2 Long and thin site.

land that might be helpful to the efficient prosecution of the works can only be achieved by negotiation with the landowners. They, in fact, may not be too pleased that the new works are taking place. Any compensation needs to be negotiated in time for inclusion in the tender sum. Access to long and thin sites can only normally be effected from the existing road pattern

– and then only when the levels and the general geography of the site allows (Figure 2.2).

Gas or fuel grid pipelines

Such pipelines only require temporary land occupation. Contractually, a wayleave will have been negotiated by the client. The wayleave will have been specified as to width and the first requirement will be the establishment of fencing to stop cattle and the public straying into the area of work. In establishing the width to be fixed between fences, due account needs to be taken of the digging and dumping width (material has to be kept on site for backfilling) plus the access width for pipe delivery and the plant used for handling the pipe into the trench (Figure 2.3).

Once the pipeline has been installed and backfilled and the land restored, fencing will normally be removed to allow the site to revert to its previous use. With gas pipelines, markers are usually sited at hedgerow lines to indicate the line of the pipeline. Where booster pumps or other related equipment is installed, fencing is needed with a permanent access for gas officials to inspect and maintain the equipment.

Planning and method selection

The plant and method selection on long and thin sites must take account of the particular demands of this type of site, especially recognizing the need to travel over previously completed work. Sequencing drainage, in particular, becomes an important factor to keep the site as dry as possible for overall continuity.

Planning and method selection is also much affected by features and their location. If a major bridge on a motorway can be constructed in advance of the road works, the type of plant used in the muckshifting may well be quite different from that needed if the bridge was not already there. The difference in cost between the two methods may also be very great.

An example of how the long and thin site planning can be affected by the features of this type of site is shown in Figure 2.4. A railway widening operation is taking place to provide four tracks instead of the original two. On the left hand side, in the diagram, the railway boundary is at the top of the cutting batter. On the right hand side, there is an existing berm, some 10 m in width before a second batter to the boundary fence. The new works comprise two new retaining walls as shown, to be constructed in mass concrete, cutting into the bottom of the batter on each side of the line. The sequence of work determined is: drive sheet piling into bottom end of batter to support excavation, acting as cantilever; excavate for wall and its foundation; pour

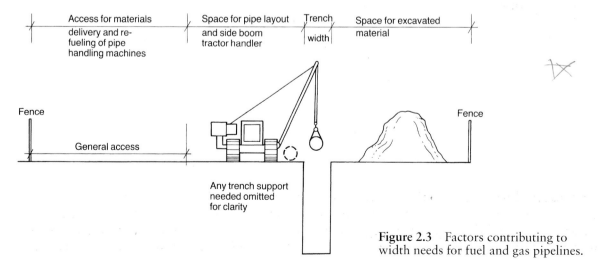

Figure 2.3 Factors contributing to width needs for fuel and gas pipelines.

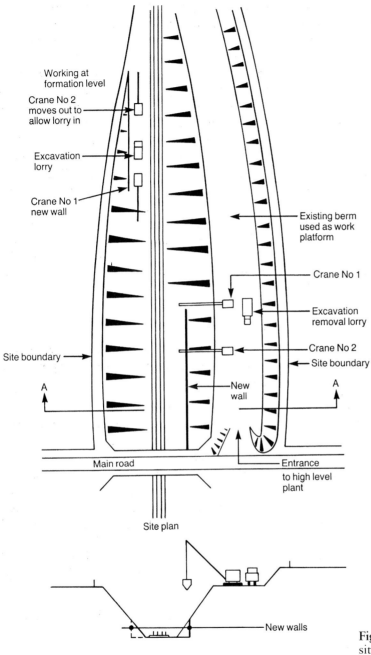

Figure 2.4 Influence of long and thin sites and their boundaries on method selection.

concrete to foundation; erect formwork to wall; pour concrete; strike formwork; at appropriate time extract sheet piling and backfill behind wall with weak concrete to form drainage layer.

On the right hand side, the cranage (excavator conversions) needed can operate from the existing berm, with enough room left behind to allow surplus excavated material to be grabbed by the first crane and loaded to vehicles for disposal. This can be done at the same time as the second crane is handling formwork and pouring concrete to the walls. Alternatively, the pouring crane can be extracting piling. In other words, all activities can be done without hindrance one to the other.

On the other side of the line there is a totally different situation. There is no room at the top to deploy the chosen plant. As a result, the only way to get access to the same sequence of events as on the right hand side is to find an access down the line where the cutting disappears, and work inwards into the cutting. It should be only too clear that the forward crane driving sheet piling and grabbing out the excavation will baulk the second machine while excavating as lorries have to travel up the formation to be filled with excavated material. It is only when the leading machine is driving piles that the second can be handling formwork and concrete. At other times this second crane has to travel back down the formation to the access entrance.

To conclude this example, the actual site costs showed that the wall on the left hand side cost twice that on the right hand side to construct.

Management and control of long and thin sites

Management and control of long and thin sites pose special problems. To get from one part of the site to another may involve a long journey off site until formations are established and stabilized. The use of radio communication between the site manager and key personnel is essential.

Restricted sites

Of the three basic site types, restricted ones call for the greatest care in planning. Access is always likely to be a problem. While plant siting needs particular care – not only to ensure proper coverage by cranes, for example, but also to make sure that materials can be delivered to the

handling equipment for onward movement. It also has to be remembered that plant readily installed may not be so easy to remove once the permanent works have been completed.

Such sites are almost invariably in town centre re-developments. Here the client needs to achieve maximum rental income from an expensive piece of land. To a lesser degree, the same circumstances exist when industrial plant enlargements are called for within the confines of an otherwise operational unit. Reconstruction of such things as railway stations are further examples of the restricted site situation.

A further difficulty that can impinge here is the deep basement. City centre sites invariably involve basement construction, which usually will occupy the bulk, if not the whole, of the site.

Planning rules

A disciplined approach to planning restricted sites is needed at all times. The principles to be followed can be stated as:

1. Study the site in depth and assess the influence of boundaries, adjacent buildings and any obstructions (Figure 2.5);
2. The restrictions of the site itself;
3. Are there any features of the site that can be exploited in the planning concept. For example, are there any unexcavated areas that have nothing built on them in the final result? Could they be used for unloading or storage of materials – or site hutting?
4. What factors relate to access and unloading, including one way traffic systems?

In analysing the methods that can be used remember:

- Sequence of work is crucial – ability to remove plant on completion may determine the final choice of plant and method (Figure 2.6).
- Access must be maintained at all times – particularly in feeding hoists and cranes. In so

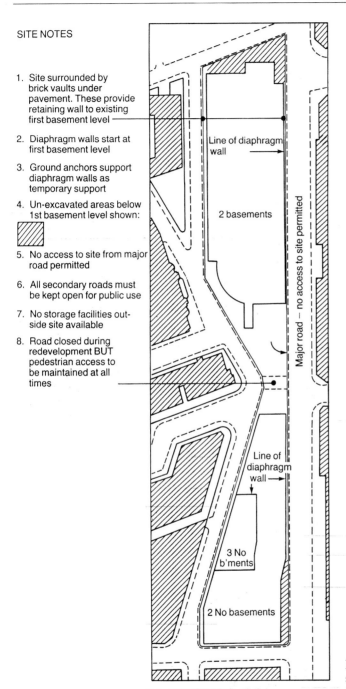

SITE NOTES

1. Site surrounded by brick vaults under pavement. These provide retaining wall to existing first basement level

2. Diaphragm walls start at first basement level

3. Ground anchors support diaphragm walls as temporary support

4. Un-excavated areas below 1st basement level shown:

5. No access to site from major road permitted

6. All secondary roads must be kept open for public use

7. No storage facilities outside site available

8. Road closed during redevelopment BUT pedestrian access to be maintained at all times

Line of diaphragm wall

2 basements

Major road – no access to site permitted

Line of diaphragm wall

3 No b'ments

2 No basements

Figure 2.5 Key factors relating to a restricted site study.

doing, the sequence of the works may be influenced.

- Method and sequence needs to be tested by the production of stage drawings to ensure that the sequence works. (That is, produce drawings of the stage of new construction reached over fixed periods of time. Four- to six-weekly intervals are those most often

Figure 2.6   Restricted site with two tower cranes. Foreground crane has room for dismantling. Rear crane has not. Planning has allowed for rear crane to be lower than the foreground one and within its operating radius. Front crane can be used to both erect and dismantle the rear crane.

Figure 2.7   A restricted site where the golden rules have not been followed. The worst boundary has not been secured before starting the main construction. The result is that what should have been continuity over the main structural area, is now split into two smaller areas to maintain access to the worst boundary completion. Further cost will be involved when the infill areas have to be constructed as further fragmentation of construction will arise.

chosen.) A good deal will depend on the complexity of what has to be built. The construction stage can then be visually tested against access and storage ability at that particular time.

Golden rules

Two golden rules should always be followed on restricted sites.

1. Always secure the most difficult boundary first (Figure 2.7).
2. Re-create space at or near ground level as soon as possible.

Temporary works

As most redevelopment work involves basement construction or upholding of some kind, temporary works and their solution can become a major element in the planning of the method and construction sequence of below ground

activities. In this area the relationship of any upholding works to the permanent works is of major importance. Temporary works will also have a major influence on time at this stage. They need careful assessment with time in mind as well as the purely technical solution.

Modern construction technology is increasingly making it possible to employ permanent construction which can function, in whole or in part, as the temporary works as well. Diaphragm walling, secant and contiguous piling installed before any excavation takes place are cases in point. How they fulfil a temporary works role is described in detail in Chapter 10.

Where such dual function is adopted, the contractural implications, with regard to responsibility needs careful attention – particularly the interface area between diaphragm or other specified walling designed by the Engineer and ground anchors adopted as a temporary support by the contractor until the permanent

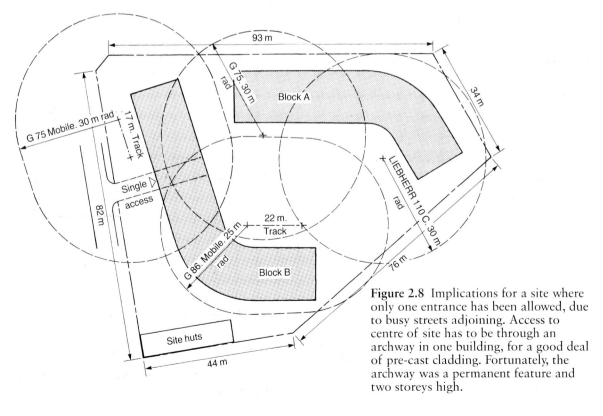

Figure 2.8 Implications for a site where only one entrance has been allowed, due to busy streets adjoining. Access to centre of site has to be through an archway in one building, for a good deal of pre-cast cladding. Fortunately, the archway was a permanent feature and two storeys high.

works are completed. This whole area is dealt with in some detail in a publication by the present author[1]. It will also be mentioned in simplified form in Chapter 10.

Individual sites

Every site, no matter to which type it belongs, will have characteristics of its own. These in turn may well impose additional restrictions on plant and method choice. Such characteristics are:

- Character and geography of the site;
- The effect (if any) of what has to be built;
- Obstructions – above, on or below ground;
- Access capability to desired plant locations;
- Noise limitations under the Control of Pollution Act[2].

When analysing a particular site, therefore, questions must be asked and solutions sought in relation to all the items listed.

Character and geography of the site
The following items need to be considered:

- Geological conditions – clay, rock, sand and gravel, silt, bog, etc.;
- Natural drainage of the site. Will this be interrupted by the new construction? Or do the permanent works allow for this?
- Slope characteristics – gentle, steep, etc. How will the movement of plant and access be affected?
- How does access on site relate to new construction?
- Are there any special restrictions on access to the site from the public highway?

Answers to all these items should come from the site inspection form.

Effect (if any) of what has to be built
- Erection of plant may be very simple early in the contract, but can it easily be removed when what has to be built has been completed?

- Does the planning have to allow for some parts of the permanent work being left out temporarily, so that long delivery building services equipment can be lifted into place at a later stage?
- Are there access problems feeding or unloading materials and equipment into the structure?
- What are the safety problems in relation to the erection of the structure? How can they be overcome? And at what cost?

Obstructions – above, on or below ground level
Some sites will have no problems in this respect. Others may contain obstructions which may well influence the choice of method and/or the location and size of plant.

Any Site Inspection needs to establish, early on, the nature of any obstructions and the questions asked – will they be a permanent feature, or will they be removed or diverted at a stated time in the contract period?

Above ground obstructions can be:
- Overhead power lines;
- Overhead pipes or services – particularly in relation to new work or alterations in existing industrial or commercial premises;
- Trees which have to be preserved at all times.

At ground level items include:
- Buildings or facilities already existing on site which have to be retained;
- Trees – arguably in this category as well as above ground;
- Maintenance of access for others;
- Unsafe areas – ducts, soft ground, badly filled-in old basements etc.

Below ground level. Apart from the obvious:
- Gas, water, electricity and telephone services;
- Existing drainage of all kinds.

Consideration has to be given to less obvious things such as:

- Specialist services in industrial locations – trade effluent pipes, compressed air lines, or other specialist services.
- Requirements of the Ancient Monuments and Archaeological Areas Act[3]. In areas designated as of potential archaeological interest any finds are protected by the above legislation. The Act provides time for investigation and may have a serious effect on contract time.
- In London, the City of London (St Pauls Cathedral Preservation) Act 1935[4], imposes specific conditions relating to work below ground in a designated area around St Paul's Cathedral. In any location, the planning engineer needs to check to see if any Acts of Parliament or local by-laws may impinge on the method planning – **at the tender stage!**

Boundary conditions

These can affect plant and method selection in a number of ways:

1. **Offloading** – can vehicles be parked in the street for offloading? If not, can provision be made on site for delivery vehicle parking? If neither alternative is tenable, consultation with the Traffic Branch of the local police will be needed to find an acceptable solution. In such cases the police are usually understanding and helpful – provided they are consulted beforehand.
2. **Plant overswing** – if tower or other cranes have to swing outside the site boundary, a number of matters have to be considered:
   (a) Overswinging adjacent property. Tested in the courts[5], this constitutes a trespass, unless prior consent has been given by the adjoining owner
   (b) Overswinging the Public Highway. In England and Wales agreement is necessary with the highway authority, but in Portsmouth a specific licence is needed as well.

In Scotland, Section 26 of the Roads (Scotland) Act 1970[6] applies. Application for permission to overswing has to be made to the highway authority. Before agreeing, they may require written agreement to such action from all statutory undertakers with plant or services under the road or pavement. In addition, they will require a certificate stating the Code of Practice under which the crane was designed.

   (c) Overswinging railway lines. If the overswing involves Railtrack plc, London Underground or other railway companies in the United Kingdom, consultation will need to take place prior to any tender submission to establish specific safety requirements. In general terms, under the various Railways Acts, it is forbidden for any crane or other item of plant to overswing railway property unless special agreements have been reached. Such agreement is unlikely to be given in respect of heavily trafficked lines. Stability calculations are normally required and the railway authority may demand greater precautions than the makers factor of safety allows. Increased costs invariably arise. Such matters must be resolved at the tender stage.

   The company concerned will normally provide guidance documents on request. By way of example, Railtrack plc have a formal guidance document[7] which they issue to all contractors working adjacent to their property. London Underground merely have a set of guidance notes[8]. Where other rail lines are involved they should be approached for documentation or guidance, i.e. Newcastle Metro, Docklands Light Railway, Glasgow underground and so on.

   (d) Where overswing has been agreed. At this stage, the height of adjacent buildings will need careful evaluation. Cases are not uncommon where tower cranes have to be erected to a greater height than would be needed for the site in question – merely to clear adjacent structures, radio and television aerials, tall trees, etc.

(e) Works close to airports. Contact with flying control at the airport is essential. They will specify maximum heights for crane jibs or piling rigs on or close to approach and glide paths. Red lights on jibs may be required at night.

Access to plant

With any piece of plant selected, it is clearly important that access to it can be maintained at all stages of the construction sequence. Siting of plant must take into account the location of the site entry and offloading areas. The construction sequence proposed needs to ensure that at no time access routes get cut off or prevented from being used to feed hoists or other plant envisaged.

Noise limitations

Under the Health and Safety at Work etc. Act 1974[9]. Section 2 (1), employers are required to protect operatives from excessive noise.

The Control of Pollution Act 1974[2], by contrast, in Sections 60/61 relates to the protection of the public from the noise of construction operations. Local authorities are empowered to set noise limits which must not be exceeded. It is essential that those responsible for planning the site methods are fully aware of the statutory requirements and that any methods proposed comply with local noise limitations. Failure to apply at the tender stage could have expensive repercussions later on if the plant does not comply, resulting in the need to alter the method concept before any work can carry on.

Of particular importance for the site planning of plant use, is to recognize that hospitals, schools and similar institutions which are close to the construction site may attract lower noise limits than the general level laid down or, alternatively, a different approach to the problem. For example, a hospital close to road alteration which was going to involve a lot of concrete breaking said that they would prefer the use of controlled explosive demolition at fixed times, rather than jack hammers going on all day. They believed that telling patients of loud bangs at fixed times was less stressful than continuous noise. The method proved highly successful.

Tidal incursions

Mention has already been made of water flow from outside the site affecting excavations on

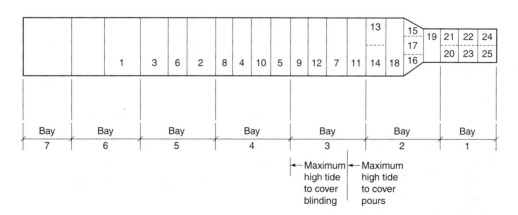

Figure 2.9 General arrangement of slipway. Concrete bays shown and extrent of tidal incursion when bay 18 was poured.

Figure 2.10 The site at high tide related to the tidal situation when concrete is to be poured.

site. A more difficult situation arises where construction can be affected by tidal conditions outside the site actually washing into the site at high tides. Such situations are best illustrated by a case study of an actual site.

The contract involved the construction of a ship building slipway. Prefabricated sections of ships were designed to be pre-assembled in a large assembly hall and then moved out to the slipway for joining together. Figure 2.9 shows the dimensions. The length was 233.5 m; width 30 m and thickness 1 m. 5990 m³ of concrete had to be poured. The figure also shows the bays that were affected by tidal encroachment. Figure 2.10 shows the site at high tide.

The original planning concept was to pump all concrete into the bays not affected by the tide. The remainder to be poured using under water concrete skips[10]. A more careful study was made when the contract was won, in relation to the tidal situation. This showed there was a high risk of the concrete being contami-

nated by sludge regularly washed in by the tide onto the blinding. As a result it was decided to reduce the size of of the lowest pours and pump the concrete into position in between tides after using high pressure hoses to wash out any sludge left by the preceding tide.

A careful study of the tide tables, as illustrated by Figure 2.11 for bay number 18, showed what time was available for all items – formwork, steel fixing and the actual concreting. As will be apparent, many hours were lost for production between 0800 and 1630, the winter working hours at the time – January to March. Overtime was not permitted.

The actual pouring time was kept to three hours for the tidal pours to allow $1\frac{1}{2}$ to 3 hours before the concrete became submerged. Pouring always started from the lowest point and the full width of the bay to give the greatest time for the pour. As previously mentioned, bays were washed by high pressure water jet. Only at the last pours, 22–25, was it necessary to use a detergent in

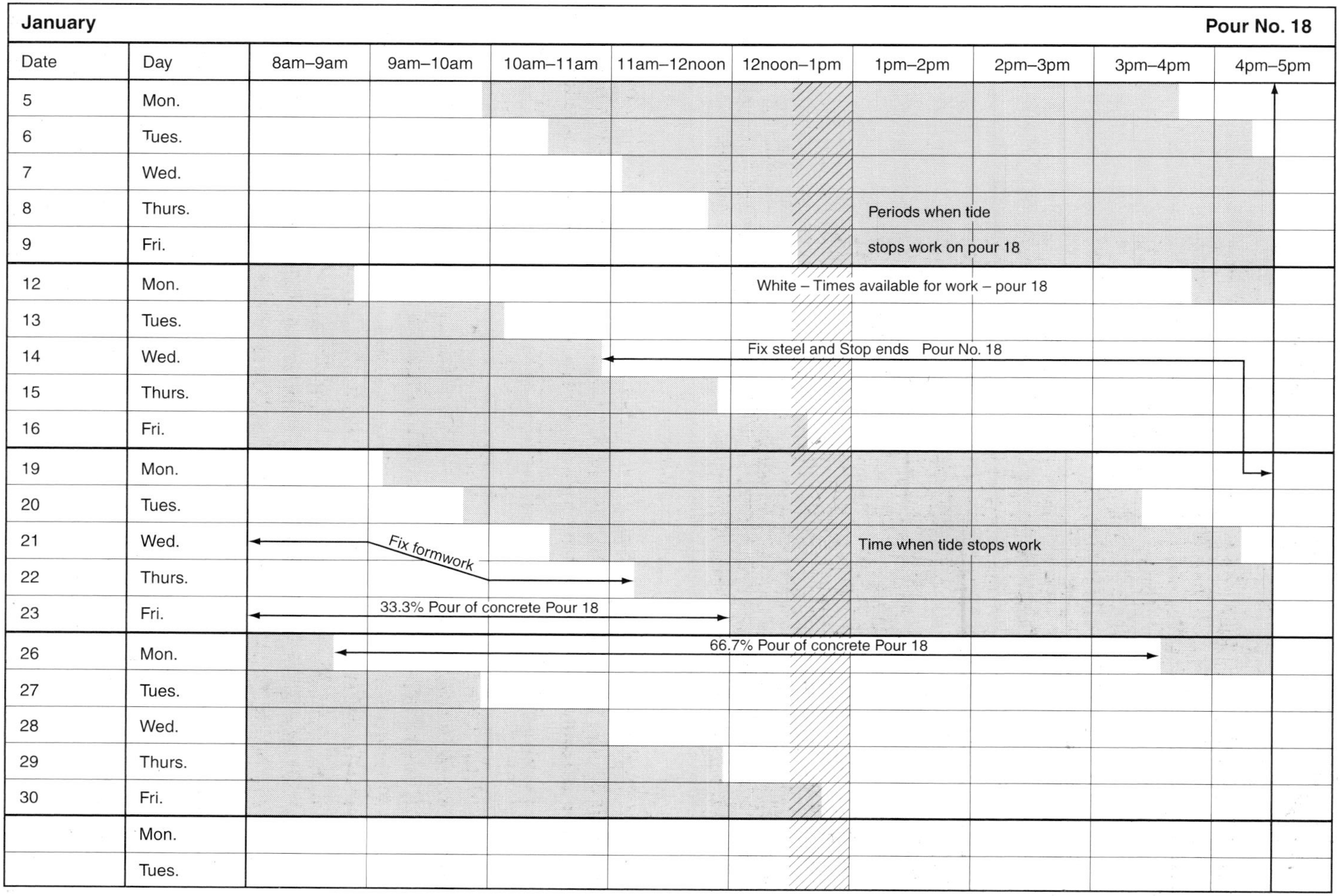

January		8am–9am	9am–10am	10am–11am	11am–12noon	12noon–1pm	1pm–2pm	2pm–3pm	3pm–4pm	4pm–5pm
Date	Day									
5	Mon.									
6	Tues.									
7	Wed.									
8	Thurs.									
9	Fri.									
12	Mon.									
13	Tues.									
14	Wed.									
15	Thurs.									
16	Fri.									
19	Mon.									
20	Tues.									
21	Wed.									
22	Thurs.									
23	Fri.									
26	Mon.									
27	Tues.									
28	Wed.									
29	Thurs.									
30	Fri.									
	Mon.									
	Tues.									

Periods when tide stops work on pour 18

White – Times available for work – pour 18

Fix steel and Stop ends   Pour No. 18

Time when tide stops work

Fix formwork

33.3% Pour of concrete Pour 18

66.7% Pour of concrete Pour 18

Figure 2.11 Tide table diagram at time of work on pour 18, showing periods when work can take place.

washing out. This was due to oil spillage from a tanker terminal next door to the shipyard, being left behind by the previous ebb tide.

References

1. Illingworth, J.R. (1987) *Temporary Works – their role in construction.* Thomas Telford, London. Chapter 8.
2. *Control of Pollution Act 1974.* HMSO, London Sections 60 & 61.
3. *Ancient Monuments and Achaeological Areas Act 1979.* HMSO, London.
4. *City of London (St Pauls Cathedral Preservation) Act 1935.* HMSO, London.
5. *Woolerton & Wilson v Richard Costain (Midlands) Ltd.* (1970), 1 All England Law Reports, p 483.
6. *Roads (Scotland) Act 1970.* HMSO London.
7. Railtrack plc (1978) *Notes for guidance of developers and others responsible for construction work adjacent to the Board's railway.* Civil Engineering Department Handbook 36. Railtrack plc, London.
8. London Underground, Department of Civil Engineering. (1984) *Notes and special conditions for work affecting London Underground.* LU, London.
9. *Health & Safety at Work etc Act 1974.* Her Majesty's Stationery Office, London.
10. Concrete Society. (1990) *Technical Report No. 035 Under Water Concreting.* Update of Technical Report No. 003. Concrete Society, Slough.

Temporary works, their role; association with plant and equipment

Temporary works are an essential part of the construction process. No construction is possible without some form of temporary works being involved. It is essential, therefore, to have a clear understanding of what constitutes temporary works, together with the planning and cost implications that arise.

Temporary works can be defined as 'any temporary construction necessary to assist the execution of the permanent works and which will be removed from the site on completion'. The purist will probably produce exceptions to this definition, but they are few in number and should not be allowed to complicate the above definition. Exceptions will be dealt with as they crop up in the following text.

The scope of temporary works

The use and type of temporary works covers a wide field. Some idea of the coverage can be gained from Table 3.1, together with the reasons why they are necessary.

While the coverage of Table 3.1 is not necessarily complete, it will be clear that the extent of temporary works use on any contract is likely to be a significant part of the activities involved. It also follows that temporary works will represent a significant part of the tender sum in relation to the overall price submitted. Unless, therefore, an adequate study is made of the temporary works involved when tendering and appropriate methods decided upon and suitably priced, there will be a danger that they are underpriced in the tender submission. As the majority, if not all, of temporary works form part of site safety provisions, underpricing can lead to attempts to save money by taking short cuts, with possibly fatal consequences. Assessment of temporary works must ensure that enough data and designs are available for the estimator to price properly.

Table 3.1 The general scope of temporary works on site

Reason for need	Temporary works items involved
The site and its boundaries	Provision of temporary access – roads, bridges, temporary support to adjacent buildings – shoring and waterproofing. Protective measures when plant required to travel under HT lines or work adjacent. Protection of the public adjacent to the works or where rights of way have to be maintained on the site. Safety measures when working adjacent to railways. Hoardings and fencing – traffic management needs.
Temporary works associated with plant use	Provision of crane tracks and bases. Provision of hoist foundations. Temporary anchorages to structure of cranes and hoists. Foundations and service needs for mixing plants. Power supply for cranes, hoists, bar benders etc.
Safe places of work and access to them	Scaffolding. Required by law and to the standards laid down by regulation. In providing scaffolding, public safety must be considered – netting to stop falling materials, safe access under scaffolds which have to be erected over a pavement, etc.
Work in excavations	The law requires that all excavations more than 1.2 m deep must be adequately supported[1]. In addition, rescue equipment and testing facilities are necessary where there is risk of foul atmospheres in the excavation.
In situ concrete work	Provision of formwork and falsework adequate to resist the concrete pressures arising and the loads to be supported until the concrete is self supporting.
Erection of structural frame	Establishing special lifting gear, methods of work to achieve safe access and temporary supports to provide stability to structural members while in the process of erection.
The site set-up	Provision of site offices, toilets, storage facilities, first aid rooms, canteens and drying facilities and appropriate services. Where overseas work is concerned, camps and other facilities may also be needed.

This latter point assumes further importance if significant changes are made to the working drawings and specification, compared to those provided at the time of tender. Unless adequate records exist of what was priced for in the tender, the formulation of a claim for additional or changed work is very difficult. The construction planner is usually best placed in this situation to advise the quantity surveyor on site as to what changes have taken place.

Temporary works related to construction activity

The way in which the various types of temporary works spread themselves amongst the main elements of construction activity is best illustrated diagrammatically. Figure 3.1 shows a flow diagram for a major building structure. To each activity are listed the likely temporary works items that will be necessary. While the cost and extent will vary according to the contract's nature, Figure 3.1 clearly illustrates the way in which temporary works spread over the whole construction sequence.

In the Introduction it was pointed out that construction only involved two fundamental activities: the handling of materials and their fixing in the final location. Even fixing is a handling operation. At the same time handling is an amalgam of plant and labour. For construction to take place, temporary works are necessary to

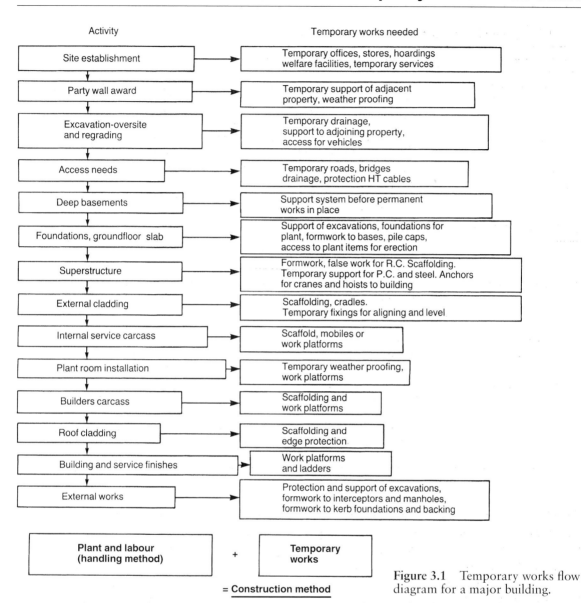

Figure 3.1 Temporary works flow diagram for a major building.

enable plant and labour to perform their duties both efficiently and safely. If now we amalgamate handling methods with temporary works we have:

Materials handling + Temporary works
(Plant and Labour) = Construction method

With this general introduction in mind, it is now appropriate to consider the role of the contract in relation to temporary works.

Temporary works and the contract

In the United Kingdom, three forms of basic contract are normally used.

1. Standard Form of Building Contract – produced by the Joint Contracts Tribunal. Current Edition JCT 80.
2. ICE Form of Contract. 6th Edition: 1991.

37

Produced by the Institution of Civil Engineers.

3. GC/Works/1. Used by Government Agencies for building and civil engineering works, e.g. Property Services Agency.

Within the above, responsibility for temporary works is the responsibility of the contractor (except in certain circumstances in the ICE form), though the wording varies a great deal. It is desirable to consider each in turn.

Standard Form of Building Contract (JCT 80)

This form of contract has many variations to suit particular applications. Temporary works as such are not mentioned in the contract. The responsibility for them is inferred only in clause 2.1. This contains the basic need for the contractor to carry out and complete the works in accordance with the contract documents. In other words, to do so the contractor will provide all necessary plant and temporary works as may be necessary.

ICE Form of Contract (6th edition 1991)

This contract has a good deal to say about temporary works. In particular, clause 8 spells out the contractor's responsibility:

8.(1) The Contractor shall subject to the provisions of the Contract construct complete and maintain the Works and provide all labour materials Construction Plant Temporary Works transport to and from and in or about the Site and everything whether of a temporary or permanent nature required in and for such construction completion and maintenance so far as the necessity for providing the same is specified in or reasonably to be inferred from the Contract.

(2) The Contractor shall take full responsibility for the adequacy stability and safety of all site operations and methods of construction provided that the Contractor shall not be responsible for the design or specification of the Permanent Works (except as may be expressly provided in the Contract) or of any Temporary Works designed by the Engineer.

Of note here is the point in clause 8.(1) relating to the provision of 'everything whether of a temporary or permanent nature ...' This is an example of a variation from the definition of temporary works already given. In the main it refers to such things as permanent formwork – that is where formwork is specified to be left in place. A frequent example is in relation to formwork to bridge decks.

GC/Works/1

Clause 26(1) is specific in making the contractor responsible for all things provided by the contractor for the completion of the works, yet not incorporated in the works. While temporary works as such are not specifically mentioned, they are clearly part of this clause.

Department of Transport work using the ICE Form of Contract

Where the ICE contract is used by the DoTp for such things as motorway construction, clause 8 is modified by the addition of clause 8A. The effect of this modification is to require the contractor to submit to the Engineer a certificate signed by an engineer of suitable qualifications and experience, prior to commencing any temporary works. The certificate is to certify that any erection proposals and temporary works details are satisfactory for the purpose intended. The person signing must be independent of the design team. In other words, an independant check.

Overseas work

Overseas, the most commonly used form of contract is the Conditions of Contract (International) for Works of Civil Engineering Construction (fourth edition 1998) prepared by the

Fédération Internationale des Ingénieurs-Conseils. It is more commonly known as the FIDIC form of contract. In essence, its clause 8 has similar wording to the ICE version and specifically states that the contractor is responsible for temporary works. As with the ICE form, the contractor is not responsible for any temporary works designed by the Engineer.

Related legislation

All construction attracts legislation. By its very nature, temporary works attracts the most by virtue of its role in providing safe working conditions, and the safe use of plant. The main items of legislation are contained in Appendix A together with a brief indication of their content. In addition, further legislation is included which, in certain circumstances, can affect temporary works and their planning and selection.

Notwithstanding the above, it is incumbent on those involved in construction planning and temporary works design to be fully conversant with the detailed requirements of each Act and Regulation. It is also necessary to check if any local by-laws exist that may affect decisions in relation to temporary works.

Codes and Standards

There are numerous Codes and Standards which have an effect on both temporary works and planning generally. As with legislation, they have been collected together in Appendix B for general convenience.

More detailed coverage

Within the confines of a book on planning, it is not possible to cover temporary works in great detail. Nevertheless, their role in the planning process needs to be fully understood. To this end, the remainder of this chapter deals with the role of temporary works associated with the use of construction plant. In all such cases, design

by persons competent in this field is an essential requirement. In the following chapter, the remaining three areas of temporary works most commonly dealt with are treated separately. The reason for this action is that scaffolding, formwork and falsework and the support of excavations are capable of resolution by individual design or by what are known as standard solutions. Those involved in construction method planning need to have a very clear appreciation of how and when each option should be used.

Those seeking more detailed information should study the relevant references included in the text and a book *Temporary Works – their role in construction*[2].

Plant-associated temporary works

Section 6 of the Health and Safety at Work etc. Act[3] covers the general duties of manufacturers etc. as regards articles and substances for use at work. Within this phraseology, the Act covers construction plant. For example, Section 6 (1) states:

> It shall be the duty of any person who designs, manufactures, imports or supplies any article for use at work
>
> (a) to ensure, so far as is reasonably practicable, that the article is so designed and constructed as to be safe and without risks to health when properly used;
>
> (b) to carry out or arrange for the carrying out of such testing and examination as may be necessary for the performance of the duty imposed upon him by the preceding paragraph;
>
> (c) to take such steps as are necessary to secure that there will be available in connection with the use of the article at work adequate information about the use for which it is designed and has been tested, and about any conditions necessary to ensure that, when put to that use, it will be safe and without risks to health.

In addition, Section 6(3) states: 'It shall be the duty of any person who erects or installs any article for use at work in any premises where that article is to be used by persons at work to ensure, so far as is reasonably practicable, that nothing about the way in which it is erected or installed makes it unsafe or a risk to health when properly used.'

Two important points arise when complying with Section 6. In the first case, the supplier of the plant or equipment is legally obliged to have available adequate information about the correct use for which it was designed. Note, at this point, that the operative word is 'available'. The onus is on the erector or the user to make sure that all the necessary instructions have been received. Second, there must be adequate information available about any conditions necessary to ensure that, when put to that use, the plant or equipment will be safe and without risks to health. This latter point covers those items for which the user must take responsibility. For example, the user of a mobile crane is responsible for seeing that adequate foundations are provided beneath the crane outriggers. Clearly, the crane designer cannot always foresee the type of ground on which the crane will operate. The provision of such foundations is clearly an item of temporary works and needs to be allowed for in pricing.

In either case responsibilities for selection of the plant and the need for special conditions for safe use rest with the user. In selecting plant at the tender stage, the method planning must take account, not only of the specification required, but also the special conditions related to use. Such conditions, in almost all cases, relate to the provision of temporary works which are necessary to ensure safe use of the equipment. For cranes, these include crane tracks and bases, anchorages to the structure, adequate foundations for outriggers, to name the most important. In the case of hoists, the same list applies, although the use of outriggers for hoists is unlikely these days. Foundations for large site

batching plants are another example of temporary works related to plant. All these matters are covered in some detail in the book on temporary works already mentioned.

From the purely method planning point of view, requirements and cost need to be established at the tender stage, so that adequate allowance can be included in the tender sum.

In establishing plant requirements and associated temporary works, foundations, in particular, need careful consideration. They are the rock that provides safe operating conditions, for the plant above. The planning engineer needs to appreciate that such foundations must relate to the ground conditions on an individual site. At the same time, where anchorages to the structure are needed by items of plant for their safe operation, the structural designer has to be consulted and his approval given. A check of the boundary situation is also needed to see if adjoining owners may require improved overturning stability for tall plant in addition to that given by the maker. See Chapter 2 about boundary conditions.

Unless adequately qualified and experienced, those doing the planning will need the services of a temporary works designer of suitable experience to produce individual designs for pricing purposes **at the tender stage**. Failure adequately to do so can lead to inadequate allowance for such temporary works, which, if the contract is won, brings an immediate loss situation. At the same time, the temporary works designer and the method planner need to be cost conscious and, within the requirements of site safety, try to develop solutions that minimize cost as far as possible.

Minimizing cost

The following examples of how cost savings can be achieved without imparing essential safety needs illustrate the way in which the planner and the temporary works designer need to be thinking.

A central redevelopment site needs five tower cranes to provide overall coverage. As there are two basements covering the whole site, the cranes have to be static, to allow the structure to be erected around them (Figure 3.2). Thus:

Five static bases needed. Average base size 7 m × 7 m × 2 m deep
Volume of each base = 7 × 7 × 2 = 98 m³ say 100 m³

Components of cost: excavate for base, support excavation, blinding to base and fix reinforcement, formwork to base edges. Followed by, temporary support to base section of mast to be cast in or consumable part of recoupable base (Figure 3.3) pour concrete, strike formwork and backfill.

Allow, for simplicity, overall rate to cover work listed above. Say £80 per cubic metre. Then:

Cost of installing base is £80 × 100 × 5 = £40 000.

While the above covers installation, what happens when the base is no longer needed? If insufficient thought has been given to the location of each base, they may end up in the way of permanent construction and have to be broken out. Breaking out old concrete (reinforced) usually costs almost double the total cost of installation, thus:

100 m³ of concrete has to be broken out at £150m³ = £15 000
As five bases are involved, we have 5 × £15 000 = £75 000
Therefore, total cost of static bases and their removal will be: 40 000 + 75 000 = £115 000.

Such sums are considerable and – in most cases – with careful planning and the agreement of the structural engineer, can be considerably reduced. Figure 3.4 illustrates, from an actual site, how the temporary works costs were dramatically reduced.

The upper surface of the crane base has been

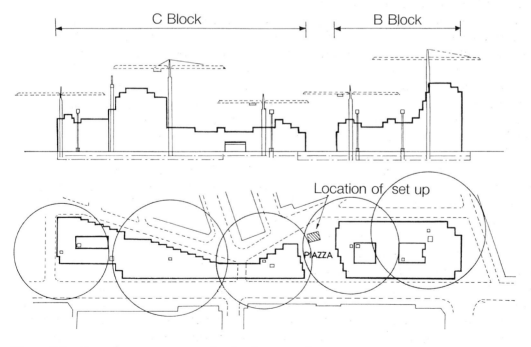

Figure 3.2 Crane layout on a central redevelopment site.

Figure 3.3 Crane base construction. Note base section of tower lying on its side prior to positioning in base before concrete is placed.

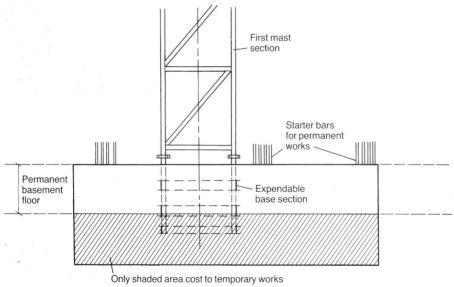

First mast section

Starter bars for permanent works

Permanent basement floor

Expendable base section

Only shaded area cost to temporary works

Figure 3.4 Use of static base as part of permanent works.

sited level with the top surface of the sub-basement floor. This floor was 1 m thick, providing a continuous thin raft over piles below. As illustrated in the diagram, the top 1 m of the crane base eventually became part of the raft and as such a measured item of work. At the same time, the base size was such that it extended under the location of six permanent

structure columns. It was therefore arranged to incorporate the starter bars for these columns in the base as well. By adopting this planning method, the base cost was reduced by nearly half and no cost was involved in removing it on completion of the crane's activity. A saving against the original concept of £95 000!

A second example, again for tower cranes,

but this time in relation to rail tracks, empha-sizes the point. (Figure 3.5.) In this case, the site was not level and for the crane to operate safely, two concrete beams were provided below the rail track assembly to provide a level base for it to sit on. Examination of the beam will show that it is broken into sections by inserting fibre board dividers when cast. At the same time, steel loops were cast in at one end of each beam section. On completion of the crane's work the track components can be lifted off in sections and returned to the depot, while the two beams can be lifted up in the preformed sections straight onto transport for disposal elsewhere. No breaking out is required.

Perhaps the ultimate in economy in providing a levelling system for a crane track is illustrated in Figures 3.6 and 3.7. Here the track is sup-ported on precast concrete pads, brought up to the correct level by setting in a sand cement mix. Note the anchor slots cast into the pads. These provide anchorages from the track waybeam to the pads. In turn, the actual track is perma-

nently fixed to the waybeams in 7 m long lengths and braced between the pairs of rails. On com-pletion of use, every component can be lifted and taken to store for future use – with the exception of the levelling compo.

Figure 3.6 Waybeam and integrated rail supported by precast concrete pads.

Figure 3.5 Minimizing cost of concrete levelling beams under crane tracks.

Figure 3.7 As 3.6, but showing sand cement packing under the pads to make up levels.

Boundary conditions

Plant associated temporary works can also be needed as a result of boundary conditions, especially when working near railways.

When the site runs alongside a railway, the railway company will not only refuse permission for crane jibs to overswing, but will also require better stability conditions against overturning than the factor of safety used in the crane design. This second factor will apply, not only to cranes, but any other tall plant – for example, piling rigs as well. When such adjacent situations arise, it is important that those doing the pre-tender planning check with the railway company in question what their requirements might be. Railtrack have a publication setting out their rules[4], and London Underground have a set of guidance notes giving their rules to be followed[5]. It is important that contact is made with the railway in question at the time of tender. Increasing stability will always cause additional expense, while having to shorten a jib may slow down construction progress or demand the installation of an additional crane, with increased cost. Whatever the restrictions may be, they must be identified and allowed for at the tender stage.

Case study

A static based tower crane was provided with a shortened jib to avoid overswing of a London Underground line. In addition, the railway company wanted extra stability against overturning. The solution is illustrated in Figure 3.8. Behind the tower crane, in relation to the railway, a second tower was erected but to a lesser height, also on a static base. Between the tower crane and the second tower, a pair of wire cables were loosely installed, to allow for the normal sway of the crane tower. If for any reason the crane began to topple, the cables would begin to tighten. Any further movement in the direction of the lines would be arrested and the crane swing left or right, back into the site.

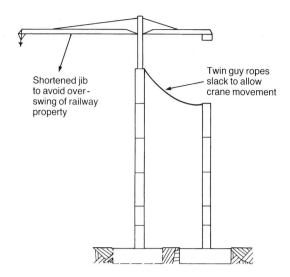

Figure 3.8 A method for increasing the stability of a tower crane working adjacent to railway lines.

As has already been seen, static bases are expensive, while a second tower would attract additional hire charges.

While the preceding part of this chapter is by no means comprehensive in relation to plant associated temporary works, the way that they should be examined will be clear: (a) adequate assessment at the tender stage; (b) adequate pricing at the tender stage; (c) always looking for cost effective solutions that can save money with safety against competition.

In the following chapter, where both standard and designed solutions become available, knowledge of the dividing line between the two is paramount for the construction planner and/or the temporary works specialist, while the same attitude to minimizing cost will apply in either case.

References

1. *Construction (Health, Safety and Welfare) Regulations 1996*. The Stationery Office, London.
2. Illingworth, J. R. (1987). *Temporary Works – their role in construction*. Thomas Telford, London.
3. *Health and Safety at Work etc Act 1974*. HMSO London.

4. Railtrack plc. (1978) *Notes for guidance of developers and others responsible for construction work adjacent to the Board's operational railway.* Civil Engineering Department Handbook 36. Railtrack plc.

5. London Underground, Department of Civil Engineering (1984) *Notes and special conditions for work affecting London Underground.* LU, London.

Chapter four

Temporary works: scaffolding, formwork and falsework, support of excavations

In the previous chapter, the need for individual design of plant associated temporary works was emphasized. By contrast, the three other main areas of temporary works use – scaffolding, formwork and falsework and the support of excavations – can have either of two approaches to their solution. These are by standard solutions or by individual design, depending on the circumstances involved. Such alternative approaches are recognized and defined in Codes of Practice in the case of scaffolding and falsework, and in definitive documents in the case of formwork and support of excavations.

Standard solutions

These may loosely be defined as solutions to a temporary works problem by reference to tables previously prepared for the likely variety of loadings that could be met in situations amenable to their use. Such tables may be available from a number of sources:

- in house tabulations showing loads and spacing of materials used on a standardized basis;
- authoritative guidance documents resulting from research and testing;
- tables provided by manufacturers marketing proprietary equipment tested to comply with the requirements of Section 6 of the Health and Safety at Work etc. Act[1].

All proprietary data is normally based on worst case loadings for a given material content and spacing.

Use of standard solutions

The value of standard solutions, in situations appropriate to their use, to the construction planner will be obvious. At the tender stage, in these circumstances, requirements in relation to temporary works can be assessed with considerable accuracy. Specialized knowledge is not needed as the solutions can be read off from the specific tabulations available. This, in turn, enables estimators to price the items more accurately.

The further spin-off arises at the construction stage. Standard solution tables or charts provide site management with the means of checking the work of subcontractors in known standard solution situations before it is too late (Figure 4.1).

Designed solutions

Designed solutions become necessary for all situations where standard solutions are inappropriate, as defined in the following sections. Persons suitably qualified must be used in these situations. In the previous chapter, it was noted that all temporary works in relation to construction plant must be dealt with by individual design.

Divisions between the two approaches

In the three areas discussed in this chapter, scaffolding, formwork and falsework, and the support of excavations, the options for standard solutions inevitably take different paths, with the dividing line between standard and designed solutions being variable. In consequence, each of the three divisions is dealt with separately.

SCAFFOLDING

Scaffolding can truly be called the maid-of-all-work in construction. It is such a commonplace element of the construction process that it tends to be taken for granted. In fact, it is a crucial item in ensuring that operatives have a safe place of work and a safe means of access to it. At the same time it is the most difficult item of temporary works to estimate accurately. Cynics are

Figure 4.1 Scaffold collapse caused by failure to follow standard details.

often heard to suggest that scaffolding can cost twice as much as the figure in the tender.

Legal requirements and codes of practice

The Health and Safety at Work etc. Act 1974[1] and the Construction (Health, Safety and Welfare) Regulations 1996[2] require that all employed persons on site must be provided with a safe place of work and safe access to it. In the case of scaffolding, the above Regulations lay down detailed requirements for safe working places and safe means to the working place. These Regulations deal with platform widths for various purposes, provision of hand rails, toe boards, prevention of materials falling off scaffolds and so on. The design of scaffolds in the structural sense, using tube and fitting methods, is dealt with in a code of practice BS 5973 : 1990[3]. The specification of the tube and fittings used is laid down in BS 1139 : 1983[4]. There are no codes of practice for the types of scaffolding commonly known as 'proprietary systems'. These are dealt with later on in this section.

Tube and fitting scaffolds

Before considering scaffolding needs in the planning of construction methods, it is necessary to be conversant with the various types of scaffolding as commonly used and the basis of their provision in the design sense.

Two fundamental configurations of scaffolding are normally used in main access and working place situations: putlog and tied independent scaffolds. Both methods are based on the use of tube and fitting methods and are covered in BS 5973.

Putlog scaffolds are those which have one line of standards to support the outside of the scaffold deck, while the inside inside edge is supported from the wall being built (Figure 4.2). To achieve the support from the wall a 'putlog' is used. This consists of a scaffold tube transom member with a flattened end to rest on brickwork, masonry or block walling. To improve general stability, it is customary to allow the standards to lean slightly inwards when erecting. The limitations of this method will be only too clear. It is only suitable for use on load-bearing walling suitable to accept the putlog ends.

Independent tied scaffolds, by contrast, have two rows of standards supporting the working deck. In this case the transoms are not built into the structure. The whole scaffold structure is independent from what is being built but for stability has to be tied into the new building.

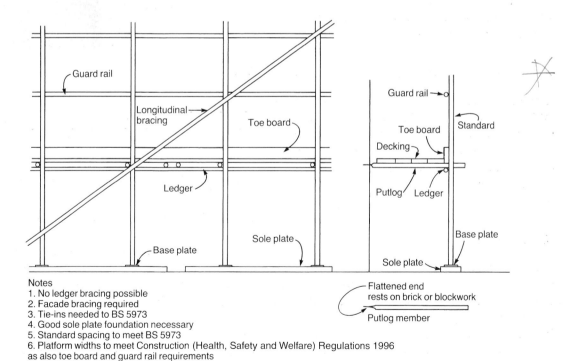

Notes
1. No ledger bracing possible
2. Facade bracing required
3. Tie-ins needed to BS 5973
4. Good sole plate foundation necessary
5. Standard spacing to meet BS 5973
6. Platform widths to meet Construction (Health, Safety and Welfare) Regulations 1996 as also toe board and guard rail requirements

Figure 4.2 Details and nomenclature for putlog scaffolds.

Figure 4.3 shows the components and their nomenclature, and Figure 4.4 the method in use.

In addition to the types mentioned above, which are the main common user scaffolds, three further types have a more specialized use as places of work.

Free-standing scaffolds are usually in the form of movable towers, although not necessarily so. They are required to be stable against overturning without any attachment to any other structure. If necessary, they may need guy or raking strut support. Such scaffolds are covered by BS 1139:Part 3. 1983[4].

Slung scaffolds are those hanging from a structure overhead but incapable of being moved sideways or raised or lowered; BS 5973 : 1990[3].

Suspended scaffolds comprise working platforms suspended on wire ropes from a suitable structure above, such that they can be raised or lowered, but not moved sideways. A special BS covers such scaffolds, BS 5974: 1990[5]. This BS covers both manual and power-operated equipment.

Proprietary scaffolding systems

Proprietary systems differ from tube and fitting scaffolds in two main ways. They are designed to avoid the use of loose fittings and are modular in concept. As previously stated, these systems are not covered by the tube and fitting code. Nor are they covered by a specific code for modular scaffolds. Instead, their load carrying capacity and structural stability work on a deemed to satisfy basis after extensive testing to validate the performance figures supplied by the maker in question.

While numerous makes are available, two main principles are usually involved. In the first,

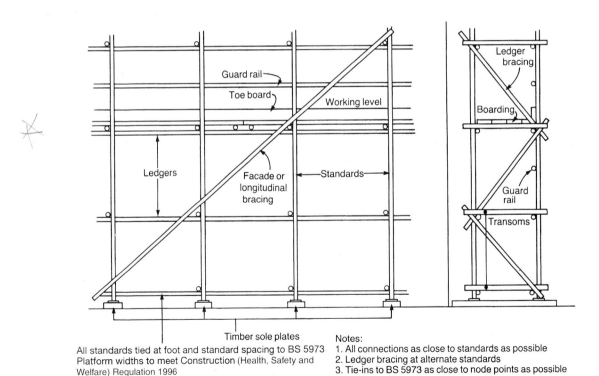

All standards tied at foot and standard spacing to BS 5973
Platform widths to meet Construction (Health, Safety and Welfare) Regulation 1996

Notes:
1. All connections as close to standards as possible
2. Ledger bracing at alternate standards
3. Tie-ins to BS 5973 as close to node points as possible

Figure 4.3   Details and nomenclature for independent tied scaffolds.

Figure 4.4 Independent tied scaffold in use. Note ledger and facade bracing, bottom of standards tied with ledgers and toe boards at working levels.

modular ledger and transom units have end fittings which drop into so called 'V' pressings on the standards and are then secured by driving home captive wedges on the ends of the ledger or transom (Figure 4.5). In the second case, modular transoms and ledgers are used but with ends that drop into a ring unit fixed to the standard. Once in place, a cup shaped element drops down the standard and, on knocking sideways, locks into place, all the ledgers and transoms at that particular point (Figure 4.6). In using either system, the modular nature is completed by the use of modular scaffold boards as well. To be able to span between standards and transoms without further support, such boards are usually 62 mm in thickness.

Value of proprietary scaffolds

In operational terms, the value of such systems can be summarized as follows.

- Both basic connection methods provide a more rigid connection to the standards than with tube and fittings.
- With all horizontal and vertical elements made in fixed modules the scaffold sets itself out. If dimensions are wrong, the pieces will not fit together.
- It is easier and quicker to train a competent scaffolder.
- Few loose fittings are needed.
- Because no couplings have to be set out or done up, erection is quicker than with tube methods.
- Losses are greatly minimized because of the size of the standard components.

Against these advantages, the capital cost is greater than with tube and fittings.

That proprietary systems have found such wide favour with contractors makes it clear that, in situations suited to their use, overall financial advantage has been found to exist. This does not

Figure 4.5 Details of Kwikstage scaffold shown in use. V pressings are clearly seen.

Figure 4.6 Details of Cuplock scaffold connection. (Scaffolding (Great Britain) Ltd).

mean that tube and fitting scaffolds have had their day. They are far more flexible to awkward situations and dimensions which do not fit the modules of proprietary methods. Each type has a role to play in providing safe access and working places.

When attempting to compare one type against another it is necessary to allow something for the losses of fittings likely to arise with tube and fitting assemblies. Before making a decision on the purchase of a modular system for use on multi-storey construction, a contractor carried out an audit of the annual losses of fittings in its tube and fitting stock. It was found that the average annual loss of fittings was 25%. In other words, the complete fitting stock was being renewed every four years.

A further factor to bear in mind when doing cost comparisons is that, given the long life of scaffolding tubes, in spite of continual abuse on site, it is the labour cost of erecting and dismantling that is the major element in any given use situation.

Division between designed and standard solutions

BS 5973 : 1993, and previous editions, recognizes the principle that standard solutions can be used for the design of scaffolding using tube and fitting methods, provided that certain conditions are met: 'Unsheeted access and working scaffolds may be constructed up to a height of 50 m without calculations provided that they are constructed in accordance with the recommendations of sections two and three (of this code), and that they do not carry loads nor have greater bay lengths than those given in Table 1.' A further recommendation is that they are not subjected to loading of materials by mechanical means, such as by rough terrain fork lift trucks (BS 5973 Paragraph 8.5.1.)

The code requires any other form of tube and fitting scaffold to be the subject of design and calculation. Section five of the code details special scaffolds outside the usual access types, which must be the subject of calculation. Section six gives technical data for calculations of the height of scaffolds while appendix B gives worked examples of calculations.

Consideration of the division between standard solutions and designed solutions will make it clear that most normal access scaffolds, not exceeding 50 m in height, come into the standard solution category.

By comparison, proprietary scaffold systems all come into the standard solution category as laid down by the maker unless modified by additional material to meet special circumstances. In such cases, any modified approach needs to be designed by the firm making the equipment. The loading limits in modified or specially designed cases will be given by the firm in question.

Compliance with the Construction (Health, Safety and Welfare) Regulations 1996[2] is, of course, mandatory, regardless of designed or standard solution use.

Assessing scaffolding needs

Today, most scaffolding requirements are provided by specialist scaffolding firms. In order to obtain a competitive price, specialist firms need to be given as accurate a picture as possible of the contractor's requirements, together with the programmed time for each item. Unless the tender planning goes into as much detail as possible, the price will be based on the main contractor's generalized guess. In such cases, which are all too frequent, scaffolding becomes a built-in loss before the contract starts.

The importance of getting the requirements correctly assessed becomes more clear by examining the usual method of tender by the scaffolding subcontractor. The submitted price will usually cover three situations:

1. provision of scaffolding as described by the main or specialist contractor who is responsible for its provision;
2. the extra over rate for the hire of the scaffold if the contract period is overrun, as a rate per week;
3. hourly labour rates for alterations to the scaffold after the initial erection.

To minimize extra charges in one or more of the above items, a detailed assessment of scaffold needs is required in the planning stage prior to tender and proper management control during the contract.

Scaffolding specification

To achieve this desirable situation, the assessment of scaffold needs should be broken down into as many individual components as possible and collected into a specification of scaffolding needs. For example, if ladder towers are needed, specify the requirements for one tower to be priced. In the event, if more than one is needed, the additional cost is merely a multiplication of the basic price. The same can apply in reverse. If three towers were thought to be needed, and it was subsequently found that two were enough, the sum to be deducted from the scaffolding firm's tender is unarguable. To cover the most common situations, many firms have standard sheets onto which dimensions can be put for a particular situation. Alternatively, standard detail drawings may have been prepared for ladder towers, hoist towers and so on.

Preparations for such scaffolding specifications are assisted by having available a check list of all the possible types of scaffolding that may be needed for both internal and external activities. Such a check list is shown in Example 4.1. While at first glance rather intimidating, it does illustrate the large number of matters that may arise when assessing the scaffolding needs for a particular contract. In practice, of course, only a limited number of items will arise on many contracts.

The importance of going to so much trouble is that scaffolding firms are not unknown to make substantial profits from contractors who

---

Example 4.1 Checklist for scaffolding requirements

1. **External elevations**
   Level of working lifts
   Number of working lifts
   Width of working platform (4 or 5 boards)
   Board inside inner standard, or hop-up brackets
   Toe boards and clips
   Type – independent or putlog
   Bridging where required for access
   Adaptions for specialists trades:
       Bricklayers
       Window fixing/glazier
       Roofing
       Cladding/welders
       Painter
   Number of adaptions
   Masons or heavy duty scaffold specified
   Tube and fittings to be galvanized (no rusting)
   Sub-structure is adequate to support scaffold – levelling or making up for sleepers required?
   Adequate fixing back to structure – tie type and who installs
   Ladder access – towers or inside scaffold, supply of ladders
2. **Scaffold to roof level structures**
   As above plus
   Scaffold walkways across roofs required

---

3. **Hoist towers**
   Hoist towers, including mesh and run offs to levels. Fix guides, runners, and gates, dismantled with main scaffold, or later
4. **Rubbish chutes**
   Number, tie ins and platforms, who supplies erects and dismantles
5. **Loading towers**
   Capacity, size, and number
   How many lifts
   Dismantled with main scaffold, or left behind
6. **Internal wall and partition scaffold**
   No. of lifts
   Length of scaffold, no. of erect and dismantle
   Width of scaffold
   Type (independent, putlog for bricklayers, trestle and board)
7. **Birdcage access scaffold**
   Bricklayers
   Services
   Ceiling fixer
   Plasterer
   Painter
   } fully boarded, requirement for scaffold
   Access scaffold for structural work – not included in formwork or falsework
8. **Mobile access towers**
   Number of, steel or aluminium
   Size of platform, height, together with intermediate
   Platforms
   Castors, winching facilities
   Ladders and guardrails
9. **Access scaffold – trestle and board**
   Number of, type
   Youngman type staging required
   Who moves?
10. **Handrails and guardrails**
    Guard rails and boards to floor edges as structure proceeds
    All openings in floor slabs
    Guard rails at roof levels
    Stairs, construction and temporary handrails
11. **Lift shafts**
    Size of platform, adaptions, scaffold lifts
    Position of standards
    Dual construction/lift engineers use
    Guard rails to openings
12. **Protective fans**
    Size, position
    Type of fan material

Tie-in details
Timing of requirement with respect to main scaffold
Nets – who supplies
13. **Other specialist requirements**
Special attendance for subcontractor
Scaffold, 'raking' or 'flying shores' required. (Design in consultation with temporary works)
14. **Protection requirements**
Extent of debris netting needs
Use of reinforced polythene sheeting for weather protection
Corrugated sheeting for temporary roof covering
Brick guards on external scaffolds

EXTERNAL WORKS
Manholes
Pump chambers
Transformer sub-stations, etc.
Retaining walls
Gantries for storage and office accommodation
Sign board support, hoardings etc.

have not done their homework and eventually have to pay for endless alterations to get the job completed.

A continuation of the above leads directly to the stage where site management takes over from the planning activity. Those who are responsible for the operation of scaffolding on site are frequently careless in keeping track of how the contract scaffold time is relating to work progress for which the scaffold is needed. Any overrun of the contract activity will result in extra sums having to be paid for overrunning the scaffold contract period. It is the author's view that scaffolding should be time planned in the same way as operational activities, resulting in a broken-down scaffolding programme. Such a programme should show time periods and the cost per week or month for each separate item or building. Figure 4.7 shows a programme of this type taken from an actual site. While regarded as a management tool, it should not be displayed as a contract programme. Rather, it should be kept in the site manager's drawer as information privy to the

contractor. If the construction period shows signs of falling behind, management can immediately assess the extra cost being incurred solely by the scaffolding as an extra over any unit loss for the work being done.

Forms of quotation

In recent years a *Model form of quotation for the hire, erection and dismantling of scaffolding* has been approved and recommended by the National Access and Scaffolding Confederation and the Construction Confederation. Attached to this document is a schedule of safety conditions. This schedule makes clear a number of matters in relation to responsibility.

1. The scaffold owner shall declare the loads that are brought down by each standard. The hirer will be responsible for adequate foundations and this information on loading of standards is necessary for him to provide the necessary support.

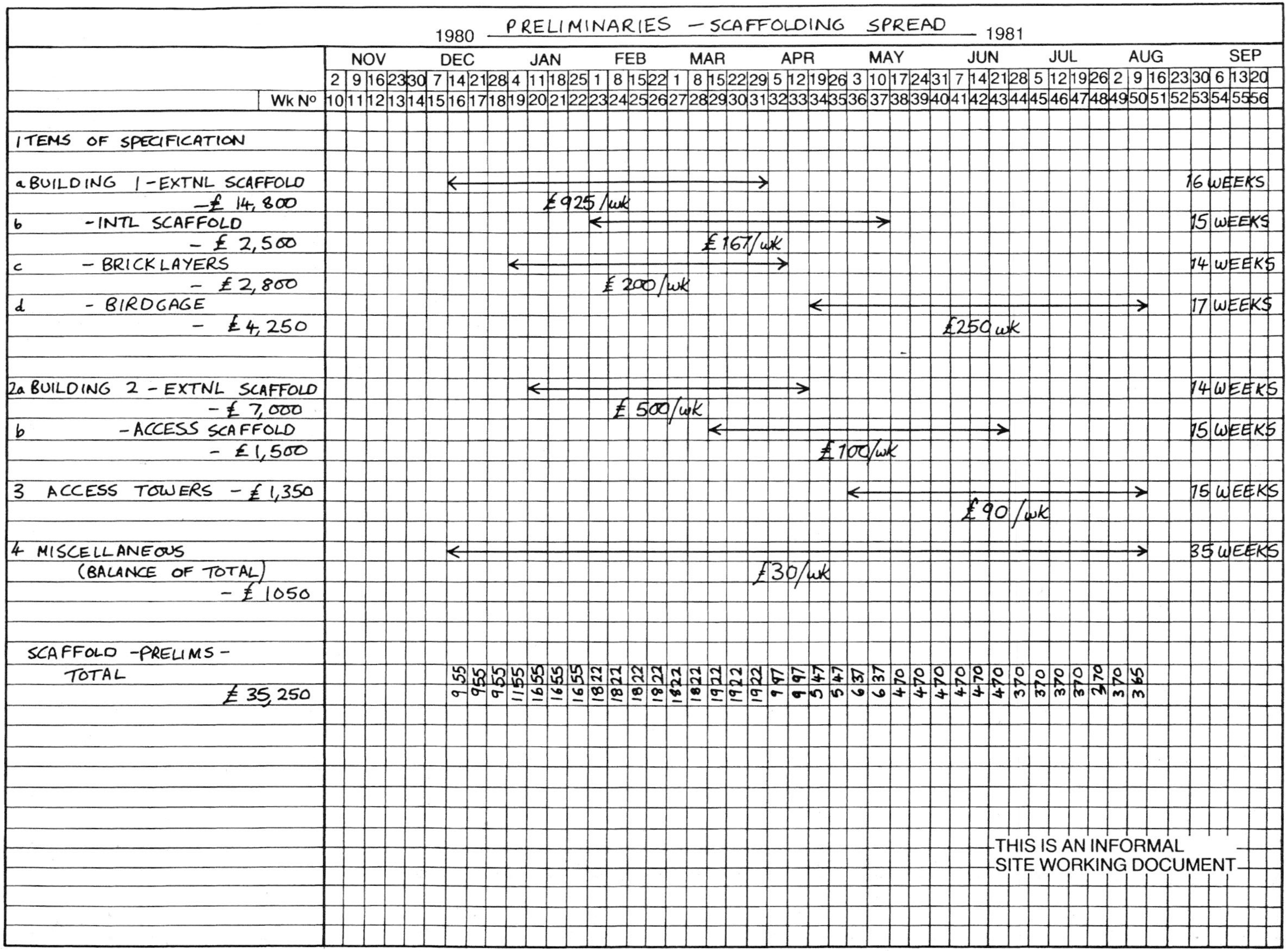
PRELIMINARIES – SCAFFOLDING SPREAD
1980
1981
NOV
DEC
JAN
FEB
MAR
APR
MAY
JUN
JUL
AUG
SEP
Wk No
ITEMS OF SPECIFICATION
a BUILDING 1 - EXTNL SCAFFOLD
- £ 14,800
£925/wk
16 WEEKS
b - INTL SCAFFOLD
- £ 2,500
£167/wk
15 WEEKS
c - BRICKLAYERS
- £ 2,800
£ 200/wk
14 WEEKS
d - BIRDCAGE
- £ 4,250
£250 wk
17 WEEKS
2a BUILDING 2 - EXTNL SCAFFOLD
- £ 7,000
£ 500/wk
14 WEEKS
b - ACCESS SCAFFOLD
- £ 1,500
£100/wk
15 WEEKS
3 ACCESS TOWERS - £ 1,350
£90/wk
15 WEEKS
4 MISCELLANEOUS
(BALANCE OF TOTAL)
- £ 1050
£30/wk
35 WEEKS
SCAFFOLD - PRELIMS - TOTAL
£ 35,250
THIS IS AN INFORMAL SITE WORKING DOCUMENT

Figure 4.7 Scaffolding programme, showing breakdown of time and cost for each area of use.

MODEL FORM OF QUOTATION FOR THE HIRE, ERECTION AND DISMANTLING OF SCAFFOLDING

(Owners)   From: Name _____

Address: _____ Our Ref: _____

_____ Your Ref: _____

(Hirers)   To: Name _____ CONTRACT QUOTATION _____

Address _____

_____ Date _____

Dear Sirs,

(Job Identification)

We thank you for your enquiry dated _____ and have pleasure in submitting the following Quotation for the hire, delivery, erection, dismantling and return of scaffolding. This Quotation is subject to the Model Conditions of Contract for Hire, Erection and Dismantling of Scaffolding, which are deemed to be incorporated in our Quotation, and to the other conditions overleaf.

Yours faithfully,

1. LOCATION

The Hirers shall ensure that the ground is cleared and ready for the erection of equipment by the date on which erection is to begin.

2. DESCRIPTION AND SPECIFICATION OF THE WORK

Including full details of type of materials and uses of scaffolding. For loading specifications see Clause A of the Safety Conditions.

3. LADDER ACCESS

4. MINIMUM PERIOD OF HIRE

_____ weeks, commencing _____

and terminating _____

Figure 4.8a

58

SAFETY CONDITIONS

A. LOADING SCAFFOLDING[g]

In order to comply with Clause 2 of the Quotation the Scaffold Structure will be constructed on a

_____ (m) (_____ ft) x _____ m (_____ ft) nominal grid and is designed to support safely

_____ No. working platforms with a distributed load of _____ kN/m²

(_____ lb per ft²) per working platform. This loading MUST NOT be exceeded

B. LOADING – SUPPORT SCAFFOLD/FALSEWORK[g]

In order to comply with Clause 2 of the Quotation the Scaffold Structure will be constructed on:

*a _____ m(ft) x _____ m(ft) grid

*as shown on drawing No. _____

It is designed to support superimposed vertical loads[h]

On each standard of _____ tonnes

As figured on Drawing No. _____

The loadings stated above MUST NOT be exceeded

Without limiting in any way the Owners obligations and liabilities for properly designing the support scaffold/falsework structure, horizontal loading due to wind, side surge and inclined forces have been allowed for:

(a)   Wind _____

(b)   Side Surge _____

(c)   Inclined forces _____

(g) Safety Conditions A and B are alternatives, delete one as appropriate.
(h) Delete one of the following two lines and complete the other.

C. ELECTRICITY CABLES

The Hirer is responsible for ensuring that any electricity cables in the vicinity of the Scaffold Structure have been isolated or insulated to the satisfaction of the local Electricity Board before commencement of erection.

D. WORKING PLATFORMS

If components of working platforms (e.g. scaffold boards, hand rails or ladders) are moved as the Work proceeds, after the Owners shall first have placed them in position, it is the responsibility of the Hirers or of the employers of labour using the Scaffold Structure to ensure that it complies with the Regulations at all times. Scaffold boards must not be used for any other purpose than as part of the Scaffold Structure.

E. FOUNDATIONS

Unless otherwise agreed to the contrary it is the responsibility of the Hirers to ensure that the ground and/or foundations provided for the Scaffold Structure is/are adequate to support the loads to be applied. The Owners shall supply to the Hirers adequate information in respect of any loads imposed on the foundations to enable this to be done. Prior to erection, the Owner will inspect the site and surfaces available before carrying out their work and will inform the Hirer of the loading to be applied and any apparent problem. Subject to this, the Hirer is responsible for any failure of the Owner's work due to a settlement of the subsoil or other defects in the foundations or base which would not be reasonably apparent to the Owner. The Hirer relies on the skill, care and expertise of the Owner in the undertaking of such inspection and advice given as a result of making the same.

Figure 4.8b

F. ANCHORAGES

It is the responsibility of the Hirers to ensure that sufficient anchorage points suitable for a positive tie into the structure are made available either in accordance with B.S.I. Code of Practice No.5973, as from time to time amended, or to a method agreed between the Hirers and the Owners. The Owners shall be responsible for ensuring an adequate number of such anchorages and their effectiveness.

The Hirers shall be responsible for ensuring that sufficient anchorage points are available during the period of the scaffold erection, use and dismantling.

G. TIES AND BRACINGS

Adequate ties and bracings will be supplied by the Owners. Ties and bracings shall not be removed or interfered with by the Hirers. If it is necessary to alter ties or bracings, this must be carried out by the Owners or by a qualified scaffolder with the agreement of the Owners.

H. CLADDING

No tarpaulins or other types of sheeting are to be fixed to the Scaffold Structure without the prior written consent of the Owners.

I. HANDING OVER CERTIFICATE

(a)   Upon completion of the erection of the Scaffold Structure the Owners shall upon request issue to the Hirers a Handing Over Certificate stating that the Scaffold Structure has been supplied and erected in accordance with the Contract and complies with all Regulations including any design requirements governing the design erection adequacy stability and safety of Scaffold Structures.

(b)   Where the Contract provides for the handing over of sections of a Scaffold Structure such certificate as is referred to in Clause I(a) above shall be issued with reference to such sections upon their dates of completion.

Provided that where a Scaffold Structure has been erected in accordance with the design of the Hirers the issue of a Handing Over Certificate pursuant to these Conditions shall not relieve the Hirers of any responsibility with regard to design. Nevertheless in such a case the Owners shall be responsible for ensuring that the Scaffold Structure is erected in accordance with the Hirers' design.

J. STATUTORY INSPECTION

Whereas the Owners will ensure that the Scaffold Structure is soundly and adequately constructed for the purpose requested by the Hirers and that when constructed it will comply with any applicable Regulation, the Owners cannot undertake to carry out the statutory inspection of the Scaffold Structure or to sign the Register of Scaffold Inspection (Form F91)(Part 1) Section A. Under Regulations these matters are the responsibility of the employers of labour using the Scaffold Structure.

Figure 4.8b continued

Figure 4.8   Model form of quotation for the hire, erection and dismantling of scaffolding. (National Access and Scaffolding Confederation and the Construction Federation).

2. The hirer is responsible for the provision of sufficient anchorage points for a positive tie into the structure, in accordance with BS 5973 or as may be agreed between the parties.
3. Those erecting the scaffold will ensure that the scaffolding is soundly and adequately constructed and complies with appropriate Regulations on handing over.

4. The legal requirements calling for regular inspections of the scaffold thereafter rest with site management.

Those parts of the document relative to site planning are reproduced in Figure 4.8. A companion document, *Model conditions of contract for the hire, erection and dismantling of scaffolding*[6] should be used in association with the above.

From a planning point of view, therefore, it is not only a question of adequately assessing scaffold needs in the most cost effective way, but ensuring that adequate design and monies have been allowed for the provision of suitable foundations and anchorage points. For those who wish to look at scaffolding in more detail, both from the operational and design point of view, two books[7] [8], in addition to the Code of Practice, will be most helpful.

Planning for economy

Much has already been made of the need for construction method planning to be alive to changing methods which may save money. Not only does the method planner need to keep abreast of new developments, but also be capable of valid cost comparison methods to validate a particular judgement. The methods of comparing alternative solutions is dealt with, in detail, in Chapter 7. However, with certain aspects of scaffolding facing growing competition from hydraulic work platforms, it is appropriate to show, Example 4.2, the detail that needs to be gone into if a truly valid comparison is to be achieved when comparing short term

scaffolds with the alternative of using mobile hydraulic work platforms, or other mechanical access systems. While the comparison method is shown as if the competing costs are worked out from first principles, the method is still valid if, in either case, subcontract prices are used.

Economical answers can also be achieved by more thoughtful consideration of alternative scaffolding concepts. Figure 4.9 illustrates a mobile scaffold for bricklayers when cladding a warehouse building. Instead of the obvious solution of scaffolding down the whole of one side of the building, the mobile scaffolding shown only covers two of the bays. The scaffolding sat on castors, roughly levelled, and at one end carried a hoist for raising the bricks and mortar. The saving over a full scaffolding was considerable.

A final example in this context, is the use of scissor-lift equipment for short term use instead of scaffolding (Figures 4.10 and 4.11). Here, the equipment is not only a work platform but a means of lifting materials as well, in this case handling the internal insulation sheeting of an industrial structure. In planning such an alternative method, it is crucial that stability of the equipment is achieved. The design will provide stable working conditions on hard level ground – e.g. a concrete factory floor. If used outside on

Figure 4.9 Mobile scaffold for brick cladding to a warehouse. Note hoist for raising mortar and bricks is included in the set-up.

Example 4.2 Format for cost comparison between scaffolding and hydraulic work platforms

COST COMPARISON FORMAT

£–p

Mechanical Access Means
Hire of equipment: £   p.w. ×   weeks =
Fuel and lubrication: £   p.w. ×   weeks =
Transportation: £   per hour ×   hours =
Operator: £   per hour ×   hour week ×   weeks =
Hire of back-up requirement (e.g. generator): £   p.w. ×   weeks =
Fuel and lubrication of back-up equipment: £   p.w. ×   weeks =
Provision of mains electricity source, plus any cabling =
Mains electricity charges =
Low pressure gas charges =
Labour – dismantle, move and re-erect (if not in S/C): manhours × £ =   _____
Total cost of mechanical means = £_____

Scaffolding or Other Static Means of Access
Hire of scaffolding etc: £   p.w. ×   weeks =
Labour – erect and strike: manhours × £   =
Labour – dismantle, move and re-erect: manhours × £   =
Transportation (including driver): £   per hour ×   hours =
Estimated losses =
Lump sum sub-contract price – materials plus labour =
Allowance for remedial work to floors etc. =   _____
Total cost of scaffolding etc. = £_____

Effective Cost of Mechanical Means
Total cost of mechanical means =
Contract time savings* – deduct   weeks' prelims × £   p.w. =
Effective cost of mechanical means = £
NOT operational time savings.

Potential Savings
Cost of scaffolding etc. =
Cost of mechanical means =
Savings by use of mechanical access means =   _____
£_____

Notes
1. Hourly labour rates are to be those representing the *overall cost to the Company of employing the operative*.
2. Where plant hire rates are quoted 'per hour', the figures to be inserted for weekly hire must obviously relate to the proposed hours worked. Fuel and lubrication calculations must likewise follow the same pattern.

Figure 4.10 Scissor life platform used for lifting as well as work platform. Ideal for short term use. (John Rusling Ltd)

Figure 4.11 Scissor lift platform in external use for cladding. Note compacted hardcore base and outriggers used for stability.

ground inadequately prepared, a dangerous situation in relation to overturning is possible. A recent fatal accident in these circumstances emphasizes the need for planning to ensure that proper allowance for a rigid formation to run on and/or outrigger use has been made.

The design and erection of the roof access and crane support at the Millennium Dome site in London (Figure 4.12) was the tallest free-standing scaffold ever built anywhere in the world (confirmed by the *Guinness Book of Records*.) The 32 m square structure, weighing 350 tonnes, sat on its own concrete slab base, and rose to over 53 m, including a 7 m high central crane podium. There were 45 miles of Kwikstage standards, shoring ties and trigger braces, comprised of some 36 000 individual components. The structure took less than four weeks to build under a full contract package. Additional beams and soldiers were used for strengthening and to ensure maximum rigidity against the potential colossal wind loadings,

Figure 4.12 World's largest free standing scaffold, the access tower at the centre of the Millennium Dome, London. (Kwikform UK Ltd).

horizontal forces and the fact that this structure carried a derrick crane at the very top. This was installed on top of its own 7 m high podium. It had 15 m jib with a 3 tonne lifting capacity. The scaffold also served an additional purpose in that it was being used for access by abseilers and installation teams working on attaching and securing the PTFE fabric roof cover.

FORMWORK AND FALSEWORK

Confusion often exists as to the difference between formwork and falsework, especially to those preparing for a career in the construction world. It is appropriate, therefore, to begin this section with authoritative definitions.

Formwork

In BS 6100, Section 6.5 [9] formwork is defined as 'A structure, usually temporary, but in some cases wholly or partly permanent, used to contain poured concrete to mould it to the required dimensions and support it until it is able to support itself. It consists, primarily, of the face con-

tact material and the bearers that directly support the face material.'

While a comprehensive glossary of terms exists, there is no code of practice for formwork, as such. There is, however, an internationally recognized definitive guide to formwork, *Formwork – a guide to good practice*, published jointly by the Concrete Society and the Institution of Structural Engineers[10]. This volume embraces all aspects of formwork, including materials, equipment, design, special formwork together with site practice and an impressive set of references and bibliography. Extensive worked examples of design are shown in the appendices. Finally, many advertisements are given at the end covering specialist services, equipment and materials related to formwork. This book[10] on formwork has also been written to be compatible with BS 5975, the code of practice for falsework.

Falsework

BS 5975[11] defines falsework as 'Any temporary structure used to support a permanent structure

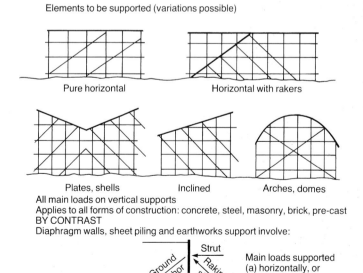

Elements to be supported (variations possible)

Pure horizontal   Horizontal with rakers

Plates, shells   Inclined   Arches, domes
All main loads on vertical supports
Applies to all forms of construction: concrete, steel, masonry, brick, pre-cast
BY CONTRAST
Diaphragm walls, sheet piling and earthworks support involve:

Strut
Ground anchor   Raking shore
Main loads supported
(a) horizontally, or
(b) at raking angle

Figure 4.13   A revised definition of falsework.

during its erection and until it becomes self supporting'. As a definition, in a code of practice for falsework, it is not entirely satisfactory as it could be read to include horizontal and raking supports. It only needs a quick study of the code of practice for falsework to realize that it is dealing with situations where the main supporting system is carrying vertical loads. A better definition, in this author's view, would be 'Falsework is any temporary structure, in which the main load bearing members are vertical, used to support a permanent structure and any associated raking elements during its erection and until it is self supporting'. Figure 4.13 illustrates the coverage of such a definition. It makes a clear distinction between the meaning of the falsework code and such items as temporary raking and horizontal support in the erection of precast elements and the support of excavations (q.v.). All such items are temporary works but not falsework.

While these two definitions provide the division between formwork and falsework, there are occasions when this situation becomes inconvenient. Formwork and falsework become combined in methods known respectively as 'table forms' and 'flying forms'. These particular varieties will be examined later in the Chapter.

Legislation

Formwork and falsework both come under the same legislation, *viz*:

- Health and Safety at Work etc Act[1]
- Factories Act 1961[12]
- The Lifting Operations and Lifting Equipment Regulations 1998[13]
- Construction (Health, Safety and Welfare) Regulations 1996[2]

For greater details of the above see the Introduction and Appendix A.

It is important to recognize that both formwork and falsework are places of work, in addition to the definitions given. It follows that the same precautions to provide a safe place of work and a safe means of access to it, as in scaffolding, apply (Figure 4.14).

Standard and designed solutions

As with scaffolding it is possible to separate the design of both items into standard and designed solutions. At this point it is desirable to consider formwork and falsework separately.

Formwork – standard solutions

Wall forms

Many proprietary types of wall forms are on the market. The makers provide tables showing the way in which the equipment has to be erected for given heights of pour and their limitations in relation to concrete pressure. The makers design is, of course, based on worst case situations. The value of proprietary systems is that they are always in panel form and can be used many times and in a wide variety of configurations (Figure 4.15).

With the growth of subcontractor formwork specialists, the use of panel systems has declined to a considerable extent. The current trend is the use of large plywood sheets, backed by timber or steel or aluminium horizontal beams, in turn supported by steel strong backs. Such equipment can be bought or hired. In difficult economic times, the availability of formwork components on hire is a big advantage to subcontractors who cannot afford high capital expenditure. An example of one such approach is illustrated in Figure 4.16. As part of the package, brackets to support working platforms, knee braces at the base to provide overturning stability and levelling devices are readily available, too.

From being the strength at the back of the wall form, the steel or aluminium soldier has been developed considerably. Manufacturers have realized that with attachments and accessories, the soldier can be made part of a structural Meccano set, giving a much wider scope of

Figure 4.14 Bridge deck using steel beams with concrete deck. No falsework needed. Note scaffolding at edges to create a safe place of work. Formwork spans between steel beams.

Figure 4.15
Example of proprietary panel wall forms. (Scaffolding Great Britain Ltd).

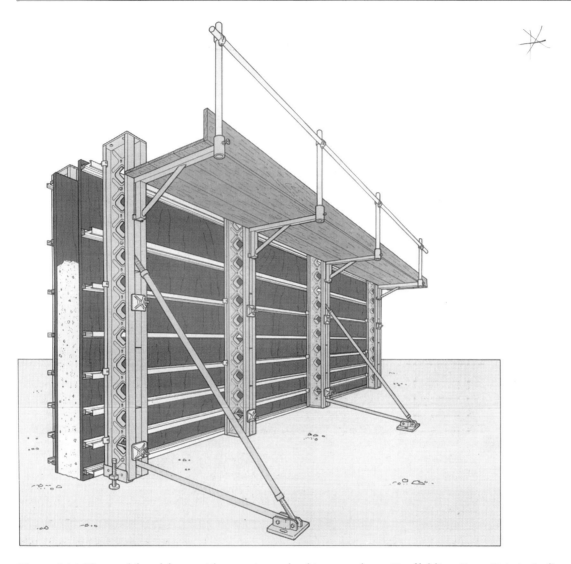

Figure 4.16 Plywood faced form with proprietary backing members. (Scaffolding Great Britain Ltd).

use. Figure 4.17 illustrates one example, the SGB Mk II soldier working in the formulation of a travelling formwork gantry. Further applications of soldiers are illustrated in Figs 4.18 and 4.19.

The application has an even wider range in use as a facade support structure (see Chapter 20). At least four major companies in the UK can provide structural solutions to a number of needs, including the support of excavations. All

work on the basis of design and supply using computer aided design methods.

Many firms sell or hire this type of formwork equipment, both in the United Kingdom and in Europe and America. The makers always provide tabulated data giving spacings for particular heights of concrete pour. The effect of this approach is to allow many variations of form composition and it reduces capital cost by using plywood as the face material with

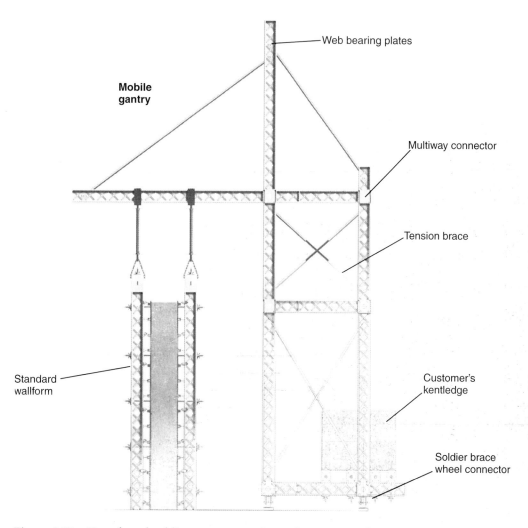

Figure 4.17 Use of steel soldiers as structural members to provide a travelling formwork system. SGB Mk II soldiers. (Scaffolding Great Britain Ltd).

the option of timber sections for the beam members as well.

A final approach to standard solution wall formwork is given by a company developing its own design tabulations for various heights of pour, whether using proprietary accessories with timber and plywood or totally plywood and timber solutions. This approach is well served by the definitive book on form design, *Formwork – a guide to good practice*[10].

Soffit forms

As with walls, proprietary systems are available, i.e. panels, and often the same as those used in walls. However, as with walls, the current trend is to use proprietary support methods (falsework) supporting the more conventional timber beams with plywood sheeting. Support methods range from the common prop to specialized support methods in either steel or aluminimum prefabricated to provide stable tower units.

Company standard designs relating to the

equipment owned can also play a part for limited use.

In general, formwork subcontractors have increasingly seen the value of one type or another of proprietary support in increasing productivity and reducing cost, while increasing the safety margins previously accepted using the simple prop.

The current support systems of this type are interconnecting aluminium frames – light, and easy to erect and dismantle. Two very similar varieties are shown in Figures 4.20 and 4.21. All such types are supplied with loading tabulations and handling methods.

A Health and Safety Executive publication, Guidance Note HS(G) 32 *Safety in falsework for in-situ beams and slabs*[14] is the definitive document for standard solutions in falsework, but has the advantage of also covering the design of the soffit and beam forms supported. Aimed at improving the competence of the smaller subcontractor, the document also has the dual function of providing the main contractor with an authorative checking document, while also giving the ability to price the contents at the tender stage planning process. Read-off designs are provided in tabulated form.

Soffit forms of standard solution type are the most numerous in use in the construction field as they arise in all floors to buildings and similar structures. The above guidance note also gives clear delineation between standard solutions and where full design by competent persons is essential. Guidance note HS(G) 32 also covers the support to the soffit forms and the provision of safe foundations. As such it will be discussed in the following section on standard solution falsework.

Formwork – designed solutions

Wall forms

Designed solutions are necessary where proprietary systems based on worst case loadings cease to be appropriate. It is not readily possible

Figure 4.18 Auto climb core formwork utilizing the structural soldier as major component. (Rapid Metal Developments Ltd).

Figure 4.19   Heavy steel forms for waste water tanks. Note the large use of soldiers.

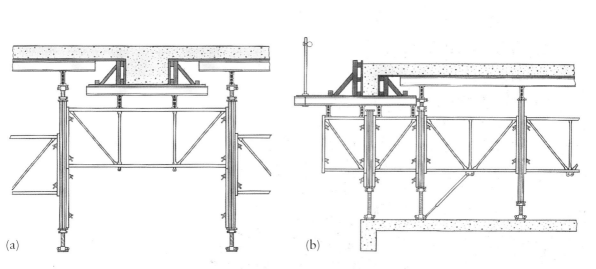

(a)   (b)

Figure 4.20   Aluminium framed falsework for use on repetitive construction. Light and easily erected and dismantled. (a) Drop beam. (b) Edge beam. (Kwikshore – Kwikform UK Ltd).

Figure 4.21 Major support structure utilizing Kwikshore frames stacked. (Kwikform UK Ltd).

to define any line between standard and designed solutions in this case. At the same time certain cases can be clearly said to be design situations, for example wall forms with battered sides, most circular formwork and situations where awkward shapes are involved and special requirements are needed to enable the forms to be struck without damage to themselves or the new concrete. Additionally, architectural concrete finishes may need the finish material to be considered as part of the structural concept of the form to minimize cost.

In considering such designs, the pressure of the wet concrete acting between the forms is dependent on a number of variables:

● The vertical height of the pour in one continuous operation;
● The rate of pour – i.e. the rate of rise of the concrete per hour in the form;
● The temperature of the concrete while placing;

● Constituents of the mix – type of cement, additives used and types of aggregates and sand.

To avoid the need for excessive calculation, Construction Industry Research and Information Association Report No 108[15] provides formulae for predicting concrete pressures. At the same time full design and practical details for good practice are given in the publication *Formwork – a guide to good practice*[10].

Soffit formwork

The design methods for soffit formwork are fully covered in *Formwork – a guide to good practice*[10]. This especially applies for sloping soffits, shell roofs, arches and staircases. As previously mentioned, soffit forms are closely associated with their supporting falsework. It is essential that the two go together in the design process. The reason is that support spacing needs to tie in with spans of support beams and the size used in the most economical way. As in the case of wall formwork, there is no defined delineation between standard and designed solutions.

Falsework

Falsework is better served than formwork in as much that clear divisions are provided between standard and designed solutions. The Code of practice for falsework, BS 5975 : 1994[11] rather loosely provides a division between the two. HSE Guidance Note HS(G) 32[14], on the other hand, provides a detailed method of assessing when standard solutions may be used and where design by persons competent to do so is necessary. The division can be more simply expressed, if not perhaps so accurately in the eyes of the pundits, by the author's version given in the box below.

Standard solutions

The falsework code provides examples of standard solutions in tabulated form, but regrettably is far from a complete guide. The publication Guidance note HS(G) 32[14], on the other hand, has been prepared to deal specifically with standard solutions in relation to beams and slabs. Read-off tables cover the full range in which standard solutions can be used, together with good practice details and procedures for safe use. It is the standard practice guide to standard solutions. Not only is it designed to improve subcon-

tractors' activities, but it also provides an authorative check list for main contractor's supervision, in the same way as it can for the soffit forms, which are included in the same document.

Designed solutions

Outside the limitations spelt out in HS(G) 32, design of falsework is essential. Here the code of practice for falsework, BS 5975[11], provides all necessary details needed for competent design in such cases. In addition, good practice, workmanship, checking of erection and dismantling and allocation of responsibilities are all set out in detail.

General

Whatever methods are used for the provision of formwork and falsework, the planner concerned must always remember that these items are a place of work and must be protected as laid down in the Construction (Health, Safety and Welfare) Regulations in addition to the sufficiency of their design to resist the loads from concrete and its placing methods.

Combined formwork and falsework

Mention was made earlier about situations where formwork and falsework can be combined into a single handleable unit. The value of such methods is that time can be saved in erection and dismantling and the labour content

---

Division between standard and designed solutions in falsework

A simplified division between standard and designed solutions can be stated as:

(a) Standard solutions may be used for the support of floors and beams where the loading is of the type normal in commercial and residential concrete construction, and where the height of support is within the range of standard telescopic props.
(b) All other situations require design specific to the circumstances by persons competent to do such work.

It should be noted that this definition does not contradict the more precise division given in Guidance note HS(G) 32

of the operation significantly reduced. To be effective, though, repetition is needed.

Table forms

Designed to be placed and struck as a single element, table forms originated in multi-storey construction to speed up the construction cycle.

Today, table forms in the original concept have been superseded both in make-up and use by the development of hydraulic table forms. Here the support structure to the form is operated up and down in the vertical plane by hydraulic rams. The fact that striking is now possible over the whole area of the form, by simultaneous lowering at every support, means that soffit forms of the trough or waffle type can be. struck as readily as flat slabs (Figure 4.22). This, in turn, allows the table unit to work on a one-plane basis, in conjunction with others, to move progressively forward over a long frontage at one level. Cranage is only needed to move the tables up to the next level if required. This type

of hydraulic table has major advantages for speed of construction of such projects as shopping malls where not more than three floors are involved, but over a considerable area. Early striking and forward movement is best achieved if the structural concept has beams between columns parallel to the line of movement, which can stay propped while the tables move forward (Figure 4.23).

The use of void formers to lighten concrete structures and to provide architecturally pleasing soffits to floor slabs or ceilings often suffered from the formers being in small units, exhibiting many joint lines on striking. This can now be overcome to a high degree by the use of purpose built trough or other shaped moulds. The moulds are formed from a twin walled extruded polypropylene sheet encasing expanded polystyrene to produce the trough system, to any desired shape. They can be tailor made to desired dimensions, especially in regard to length, so avoiding joint marks. Figure 4.24 shows the fundamental details, including the

Figure 4.22 Indumat hydraulic table forms. Ideal for soffit formwork moving horizontally over large areas – shopping centres, multistorey warehousing.

Figure 4.23 Indumat system used for greater heights on a shopping centre.

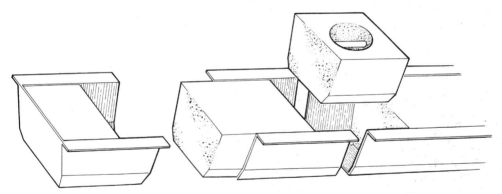

Figure 4.24 Principle of Cordek formwork panels to form trough or other shapes on floor soffits. (Cordek Ltd.)

method of striking, while Figure 4.25 shows the quality of a large trough shaped soffit.

Flying forms

The flying form is today's replacement of the table form for vertical movement. While there is some disagreement on terminology, the flying form is in growing international use as the preferred term. Figure 4.26 illustrates the main points of construction. Steel or aluminium trusses, laterally braced, support the table top. An American publication[16], quotes table tops up to 186 m^2 as being possible.

Figure 4.25 Superior soffit appearance achieved with Cordek trough moulds. (Cordek Ltd).

Figure 4.26 Support falsework in aluminium, showing connection system (Scaffolding Great Britain Ltd).

Figure 4.27 Flying form being positioned. (Scaffolding Great Britain Ltd).

The actual support method for the flying form, when in place, is by telescopic legs supported on adjustable jacks. The jacks provide fine adjustment for level and release, while the telescopic legs lower the form onto a system of rollers to move the form out of the structure for crane slings to be attached. After continuous extraction by rolling out, further slings are attached for final removal from the building and re-location (Figure 4.27). Before lowering to a new position, the telescopic legs are withdrawn and reset at predetermined levels to lower the form as nearly level as possible, leaving the final adjustment to be carried out with the replaced screw jacks.

It goes without saying, that all such forms must have connections between components which are strong enough to withstand any strains from handling.

When considering the use of table forms or flying forms, consultations with specialist formwork and falsework suppliers is most desirable.

Formwork cost factors

Formwork cost per use is based on two elements: cost of making specially or the hire rate charged by a supplier (usually monthly) and the labour and plant costs for erecting and striking. To these main costs must be added what are normally called consumables – lost parts of wall ties, built in fixings for services or other items, including lift guides, trunking etc. Such consumables will apply to both wall and soffit forms.

If the forms for walls or soffits are specially made for a contract, it is usual to write off their cost on the contract. The price per use in the bill of quantities will therefore reflect the number of uses that can be obtained on the contract. This in turn relates to whether the formwork is used in walls or soffits.

Formwork in walls

With wall formwork, striking times can be very fast if the ambient temperatures are suitable. A British Cement Association publication[17] gives up to date views on fast striking times in walls and floors. With walls the times can be very fast in what might be classed as normal British weather. In summer, pours completed in late afternoon can be struck the following morning. Even in colder times, striking can be achieved by late afternoon the following day.

The number of uses possible in a given time can therefore be considerable – provided that the repetition factor is present. The cost per use in material terms becomes the making cost divided by the number of uses. To this has to be added the fix and strike costs per use for final insertion in the bill of quantities.

The situation is somewhat different if the formwork is hired-in proprietary equipment. Here a sum per month is the usual form of charge. In this case, therefore, it is the monthly hire charge per m^2 divided by the number of uses that can be obtained in a month that provides the material element in the bill rate.

Formwork in soffits

In the case of soffit formwork, the ability to obtain high use factors is not normally compa-

rable with wall formwork. The concrete has to obtain a much higher strength before striking than in the case of walls. Because of this, the forms have to remain in place much longer. In addition, to provide the progress necessary elsewhere, large areas of soffit forms are needed (Chapter 11). In many building situations, for example, a complete floor of soffit forms is necessary to allow the best progress overall in relation to the structural frame.

Amongst the proprietary types of floor formwork there are systems designed to allow early striking of the main components, with the propping system left behind to act as dead shoring until the concrete as a whole is of sufficient strength to be completely struck out.

The only other possibility for reducing form-

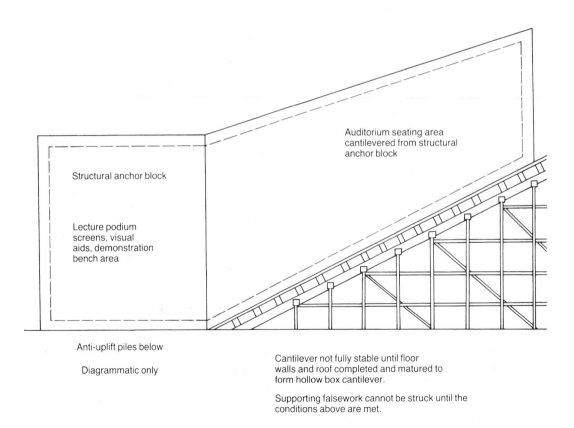

Auditorium seating area cantilevered from structural anchor block

Structural anchor block

Lecture podium screens, visual aids, demonstration bench area

Anti-uplift piles below

Diagrammatic only

Cantilever not fully stable until floor walls and roof completed and matured to form hollow box cantilever.

Supporting falsework cannot be struck until the conditions above are met.

Figure 4.28   Lecture theatre where falsework had to stay in place until walls and roof were concreted and matured.

78

work costs in floors is to consider approaching the structural engineer to change the cement specification to allow earlier striking of the formwork. That is: to be able to achieve the desired 28 day strength in considerably less time. The decision factor here is that the contractor will have to pay the extra cost of the cement. In so doing, will it provide cost savings elsewhere which more than offset the extra cement cost?

Falsework cost factors

With falsework, the ability to reduce cost is more limited than with formwork. For example, in the proprietary system described above, where the propping system is left in place while the soffit forms are released.

If, however, the cement type is changed to give high early strength, there is the ability to strike both forms and falsework at the same time in the right circumstances. For example:

Assume monthly hire rate for proprietary forms and falsework is £x per m².

If turnround of equipment is once per month (fix and strike) cost of formwork and falsework per use £x per m².

If, on the other hand, by changing the cement type we can achieve two uses per month the unit cost of the forms and their support becomes £x/2 per m² in terms of the cost of provision.

Labour costs for fix and strike will only alter to the extent that the more the repetition the better the labour force will become in relation to their output.

Whether this solution is a desirable one will depend on the method planning analysis for the structure as a whole.

Where non-hired equipment is used and the cost of provision is based on writing off on the contract, it will be the number of uses that determines the unit provision rate. Time, in assessing cost, does not come into it, except in the sense that the fix and strike performance still needs to be as efficient as possible.

Finally, in this section, it needs to be recognized that complex falsework in the designed category will normally be a one-off time which is related to all the activities that have to be performed above it. Such time will only be apparent when the overall programme for the structure has been determined. For example, a lecture theatre with sloping seating had been designed as cantilevering off heavy foundations at the lecturers end. The back end was therefore in mid-air with a void underneath. The falsework needed for the supporting floor also had to be capable of supporting concrete side walls and roof which, in the final state, partially support the stepped floor (Figure 4.28). In the event, the falsework had to remain in position for three months until all the concrete in the structure had reached the required strength.

Assessment of formwork and falsework quantities

The number of uses that can be achieved for formwork and falsework needed in a structure is a function of the quantities that have to be provided in the realization of the fastest and most cost effective construction method for the concrete works. The number of uses, therefore, depends as much on the methods adopted for the fixing of reinforcement and placing of concrete as on the items themselves. This whole aspect is dealt with in more detail in Chapter 11, where method selection for Reinforced Concrete Structures is considered.

Management control

Frequent mention has been made in relation to safety requirements when dealing with formwork and falsework. Today, when much of this type of activity is sub-let to subcontractors, there is a real need for the main or managing contractor to be competent to oversee the subcontractors' actions in both technical and safety aspects. In recent times it has become even more important where contractual requirements call for quality assurance systems from the main or managing contractor and his subcontractors. The references given in this section will give site managements a good introduction to formwork and falsework in the technical, safety and checking and supervision needs.

SUPPORT OF EXCAVATIONS

Very little new construction can take place without recourse to excavation. Some excavations may be very deep, while others could be called superficial. In between, probably the most prolific form of excavation, outside of bulk digging, is in trenches.

It is a regrettable fact of life that the collapse of excavations is a major source of accidents in the construction industry, many of which are fatal. Not surprisingly, excavation attracts a good deal of safety legislation in relation to those who manage such work and the operatives who do the work.

Legal requirements

The following Acts and Regulations have a direct bearing on the selection of methods and their operational use in practice.

1. Health and Safety at Work etc. Act 1974[1];
2. Construction (Health, Safety and Welfare Regulations 1996[2]. These require that any trench exceeding 1.2 m in depth must be adequately supported 'unless no fall or dislodgement of earth or other material so as to bury or trap a person employed or so as to strike a person employed from a height is liable to occur.' Even in rock excavations, it does not always follow that falls will not occur. Much depends on the strata and loose rock fragments left after excavation;
3. The Lifting Operations and Lifting Equipment Regulations 1998[13];
4. Control of Pollution Act 1974 – sections 60 and 61[18];

In addition to the above, other legislation relates to the consequences of excavation on the property of others.

1. Highways Act 1959[19] and, in London, the GLC General Powers Act 1966[20] deal, *inter alia*, with procedures in respect of excavations within 9 m of a highway.
2. *Halsbury's Laws of England* defines the natural right of support as 'giving every owner of land as an incident of his ownership the right to prevent such use of the neighbouring land as will withdraw the support which the neighbouring land naturally affords his land'. In this case, adjoining excavations must not affect adjacent buildings or withdraw support from the land on which they sit.
3. In such situations as in 2. above, agreements are usually made with adjoining owners, by the client's representatives, in which a schedule of requirements to protect the adjoining property during construction will be listed. Such agreements are usually known as party wall awards.

Clearly, all the above factors need to be taken into account when assessing and pricing temporary works needs at the tender stage. It is also a good policy to check with the local authority if they have any local by-laws which may apply to excavations and their support.

Standards and codes

There are no codes or standards which specifically deal with the support of excavations. BS 8004: 1986, Code of practice for Foundations [21] and BS 6031: 1981, Code of practice for Earthworks[22], both make a limited reference to the support of excavations. There are today, however, a significant number of publications which provide authoritative guidance on the safe support of excavations. There are noted as they arise in the text of this section, and listed in the references section at the end of the chapter.

Types of excavation

Bulk excavation

Bulk excavation can be described as the removal of high volumes of earth in excavating to grade for roadworks, railways, and site preparation for large industrial and commercial complexes on virgin ground. Within this area, one also needs to include open cast mining.

In all cases high capacity excavation machinery and transport is involved – excavators, scrapers, dump trucks and so on. The techniques involved for least cost operation are often complicated and do not fit readily into this text, so, apart from occasional mention in passing, this particular aspect of excavation is not dealt with here.

The excavation being covered here divides into two distinct types: trenches and wide excavations. Each type will be considered separately.

Trench excavation

Trench excavation is defined in *Timber in Excavations*[23] as 'an excavation whose length greatly exceeds its width. It may have vertical sides, which usually require strutting from side to side, or battered sides requiring no support. A trench is generally accepted as not exceeding 5 m in width.'

Trench excavation is primarily associated with the installation of the many service pipes and cables involved in modern construction of all kinds. It can also arise in connection with foundation construction (Chapter 8). As such, trench excavation is the most common form of digging requiring support to provide safe working conditions for those laying pipes and cables within the trench. It is also the most likely cause of accidents due to failure to provide adequate support.

Within this same definition one should include pier holes and small shafts. Pier holes regularly arise in foundation excavation, while small shafts arise in excavating manholes to go with trench digging, and will usually be an enlargement locally of the trench excavation.

Wide excavations

A wide excavation has been defined in *Timber in Excavations*[23] as: 'an excavation whose width exceeds 5 m. In plan its shape can be infinitely variable from an extra wide trench in open ground to an irregular shape defined by adjacent buildings as in city centre developments.'

It may be possible to support wide excavations by strutting from side to side, but single face support may be the only economically sensible approach. If the support of adjoining property is involved even greater care in providing the support system is needed. Many such wide excavations arise in deep basement construction for industrial and commercial developments. They are very often close to adjacent structures or roads.

Basic types of support

Excavations can be supported in three basic ways: double sided, single sided with raking support and simple or modified cantilever. Each method is described in turn, with its appropriate terminology and role in the support situation.

Double sided support

Figure 4.29 illustrates the key features. The support structure maintains the *status quo* between the exposed excavated faces. The sheeting on both sides collects the earth pressure on the face and transfers it *via* the walings to struts which have equal and opposite forces at each end and so maintain the equilibrium of the system. With this method it should be noted that all forces involved are horizontal.

Single sided with raking support

In this case, two types of raker can be used: raking struts inside the excavate area (Figure 4.30),

Double-sided support

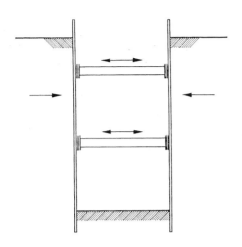

All forces horizontal

Figure 4.29 Double sided support to trenches. All forces horizontal.

or ground anchors which are installed outside the line of support (Figure 4.31).

In the first case, the sheeting method collects the earth pressure and transfers it to the walings. The walings, in turn, transfer the loading to the raking struts. Finally, the forces in the rakers have to be suitably supported by foundations which will safely distribute the loads to the ground. In this method a number of key points must be understood.

- The forces in the rakers are greater than the horizontal force from the earth pressure.
- The raking members introduce an upward vertical force into the support system as a whole.
- The whole support system will fail if the foundation block is inadequately designed in relation to the bearing capacity of the ground.
- Stability of the whole system also depends on adequate toeing in of the sheeting to resist the upward thrust induced by the vertical component of the thrust in the raking members.

Single-sided — with raking support

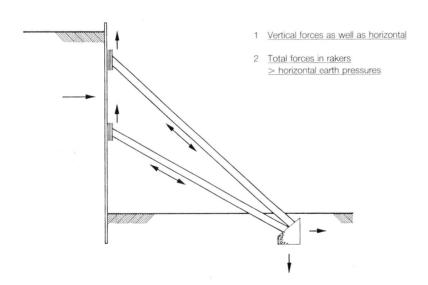

1 Vertical forces as well as horizontal

2 Total forces in rakers
   > horizontal earth pressures

Figure 4.30 Raking shore support internally. Forces induced. Note uplift present.

In the second case, the earth pressure is again collected from the sheeting system, *via* the walings, but now transferred by ground anchors into the ground behind the sheeting. The key points in this method are:

- The forces in the ground anchors are greater than the horizontal force from the earth pressure.
- The raking anchors introduce a downward force into the support system as a whole.
- The support system as a whole will fail if the ground anchors are inadequately designed and installed.
- Stability of the whole also depends on the ground conditions at the bottom of the face support being adequate to resist the vertical downward force induced by the ground anchors.

Cantilever or modified cantilever support

Figures 4.32 (a) and (b) show the essential features of this method. A sheeting material that has a suitable section modulus is required in this approach. That is, the material must have an inherent stiffness against bending. Steel sheet piling and its derivatives are the materials normally used in solutions of this type. While cantilever methods can be used as a support it is necessary to recognize the limitations they impose.

1. All cantilevers deflect under load unless they are hopelessly uneconomic in cross section.

CANTILEVER SYSTEMS

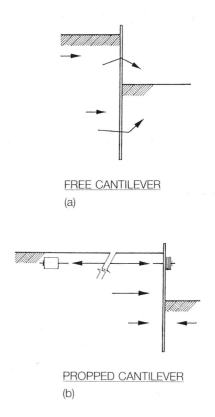

FREE CANTILEVER

(a)

PROPPED CANTILEVER

(b)

Single-sided support

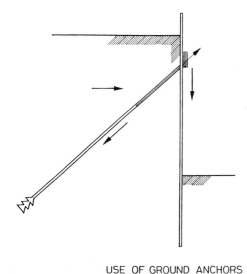

USE OF GROUND ANCHORS

Figure 4.31 Ground anchors externally. Forces induced. Note downward force on system.

Figure 4.32 (a) Essential features of cantilever support (b) Propped cantilevers.

In consequence, pure cantilevers can only be used in locations where there is no risk of settlement behind the sheeting endangering the stability of adjoining buildings or other installations. In practice, this means the choice for use is limited to open areas where a small degree of deflection, causing settlement behind the sheeting, would cause no problems.

2. The problems in (1) above can be overcome if the propped cantilever concept is adopted as shown in Figure 4.32 (b). For success, it is essential that suitable ground conditions exist for an adequate anchorage block at an equally suitable distance behind the sheeting support line.

Standard and designed solutions

As previously mentioned, the support of excavations can be divided into methods needing either standard solutions or methods calling for purpose design by those competent to do so.

Re-examination of the three basic types of support makes it clear that only double sided supports, together with shallow pits and shafts, are suitable for the use of standard solutions. All loads are horizontal and a *status quo* situation exists. With raking supports, vertical loads are introduced and knowledge of ground conditions and soil strengths are essential together with the necessary design knowledge if safe solutions are to be achieved. In the case of the use of ground anchors, their installation and

---

Limitations of standard solutions in the support of excavations

There is general agreement that the following limitations on the use of standard solutions should apply in the support of excavations.

Criteria
Standard solutions should only be used for the support of excavations:
(a) where open cut not exceeding 6 m in depth is feasible;
(b) with double sided support, in non-water-bearing ground, to an excavation not exceeding 6 m deep;
(c) in shallow pits not exceeding 6 m deep;
(d) in water bearing ground, where water problems have been eliminated by other means (e.g. well-pointing), within the limitations of (b) and (c).

Procedures
In the above situations the procedures below should be followed.
1. When deciding the batter of an open cut excavation, proper account must be taken of the material and its characteristics and the safe slopes recommended in either *Trenching practice*[24] or *Timber in excavations*[23].
2. Do not assume that excavation in rock is necessarily stable. Look for sloping strata, fissures and loose material after blasting. Support unless absolutely sure that the material is stable.
3. Supervision should make sure that persons erecting or removing supports have been adequately instructed on the methods to be used.
4. Where proprietary methods are being used, the procedures must be strictly in accordance with the manufacturers' instructions.
5. All relevant legislation must be complied with.

design is a specialist situation, requiring specialist equipment and knowledge. The same sort of knowledge is also needed where cantilevers are involved.

Standard solutions

Criteria for the use of standard solutions need spelling out if accidents are to be avoided. At the same time, standard solutions need to be limited in relation to the depth to which they can be used. The recommended limitations on the use of standard solutions, whether by in-company tables or those provided by the makers of proprietary support methods, are given in the box above. Note that the depth limitation is kept to 6 m.

Two publications provide valuable data for standard solution assessment. *Trenching Practice*[24] specifically relates to trenching support and covers tables for the support needed for various types of ground, the materials needed, good practice in installation and backfilling and removal of the suporting material. In *Timber in Excavations*[23], standard solutions are covered in much the same way as *Trenching Practice*. Additionally, however, shafts and pits and small headings are dealt with, together with a worked example and construction details of the most widely used support method for wide excavations. While it has already been emphasized that wide single sided excavations need design of the support by competent persons, the worked example referred to above will allow the method planner to get a close assessment of the material content needed in a situation where such an approach is required, at the tender stage.

Proprietary systems

In past times, accidents in trenches frequently occurred while a support method was being installed. The Health and Safety at Work etc. Act makes it clear that a safe place of work and access to it must be provided by the employer. In response to this need proprietry methods have been increasingly developed in recent years. The available systems fall into a number of well defined groups:

- aluminium walings with hydraulic struts built into an assembly;
- shields or boxes;
- box or plate supports;
- special methods.

Aluminium walings and hydraulic struts

The purpose of this system is to allow the waling/strut assembly to be lowered into an excavated trench in which vertical timbers or steel trench sheeting have already been placed and then jacked out to the sheeting to force it tightly against the excavated faces. At no time is it necessary for anyone to enter the trench until the support is fully in place (Figure 4.33).

Figure 4.33 Power shore system Hydraulic water pressure. Aluminium walings. (Mabey Ltd).

Figure 4.34   Super shaft brace. Long span rectangular sections. (Mabey Ltd).

A variation is used for strutting large manholes or pits (Figure 4.34).

Those who supply such equipment provide tables setting out the conditions of use in a variety of ground conditions. It is important that the supplier's conditions are always followed.

Shields or drag boxes

Strictly speaking shields or drag boxes are not support methods at all (Figure 4.35). They are designed to be pulled along a trench, once excavated, by the digger to protect those working in the trench from falls of earth. As such, they do not push out to hold the sides of the excavation.

Figure 4.35   Drag box support (Mabey Ltd).

The box formation is rigid and less in width than the trench. In use it is important that the excavator has sufficient power to pull the box along. Note the cutting edge in the illustration to assist forward movement.

Box and plate lining systems

These types should not be confused with the previous method. They are designed for trench support as well as protection of operatives. Figure 4.36 illustrates the box method. The term 'box' here refers to strutted excavation support walls of a modular nature, which can be joined together vertically and ... s rigid in

... em starts ... rtical sol- ... t against ... ticals are ... down in ... 4.37). As ... s can be ... ard from

... y services ... lem as the

... ire. Their ... to mini- ... l, coupled ... possible. ... nanced by ... nting sys- ... lay pipes ... ack filling ... methods, ... rt lengths ... backfilled ... be moved ... re rate is ... e uses one ... e per unit

Figure 4.36 Box support method. Note box section being pushed into the ground by the digger bucket (Mabey Ltd).

Planning for the use of proprietary systems

To assist in the process of choosing which proprietary method would be the best in a given situation, the Construction Industry Research and Information Association has produced Technical Note No 95[25], now in its third edition, which lists all the proprietary methods currently on the market with their capabilities and specifications, together with the makers' names, addresses and telephone numbers. Once the best specification has been found for the situation in question, the supplier can be contacted for prices/hire rates for the quantity required.

Only the hydraulic waling systems lend themselves to installation between crossing services in

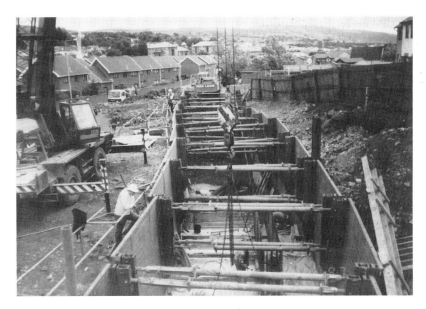

Figure 4.37 Super trench box (Mabey Ltd).

the above methods. Where services are complex and present serious difficulties for safe support of excavations through which they cross it is better to use and allow for one of the designed solutions which follow. The advantages are fully described in the text and illustrated in Figure 4.40. With services crossing an excavation it is better to be safe than sorry.

Designed solutions

As previously stated, only double sided support up to 6 m in depth, together with shallow pits and bases up to the same depth are suitable for standard solution application. Support involving rakers or ground anchors, together with cantilever systems require specific design by persons competent to do so.

Two principles of support for wide excavations form by far the most common solutions used in this context. They are (a) what is usually known as the soldier or H-pile method and (b) the use of interlocking steel sheet piling. In broad terms, each approach has its own use situation. H-piling is related to excavation support where free water is not present – or water pressure has been removed by the installation of

wellpointing. Sheet steel piling, having interlock connections between sheets, is mainly used in wet ground conditions or in free water conditions as in the case of cofferdams for bridge piers actually to be built in the waters of a river or the sea.

H-piling (soldier piling)

Figure 4.38 illustrates the main principles of the H-piling method. Universal steel column sections are pre-driven into the ground to a suitable depth. Once in place, excavation can be commenced. As the face to be supported is exposed, horizontal timbers are inserted behind the face flanges of the H-piles and wedged into place. Figure 4.39 illustrates the way in which this is done. To achieve this situation, a small degree of excavation between the H-piles is needed to accept the horizontal timber. The timbers themselves are positioned downwards, that is one under the other as excavation proceeds.

The crucial advantages of the method are:

1. The main support structure is in place before any excavation is started.

Figure 4.38 H-pile method of excavation support. Driving with drop hammer on hanging leaders.

2. The horizontal timbering can be positioned to fit the ground conditions met at the time, i.e. close timbering can be reduced if the ground is found to be stiffer than expected, to hit and miss or even wider spacing.
3. There is a retrieval capability in the method if the timbering is seen to be inadequate by only using a spacing of hit and miss or even wider spaces. Extra timbers can be installed in the gaps without difficulty.
4. Services crossing the excavation present no problems at all. The H-piles are driven between the services and the horizontal timbers inserted above and below the service in question (Figure 4.40).
5. With pre-driven piles, the ground has an arching effect between the piles (Figure 4.41). In consequence, the timber thickness that has to be used is less than that which would be calculated for a uniformly distributed earth pressure.

The system is suitable for wide or deep trenches, single sided excavations supported by raking struts or ground anchors, but is not suitable for any form of cantilever support unless conditions exist for a propped cantilever solution. The need for individual design by competent persons will be only too obvious.

World wide, this method is the most commonly used for the support of major excavations where free ground water is not involved.

The H-pile approach is capable of great versatility. The book *Temporary Works – their role in construction*[8] examines this aspect in some detail and contains case studies of use in actual contract conditions. As previously mentioned earlier in this section, the publication *Timber in Excavations*[23], gives a fully worked design example of the design of H-pile which allows the likely cost of this method of support at the tender stage by the construction planner. It does emphasize, though, that the final operational

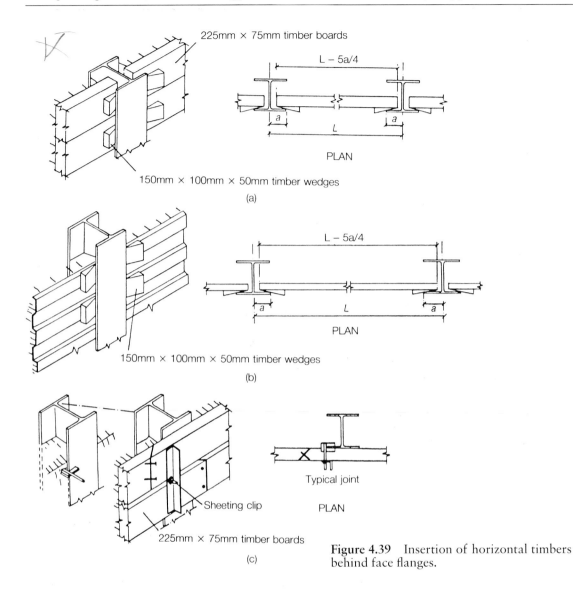

225mm × 75mm timber boards

L − 5a/4

a

L

PLAN

150mm × 100mm × 50mm timber wedges

(a)

L − 5a/4

a

L

a

PLAN

150mm × 100mm × 50mm timber wedges

(b)

Sheeting clip

225mm × 75mm timber boards

(c)

Typical joint

PLAN

Figure 4.39 Insertion of horizontal timbers behind face flanges.

design must be carried out by a competent person.

When considering use with ground anchor support, it must be remembered that if the anchors are installed outside the site boundary, it will constitute a trespass of adjoining property unless agreement has previously been achieved with all adjoining owners. Statutory undertakers also need contracting to see if they may impose minimum distances from their plant or mains of any ground anchors.

Steel sheet piling

Steel sheet piling's main advantage is its ability to retain ground with free water conditions or keep out water while building bridge piers in rivers or tidal estuaries or like situations. Such

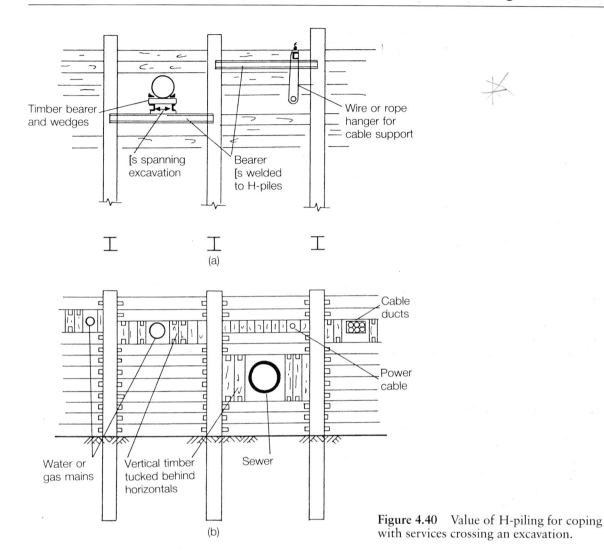

Timber bearer
and wedges

[s spanning
excavation

Bearer
[s welded
to H-piles

Wire or rope
hanger for
cable support

(a)

Cable
ducts

Power
cable

Water or
gas mains

Vertical timber
tucked behind
horizontals

Sewer

(b)

Figure 4.40 Value of H-piling for coping with services crossing an excavation.

piling is designed with interlocking sides, but it is unwise to assume that the interlocks are always watertight. Expensive caulking may be needed in some sections to provide the truly watertight conditions the situation demands.

Figure 4.42 illustrates the type of interlock frequently used and also shows the way in which steel sheet piling provides a good section modulus against bending. Because of this it is the only really satisfactory material for use in cantilever support methods.

Sheet piling also has an important role to

play where a water-bearing strata overlays an impervious one. By arranging penetration into the impervious strata water can be cut off from penetrating the impervious layer below (Figure 4.43).

Against these good qualities, there are a number of significant disadvantages.

1. The cost of driving is high for the area covered. The whole supported area has to be driven, unlike H-piling where the driven material is only the H-piles themselves.

Figure 4.41 Arching effect between H-piles.

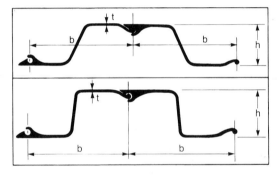

Figure 4.42 Sheet piling interlocks.

2. In general, the driving is a noisy process, though the makers of driving hammers are making strenuous efforts to cut down the level of noise to an acceptable level. This is being achieved by the use of what is known as 'impulse' or 'hydraulic' hammers. Indeed the makers claim, today, that with these new hammers, noise levels are within the limits normally imposed by local authorities under their authority under the Control of Pollution Act[18].

3. Sheet piling is much more expensive than the materials used in H-piling but is capable of many recoveries and reuse if properly installed and maintained.

4. The driving and extraction plant is more expensive than for H-piling.

In today's world, H-piling is by far the most used major support system. It is only in wet conditions that steel sheet piling really comes into its own. It also goes without saying that sheet piling must be a designed solution by competent persons. Further information on driving plant is given in *Temporary Works – their role in construction*[8].

Choice of designed method

Where designed solutions are clearly necessary, the use of H-piling will be the most economical by a considerable margin even if it is necessary to leave it in place. Sheet piling is an expensive alternative and only comes into its own where the other method cannot cope with the water situation directly, or where the cost of well-pointing plus H-piling can be shown to be greater than the sheet piling answer.

In Figure 4.44 the sheet piling shown was

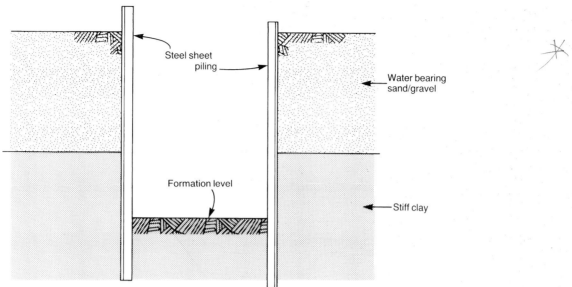

Notes:
1. Sheet piling driven into impermeable strata
2. If leaks are seen at clutches limited cauking may be needed
3. Water pressure behind piling must be allowed for in strutting calculations

Figure 4.43 Cut off effect with sheet piling.

Figure 4.44
Sheet piling is not always watertight, as explained in the text.

installed as a measured item to be left in place with a concrete retaining wall built in front of it. The idea was to improve the certainty of the wall not allowing any damp penetration. In the event, water seepage proved to be a problem and cost a great deal of money to eradicate. The cause was a two part matter, the sheet piles were not watertight, while the need to leave the raking shores in place until the concrete wall had been built required small areas of wall to be filled in at a later stage. The filling of these small areas so that the patch was watertight proved to be a very difficult task and one that involved a good deal of extra cost.

References

1. *Health and Safety at Work etc. Act 1974.* HMSO, London.
2. *Construction (Health, Safety and Welfare) Regulations 1996.* HMSO, London.
3. British Standards Institution. *Code of practice for scaffolding.* BS 5973 : 1993. BSI, London.
4. British Standards Institution. *Specification for prefabricated access and working towers.* BS 1139 : 1983: part 3, BSI London.
5. British Standards Institution. *Code of practice for temporarily installed suspended scaffolds and access equipment.* BS 5974 : 1990. BSI, London.
6. Construction Confederation and the National Access and Scaffolding Confederation. *Model conditions of contract for the hire, erection and dismantling of scaffolding.* NASC, London 1986.
7. Brand, R.E. (1975) *Falsework and access scaffolds in tubular steel.* McGraw Hill, Maidenhead.
8. Illingworth, J.R. (1987) *Temporary Works – their role in construction.* Thomas Telford, London.
9. British Standards Institution. *Glossary of building and civil engineering terms, Section 6.5 Formwork.* BS 6100 : 1987. BSI, London.
10. Concrete Society and The Institution of Structural Engineers (1986). *Formwork – a guide to good practice.* Concrete Society, London.
11. British Standards Institution. *Code of practice for falsework.* BS 5975 : 1994 BSI, London.
12. *Factories Act 1961.* HMSO, London.
13. *The Lifting Operations and Lifting Equipment Regulations 1998.* HMSO, London.
14. Health and Safety Executive. *Safety in falsework for in-situ beams and slabs.* Guidance Note HS/G 32. HMSO, London.
15. Harrison, T.A. and Clear, C. (1985) *Concrete pressures on formwork.* Construction Industry Research and Information Association Report No 108. CIRIA, London.
16. Hurd, M.K. (1979) *Formwork for concrete.* American Concrete Institute, Detroit.
17. Harrison, T.A. (1977) *The application of accelerated curing to apartment formwork systems* British Cement Association.
18. *Control of Pollution Act 1974.* Sections 60 & 61. HMSO, London.
19. *Highways Act 1959*; HMSO, London.
20. *GLC (General Powers) Act 1966.* HMSO, London.
21. British Standards Institution. *Code of practice for foundations.* BS 8004 : 1986. BSI, London.
22. British Standards Institution. *Code of practice for earthworks.* BS 6031 : 1986. BSI, London.
23. Timber Research and Development Association, (1990) *Timber in excavations*, 3rd edition. Thomas Telford, London.
24. Irvine, D.J. and Smith, R.J.H. (1983) *Trenching practice.* Report No 97. Construction Industry Research and Information Association, London.
25. Mackay E.B. (1986) *Proprietary trench support systems.* Technical Note No 95, 3rd edition. Construction Industry Research and Information Association, London.

Chapter five

Influence of design on construction cost and buildability

It is a pre-requisite for construction method planners and site management, that they should have a proper appreciation of the influence of design on construction cost, not only in relation to the actual design itself, but the detailing of items and the influence of contract specifications.

To achieve this desirable situation, four areas of knowledge are needed.

1. The breakdown of the contract sum into the major construction components.
2. Where a number of independent items combine to produce a main construction component, it is useful to know the cost breakdown of the individual items that make up the whole. For example, in a concrete structure, the cost elements for formwork, reinforcement and the concrete.
3. A technical appreciation of how design detailing and the specification may affect construction cost for all key elements billed.
4. A knowledge of building details and up to date methods of construction, and how such matters can affect the speed and quality of construction.

Each of these four areas of knowledge can now be examined in more detail.

Where the money is

The breakdown of construction cost into the main construction elements can be arrived at in a number of ways.

1. Site management can take the priced bill of quantities and individually extract the single or bulk item costs required. Once done it is better to convert these sums into percentages of the contract whole. The information is then independent to a great extent of inflationary factors, and so has future use potential. From these figures, management is made aware of the location of the main sums of money in measured work. With this information, the main areas for management supervision become clear.
2. Post contract analysis of costs and total costs will allow percentage breakdowns to compare with the tender assessments and provide more accurate data for use in the future, especially where the information provided at tender stage is slim.

95

3. From (2) above, it should also be possible to extract the man hours associated with the major items. Of considerable value for pretender planning.

Figure 5.1(a) published by quantity surveyors Davis Langdon & Everest illustrates the percentage cost breakdown for six main construction activities: substructure, superstructure, internal finishes, fitting, services and external works. These percentages are shown comparatively for eight different building types. Such information gives an immediate feel of how the contract sum is divided and where the method

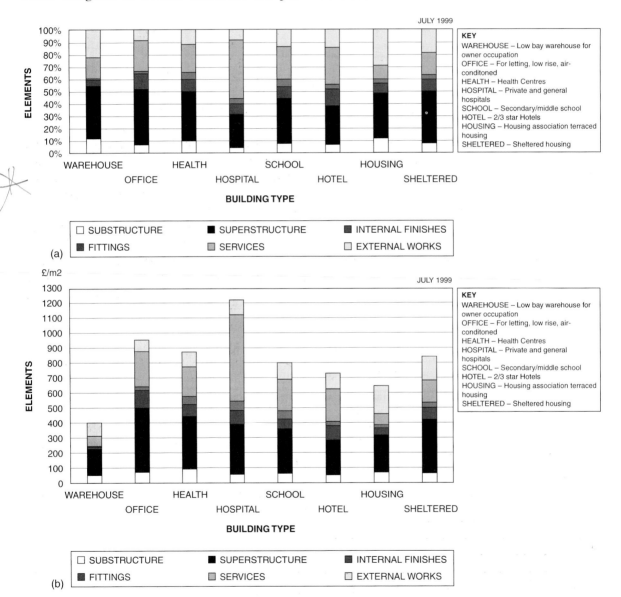

Figure 5.1 Construction cost breakdown for various building types. Statistics provided by Davis Langdon & Everest.

Table 5.1 Cost breakdown for the reinforced concrete frame of a fairly typical office building, six storeys in height; the figures include cost of materials and are expressed as a percentage of total cost.

	Columns		Beams		Slabs		Walls		Overall
Formwork	4.5 (24.4)	+	8.5 (33.0)	+	18.5 (47.7)	+	8 (53.1)	=	39.5
Reinforcement	12 (60.9)	+	14.5 (56.3)	+	3 (7.1)	+	3.5 (23.8)	=	33.0
Concrete	3 (14.7)	+	3 (10.7)	+	18 (45.2)	+	3.5 (23.1)	=	27.5
	(100.0)		(100.0)		(100.0)		(100.0)		(100.0)

planning needs to concentrate most – on areas of greatest percentage cost.

Figure 5.1(b), by contrast, illustrates the cost per m^2 for the same building types, based on July 1999 prices. While such figures change all the time, the reader can obtain a positive feel of the relative costs per m^2 of widely varying building types.

Table 5.1 is based on the same method of presentation but provides a more detailed breakdown for the cost components in a concrete superstructure. In this case, a six storey high quality city office block. The figures were obtained from the final bill of quantities.

Study of these figures provides a vivid insight into where the money is in a reinforced concrete frame. To provide even greater analysis, the figures are given in two ways: those in brackets show the percentage proportion of cost in the make up of one construction element. For example: taking the first component, columns, we see that, of its total cost, reinforcement makes up 60.9%, formwork 24.4% and the concrete the remaining 14.7%. With the second set of figures, by comparison, take a particular item, say concrete, and show the percentage cost of the concrete in each of the structural elements and add up the concrete cost as a percentage of the whole structure. Thus, each of the three components – formwork, reinforcement and concrete – add up to 100%. With more detailed breakdowns of this type, the fig-

ures given are not necessarily common to all concrete structures, which can vary considerably depending on the design concept. Nevertheless, as experience grows, records of breakdown for particular structural concepts will gradually provide a sound average for a particular structural design concept.

In planning terms, such information can be extremely useful at the pre-tender planning stage where drawing information is very incomplete, but it is known that the structural form is based on a column, beam, and floor concept with structural walls providing lateral bracing. Again, with experience, the method planner can associate the above with the tender breakdown of man hours per element abstracted from the original bill breakdown into materials, plant and labour.

Taking things a stage further, the value of this second method is that one can see at a glance where the main elements of cost for the structure lie. These pointers provide information regarding items needing the most consideration in relation to method planning and, later, management control.

Influence of design on construction cost

The contractor's ability to operate efficiently and competitively is directly affected by the design concept and the resultant detailing. To be successful, those responsible for construction method planning need to have a good grasp of design methods and detailing, in general terms, so that, in the process of deciding the methods to be adopted, they can become aware of any design details which could be the cause of problems at a later stage. By bringing such matters to light with the designer, the contractor can help to avoid errors of good practice which may become the subject of acrimonious argument later in the contract. Such action is an important part of any contract which has quality assurance requirements built in to it.

An additional factor in this area is the ability of the contractor to bring to the designer's notice any difficulties of execution and suggest alternative ways which could speed up the contract, improve quality and give the client better value for money. This latter aspect is growing in importance in management contracts and managing contracting. In either case, the main contractor or management contractor is usually employed at an early stage to advise on tailoring the design and detailing to provide, not only cost savings but also quicker construction to the quality desired.

Problems in design and detailing, which may subsequently cause poor construction to result, tend to arise in non-repetitive construction – usually in the industrial and commercial field. All such contracts are usually one-off situations and lessons learned from previous contracts are not necessarily relevant. From the contractors point of view, it is desirable to establish *aide memoire* check lists giving examples of what to look for, either at the tender stage or when working with the design team in the management role. Examples of suitable check lists are given later in this chapter.

Influence of the specification

Specifications play an important role in determining construction cost. Though mentioned in a previous chapter, they require further com-

ment here in relation to their influence on construction cost.

Many specifications used by the professional team are standard pre-printed documents which are used from job to job. As such, problems often arise as to whether their interpretation properly fits the specific contract. Experience on previous contracts with a particular architect or engineer will give a lead as to whether clauses are recognized as dated but not yet brought up to date or whether they are intended to be strictly enforced. If no experience exists, the tender construction planner will need to visit the architect's or engineer's office and ask for answers to any areas of the specification which do not relate to current established thinking. Alterations agreed will need to be confirmed in writing after the meeting.

In financial terms, the implications of clauses can have a marked effect on the cost of specific operations. For example:

- The sequence of operations stated may not be the most cost effective to the contractor.
- The requirement that industrial premises' floors are to be concreted checker board fashion is expensive and no longer recognized as the best practice (see Chapter 7 for greater detail on this).
- The potential re-use of formwork can be greatly limited if the specification does not use modern quick strike principles. Hence for a given rate of progress more formwork would be needed at extra cost (see Chapter 4, for greater detail).
- Requirements in relation to partial handover may restrict access or limit the type of plant that can be used. Such requirements may be needed in speed of completion by the client, but invariably will result in increased cost.

In Part Two, more detail will be given on the influence of the specification in construction method decisions.

Buildability

Over recent years, the term 'buildability' has become associated with the influence of design on construction cost and practicability of execution. Buildability as an area for discussion came to the fore in 1979/80 when the Construction Industry Research and Information Association decided to approach building contractors who were members of the Association, to find out what they regarded as the main problems of building practice. As a result, it was found that frequent reference was made to buildability and its problems. While no one had yet invented a definition, it was implied that building designers were not allowing clients to obtain the best possible value for money in terms of the efficiency with which the building work was carried out. This was seen as largely due to the separation of the design and construction functions.

About the same time as the CIRIA research was going on, the University of Reading prepared a study for the Royal Institute of Chartered Surveyors entitled *UK and US Construction Industries – A Comparison of Design and Construct Procedures*[1]. In this study it was stated:

> 4.52. Obviously, detail design decisions have a very high impact on cost and time. In the USA contractors have an important influence at the design stage because they have the knowledge of construction methods, actual costs and the value of time. The orthodox UK system prevents this involvement.
> 4.53. The message is clear. Detail design cannot be divorced from construction without major cost and time penalties.

The growing interest in 'buildability' was given further stimulus by a conference organized by the National Federation of Building Trades Employers (now the Building Employers Confederation) held in November 1981 entitled *Value for Money in Building – Can we learn from North America*?[2]. At this conference, 'buildability' was regarded as a key factor in better US performance than in the UK.

Continuing discussion triggered a second conference on the topic of 'Can we halve our building costs?' held in June 1982[3]. In a paper by Williams[4], the first definition of buildability was put forward: 'By buildability I mean the understanding of the most economic and efficient way of putting a building together'.

In 1983 the CIRIA work culminated in a report entitled *Buildability – an Assessment*[5]. In this document buildability was defined somewhat differently from that of Williams: 'Buildability is the extent to which the design of a building facilitates ease of construction, subject to the overall requirements for the completed building'. A key statement in the CIRIA report was: 'Good buildability leads to major cost benefits for clients, designers and builders'.

In late 1983, this author won a Chartered Institute of Building competition with a paper entitled *Buildability – tomorrow's need*[6]. In this paper, a further definition was put forward designed to relate to the definition of the construction process now put forward at the beginning of this book: 'By buildability is meant design and detailing which recognize the problems of the assembly process in achieving the desired result safely and at least cost to the client'.

While a detailed study of buildability is not possible here, its importance in relation to achieving cost benefits to the client by making the erection process as simple as possible, should at least be clear. Those who wish to become more conversant with the subject will find the references given at the end of this chapter of considerable interest.

It should also be apparent that construction method planners can play an important part in achieving good buildability, both in management types of contract and the traditional types. To be effective, however, the persons concerned need to build up a thorough knowledge of the construction process. There are no short cuts to this knowledge; experience is all.

More recently, the advent of The Construction (Design and Management) Regulations 1994[7] have refocused attention on the subject. These regulations require the integration of all parties – client, professional team, main contractor, subcontractors and any self employed labour – to work together to create the best possible health and safety conditions on site. The need for close relation should also open the door to more communication in relation to buildability, with subsequent benefit to the client.

Examples of the influence of buildability

Case 1

Figure 5.2 illustrates the cross-section of the edge beam to an office building, occurring on all floors and covering the complete perimeter. The original drawings showed the construction to be in in-situ concrete, as was the rest of the structure.

When considering the problems this section would create in relation to the formwork, the temporary works engineer concerned recognized the problems of getting the concrete into the forms and properly compacted. The cast-in inserts, in particular, not only helped to block the available concreting access, but would also make the striking of the formwork risky, in the sense that the section was very thin. Unless the concrete was very mature the risk of pulling the bottom section away from the top was very real, especially if the inserts had been too solidly fixed to the formwork prior to pouring.

As these edge beams were the fixing points for the building fenestration, great accuracy in floor to floor dimensions was essential, as well as line and level. The contractor could see great difficulty in achieving these requirements from a quality point of view as well as the time element

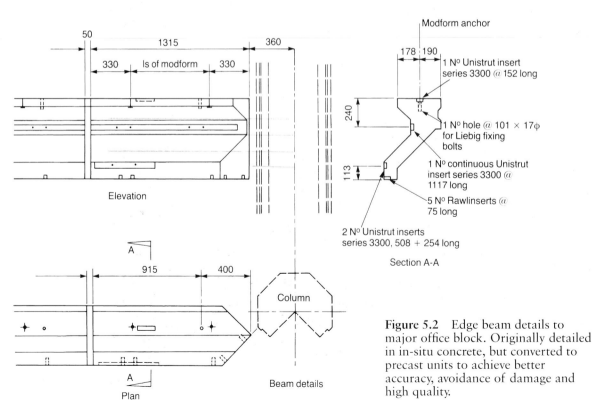

Figure 5.2 Edge beam details to major office block. Originally detailed in in-situ concrete, but converted to precast units to achieve better accuracy, avoidance of damage and high quality.

delaying the rest of the structure. The consulting engineer was approached and the problems discussed. The contractor put forward the concept of pre-casting the edge beams between columns with an in-situ connection at the columns in each case. In so doing, the edge beams could be supported on the soffit formwork and accurately lined and levelled before any concrete was poured. This approach was accepted by the engineer and the design altered. The result was a great success. Line and level achieved were excellent and subsequently the window installation went without a hitch. Figure 5.3 shows the beams in place before concreting.

While the quantity surveyor will assert that precast concrete is more expensive than in-situ, this example of buildability illustrates the need to compare costs beyond the straightforward direct comparison. An in-situ solution would have presented great problems and delayed the rest of the structural completion, thus increasing cost. Quality was guaranteed and no delays in window fixing arose – again creating financial savings. In all, this case was a good example of the value, at times, of spending money to save money elsewhere.

Case 2

In the toilet areas of an office building, the finishing details were designed in such a way that

Figure 5.3 Pre-cast edge beam units, as in Figure 5.2, prior to concreting of floor. Note also column forms specially made in steel.

the tiler had to make four separate visits to integrate with other trades. This was very costly when one considered that there were four toilet units on every floor and the building was 28 storeys in height.

Fragmenting any trade activity lowers productivity and increases cost. With very little effort it was possible to change the detailing so that the tiler could complete his work in one visit.

Case 3

Large area floors, such as one meets in warehousing or production areas, were at one time specified to be poured in checker board fashion (Figure 5.4). The idea was that working in this way would allow shrinkage to take place before the adjoining concrete was placed. The effect of this concept was to make access to the bay to be poured very difficult, especially with the introduction of ready mixed concrete as the normal means of supply. To avoid running over relatively green concrete, it was usually necessary to have a secondary handling method to reach over completed bays when depositing concrete in the 'miss' areas. All of which increased cost.

Continuing research on this type of floor

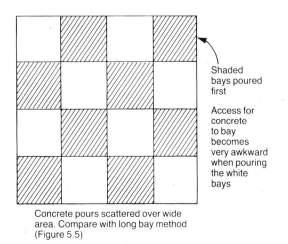

Shaded bays poured first

Access for concrete to bay becomes very awkward when pouring the white bays

Concrete pours scattered over wide area. Compare with long bay method (Figure 5.5)

Figure 5.4   Chequer board pattern of floor bays.

showed that floor construction in long continuous bays was perfectly possible, provided crack inducers at the bottom of the slab were provided at specified intervals and, when the concrete had reached a suitable strength, the bays were saw cut across the width over the location of the crack inducers. The adoption of this new approach resulted in quicker construction with cost reductions and a major improvement in level quality of the floor. The saw cut joints made the levels on both sides of the 'joint' the same – often not the case when checker board methods were used. Three publications [8][9][10] are available from the British Cement Association covering the design, good practice and construction method for such floors.

This particular case study is also an excellent example of the way in which improved buildability can create further benefits in relation to the planning of works of this type. In factory and storage buildings the external walls usually have the main workload in relation to finishes and services – for example, sheeting externally with insulation and internal lining inside. These items also associate with service feeders within or along the external walls. By using long bay construction and concreting the bays immediately adjacent to the external walls first (Figure 5.5), work can start at the earliest possible moment on the cladding, lining and service carcass. In other words a much bigger overlap in programme items is possible than in the case of checker board pouring.

Case 4

Those who detail reinforcement often have no practical experience of the problems of its assembly on site. Figure 5.6 is a case in point. Open links make all the difference between a time consuming assembly compared to the closed links detailed. This particular illustration is published by permission from a booklet published by the Concrete Society many years ago[11]. Recently re-published, the message put over is as valid today as it ever was.

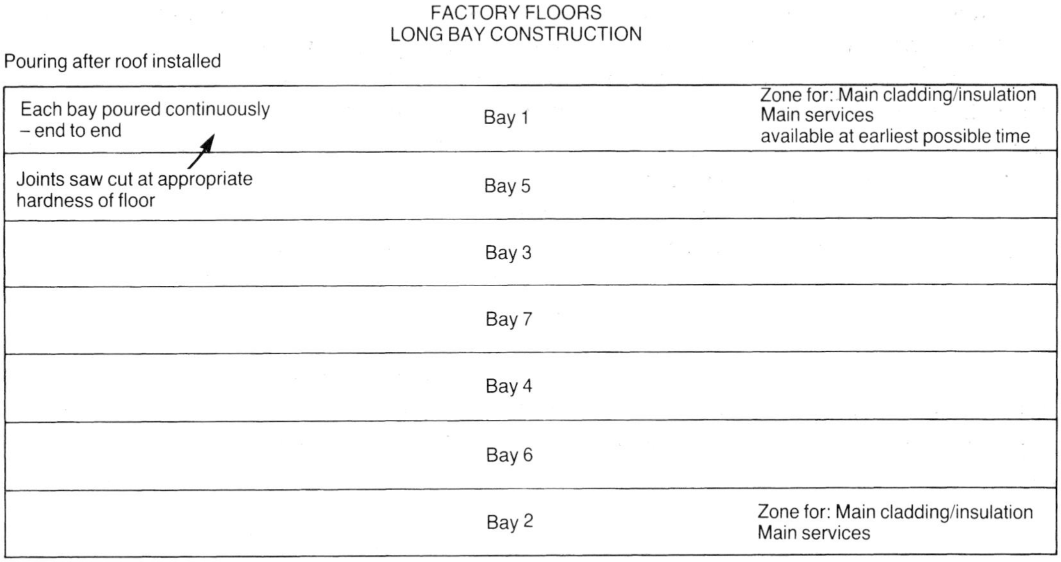

FACTORY FLOORS
LONG BAY CONSTRUCTION

Pouring after roof installed

Each bay poured continuously – end to end	Bay 1	Zone for: Main cladding/insulation Main services available at earliest possible time
Joints saw cut at appropriate hardness of floor	Bay 5	
	Bay 3	
	Bay 7	
	Bay 4	
	Bay 6	
	Bay 2	Zone for: Main cladding/insulation Main services

Figure 5.5 Value of long bay construction in overall planning benefits. For specification, methods and operational details, see references[8][9][10].

The four brief case studies above should provide the feeling of what buildability is all about. It does not contest the professional teams ideas. It provides the means of improving construction time at less cost by ensuring that detailing recognizes what makes construction more straightforward and cost effective.

Check-lists relating to design and buildability

Many problems relating to poor design or bad construction arise over and over again in industrial and commercial projects because each design is a one off situation. At the tender stage and later, when the detailed working drawings become available, those involved in method planning need carefully to examine drawings, as they become available, to see if details arise which may be the cause of future problems. If they do, discussion with the design team should be held before any construction actually takes place.

Check-lists 5.1 to 5.8 have been built up over the years from situations that have arisen on a wide spectrum of industrial and commercial contracts.

These check-lists do not attempt to deal with specialist services as these can be very complex and details are not usually available until late in the contract. They are discussed in some detail in Chapter 14 in relation to their effect on the general planning and sub-contractor control.

It is not claimed that the check-lists are necessarily complete. They provide an *aide memoire* for checking working drawings and can be added to as experience develops. The main aim, with such check-lists, is to locate bad detailing, or awkward to build details at the start and avoid acrimonious arguments at later stages as to who was responsible.

Relationship of plant to the building concept

When building construction was still a trades oriented operation and largely low rise in character, the forms of measurement were **quantity**

103

With closed links, long bars may have to be fed in from the end. This is awkward – sometimes impossible. Complex reinforcement is easier to fix if you detail open links.

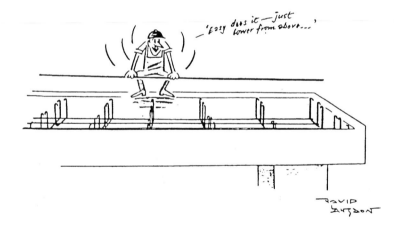

Figure 5.6 Detailing open instead of closed links simplifies and speeds up reinforcement assembly. (Concrete Society).

Check-list **5.1** Basements and foundations

1. Concrete. Is the concrete specified the correct type, with the right cement, for the ground and water conditions to be encountered?
2. Piling. Has sleeving been specified where piles are to pass through soft ground?
3. Foundations generally. Has due allowance been made in the design for differing settlements between sections of the project: with differing loadings or foundation levels?
4. Waterproof construction. Are waterbar joints continuous? Can they be installed as detailed when read in conjunction with reinforcing steel? How effective are the details for service holes and pipes passing through waterproof surfaces?
5. Damp proof membranes. Have they been detailed with proper continuity? Do they avoid potential cold spots?
6. Service entries and exits. Are they detailed in such a way that they will not break down when the building settles?
7. Sub-floor basement drainage. As detailed does it adversely affect the best construction sequence?

Check-list **5.2** Superstructure – in-situ concrete

1. Is the cover specified adequate to prevent corrosion in the prevailing circumstances?
2. Is the reinforcement detailed to allow adequate entry for vibrators and proper compaction of the concrete? Lap areas are particularly vulnerable in this respect.
3. Is the aggregate size suitable for the cross-section involved? Or to allow adequate flow through the reinforcement, or into the cover specified?
4. How is the ground slab bay size specified? Long strip construction has considerable advantages – more efficient use of plant and labour. Allows earlier installation of services in factory type structures (Figure 5.5). Large area ground floor slab is growing in use. This method needs great expertise from specialist contractors.
5. Vertical members. If inserts for ties for brickwork or partitions or curtain wall fixings are specified, would it be better to drill for such fixings later than try to cast in to the accuracy required?
6. Horizontal members. Are construction joints realistic? Large area pours for suspended slabs have been proved in practice and save the client money by reduction in stop ends, apart from speeding up the work.
7. What is the final finish to the slab? Is this compatible with the specified concrete finish?
8. Roof level. Are the falls specified realistic? How are they to be created?
9. In floors of waffle or trough section, are the side slopes adequate for easy release?

Check-list **5.3**   Superstructure – pre-cast concrete

1. Are structural joints fully detailed and capable of proper execution on site? Need for trial erection to be sure of practicality?
2. With pre-cast cladding are the fixing details in compatible, non-corroding materials?
3. Are holes and fixings in floor elements detailed and part of the floor element design?
4. Who is responsible for agreeing lifting points in vertical or horizontal components, together with inserts for temporary supports during erection?
5. In agreeing (4) above, does the lifting and positioning of units need additional reinforcement to prevent cracking over and above that for the final structural state? (See Chapter 12 for more detail.)
6. Where enough repetition exists in an in-situ concrete structure, it may pay to consider pre-cast stairs as they are likely to be more accurate and cheaper. Note: they should be designed to have starter bars and be cast into in-situ landings, not supported onto previously cast landings. (Accuracy problems between landings can arise when seating allowances become eroded).
7. Has the designer considered the problems of erection? In particular, any instability problems that may arise at certain stages in the erection? Are these made clear to the contractor?
8. Pre-cast floors and roof slabs are likely to be made with a built-in camber. What procedure is to be followed if individual units are found to have varying cambers? To achieve the desired floor level it may mean increasing the thickness of floor screeds and increasing the weight on the structure at the same time.

Check-list **5.4**   Superstructure – steelwork

1. Is partial pre-assembly a possibility? If it is, erection will be speeded up.
2. Has the designer considered the problems of erection? Do the drawings indicate any areas of instability which may arise during erection? If not, the matter should be queried.
3. If the steel has to be cased for fire protection, has dry casing been specified? Casing in concrete is expensive and time consuming.
4. If wet casing is essential for a particular situation, can agreement be reached to carry out the casing before the steel is erected? Cost will be reduced and time saved at the building completion end. See Chapter 13 for more detail on this method.
5. Most steel structures today use profiled steel decking as the support for in-situ reinforced concrete floors. Are the details clear of the end fixities needed to allow the decking to achieve full load carrying capacity without additional support?
6. Steel structures pose the most problems in relation to creating safe places of work, although the use of profiled steel decking has greatly improved this situation. Are any unusual problems on the contract in question likely to arise? A useful guide is contained in Guidance Note GS 28/3 from the Health and Safety Executive, Part 3[12].

Check-list **5.5** Building works – brickwork

1. External walls. Is the type of brick specified suitable for external work in respect of strength and porosity?
2. Are toes, cills or other supports for the external brickwork designed for effective execution? There is a long history of failure due to poorly executed items of this type.
3. Are DPCs and cavity trays properly detailed?
4. Is the correct strength of mortar specified?
5. If special bricks are needed, have they been pre-ordered if long delivery is involved?
6. Expansion joints. Are they correctly detailed and to the dimensions appropriate to the circumstances? Vertically and horizontally? Do they match the joints in the structure?
7. Are openings detailed to fit in with the brick module? If not, the brickwork price will need to allow for considerable waste. See Chapter 16.
8. If the specification calls for colour matching of bricks, the cost can grow considerably, in both materials and labour sorting. Waste levels can become extremely high unless the wrong colour bricks can be built in as commons somewhere.

Check-list **5.6** Building work – cladding and windows

1. If curtain walling is specified, it will be made to very small tolerances. The structure will be much less so. The approach to solving this problem is given in Chapter 14. None the less, check that realistic tolerances have been provided for. (After troubles at the CIS building in Manchester many years ago, the Building Research Establishment recommended that fixings for curtain walling should have an adjustment capability of +/– 50 mm in three dimensions.)
2. The supplier of curtain walling should also be made responsible for the supply and fix of windows and their glazing.
3. Are fire stops and smoke barriers properly detailed.
4. It is usually necessary to leave out sections of curtain walling to maintain access to all floors after the curtain walling is substantially complete, both from hoists and loading out platforms served by cranes. Are convenient breaks (joints) available?
5. With pre-cast concrete cladding are fixings capable of inspection after installation? Are they in compatible metals?
6. Are DPCs between PC units easy to place and check after adjacent units brought together?
7. Are PC units capable of adjustment to obtain eye-sweet lines?
8. With windows cast into the structure or built into brickwork or other cladding, are sub-frames specified or is it necessary to use dummy frames to ensure the correct opening?
9. Does the frame supplier also fix the glazing? If not should be made so.
10. Are DPCs properly detailed and cavity trays correctly located?

Check-list **5.7**   Building work – roofing

1. Flat roofs. Flat roofs are a constant source of trouble. Has the best and most up to date advice been taken? Have the anticipated expansion and contraction movements been taken into account in the detailing?
2. Are flashing details practical?
3. Is the run-off detailing adequate?
4. Are uplift anchorages adequate, both in number and fixings?

Check-list **5.8**   Building work – finishes

1. Floor. Is the sub-base suitable for the applied finish?
2. Walls. Are finishes suitable for the occupancy purpose?
3. Tiling. Can the tiling be carried out with the minimum of visits? This applies to both walls and floors. Re-visits cost a great deal of money.
4. How do finishing trades fit in with services installations? See case study in Chapter 14.

---

related. By this was meant that the unit price was built up from a gang size, its output and the materials cost of what was being built. If the amount of measured work increased or decreased, the unit price did not materially change on low rise construction, unless something more elaborate than a platform hoist was used. In these circumstances, the designer could change the quantities involved with no need to renegotiate the rates in the bill of quantities.

Since that time, with the growth of mechanization to deal with increasing height of construction and the handling of a much more wide ranging scope of material content, the situation has radically changed. The form of measurement has now become **method related**. That is

to say, the influence of the plant and methods adopted becomes significant in the pricing of the work. The assessment of the plant needed is based on the tender, understanding of what has to be done, the weights to be lifted and the radius at which the operations have to take place.

Reduction in quantity does not, in the end, allow the unit rate to remain unchanged. The plant involved will have been priced on the basis of the time on site, together with the overheads that go with getting it there, taking it away on completion and its establishment on site and subsequent dismantling and removal. Once on site, lifting plant is not cheap to remove and replace with something else, apart from the construction delay while this is being done. Reduction in quantity will have the effect of increasing cost for the client, once the contract has been let. On the other hand, if the designer alters, at a late stage, the weight of some components to be lifted, this may not be possible with the lifting equipment originally allowed for. To cope with such a situation, it may well involve hiring in a further piece of plant, on a short term basis, to deal with elements outside the scope of the original handling concept at considerable expense.

If the best value for money, as far as the client is concerned, is to be achieved, designers today need to be at least conscious of such matters and, in the design consideration, be able to assess the consequences of a decision in financial terms – or at least seek advice from someone who can compare alternative solutions.

While it would be unrealistic to expect the designer to be as knowledgeable in relation to the selection of plant and methods as the contractor, he should be in a position to assess whether what is proposed is capable of being built in as economical fashion as possible. With the growth of management contracting or the use of a managing contractor, the advice is readily available at an early stage so that the method of construction can be incorporated into the design and detailing. At the same time, structural engineers, today, are taking much more interest in construction techniques, both in relation to temporary works designed by the contractor, whether or not instabilities arise during the erection of the structure, the possibilities of using permanent works as temporary works and modifying the design. For example, to allow plant items to travel on or over permanent works in a way which will permit smaller (and cheaper) plant to be used in a handling situation. Even if the structural members need increasing in strength to achieve this latter solution, the additional costs will usually be much less than the saving achieved on the handling plant.

Example

One leading structural engineer, world renowned for his pre-cast concrete structures, always provides, with the contract drawings, his concept of the way in which the structure can be built. Indicated also is the plant for handling that was envisaged when making joints in the structure. While there is no obligation on tenderers to use the method put forward, they are at least shown that the engineer has detailed his design to be capable of erection in an economical way.

This sort of approach fits in well with the guidance provided by the Health and Safety Executive in their Guidance Note GS 28[12]. In part 1 of this document the designer's contribution to safety is laid down, especially in relation to ensuring that his design is stable at all times during erection and, if not, saying what precautions the contractor should take at critical times.

The relationship of plant to the dimensions of the structure (and potential cost) is an important aspect, particularly in relation to the cost of erection, but it must also be borne in mind that the satisfactory siting of plant depends on being able to get materials within its reach or operat-

ing loading point. This is especially true of hoists.

From the above, it should be clear that the client's best interests are served by a close association between the designer and the contractor. Within this association those involved in the method planning have a major part to play especially at the tender stage or, in management type situations, at the design stage.

References

1. University of Reading, Department of Construction Management (1979). *UK and US Construction Industries – A Comparison of Design and Construct Procedures.* Sponsored and published by The Royal Institution of Chartered Surveyors, London.
2. National Federation of Building Trades Employers (now The Building Employers Confederation) (1981) *Value for money in building – can we learn from North America?* Proceedings of conference, London.
3. *Can we halve our building costs?*, Building Economics Bureau Ltd Proceedings of conference, London 1982.
4. Williams, B. 'A resume of the NFBTE Conference and review of published data' Paper to conference (ref 3 above), put forward first definition of 'Buildability'.
5. Construction Industry Research and Information Association. (1983) *Buildability – an assessment* Report by working party, CIRIA, London.
6. Illingworth, J. R. *Buildability – tomorrow's need?* Award winning paper in Ian Murray Leslie Competition, Chartered Institute of Building. Published in *Building Technology and Management*, London, February 1984.
7. *The Construction (Design and Management) Regulations 1994.* Reprinted 1995. HMSO, London.
8. Concrete Society (1994). *Concrete ground floors.* Technical Report No. 34. Second edition. Concrete Society, Slough.
9. Concrete Society (1997). *Concrete industrial ground floors. Specification and control of surface regularity of free movement area.* Supplement to Technical Report No. 34. Concrete Society, Slough.
10. Association of Concrete Industrial Flooring Contractors and The Concrete Society (1998). Guidance on specification, mix design and production of concrete for industrial floors. *Good concrete guide.* June 1998. Concrete Society, Slough.
11. Concrete Society. *Please be practical – detailing for easier construction* Cartoons by David Langdon. Concrete Society, London 1970, Reprinted 1991.
12. Health and Safety Executive. *Guidance Note GS 28, parts 1–4. Safe erection of structures.* Part 1 1984, Part 2 1985, Part 3 1986 and Part 4 1986. HMSO, London.

Chapter six

Cost elements in tender and contract planning

Those involved in tender or contract planning need to be well versed in the cost factors likely to influence decisions relating to plant and method choice. In the preceding chapter the influence of the designer was considered and the ways in which the contractor could contribute to speeding up construction and reducing cost. In this and the following chapter the cost elements related to labour and plant and the selection of plant and methods are examined, together with the basis of adequate comparisons of cost to ensure that the most economic solution is used.

As stated in the Introduction, construction is all about the handling of materials by plant and labour to provide the completed whole. If the most efficient method is to be chosen, the method planning needs to be based on a sound understanding of labour and plant costs, the relation of plant to labour and the understanding of how alternative solutions can be compared in cost terms.

The cost of labour

Within the definition of planning, the selection of plant and methods tends to be regarded as the key item. Labour requirements, by contrast, are often seen as not nearly so important. It is often said, 'If you need a few more men/women they can always be taken on – and if you have too many, does it really matter? Or why not sub-let and let someone else have the worry?

Such a casual approach to the labour element indicates of lack of both knowledge and appreciation of the true cost of labour in construction today.

Direct or sub-let

More and more work is being sub-let by main contractors in the trades specialities, as distinct from the traditional specialist subcontractors – heating, lighting, air-conditioning, plumbing and so on. In many cases, the main contractor may not employ any direct labour at all.

In these circumstances, senior management may be heard to say that all problems related to labour are for the subcontractors. Superficially, this may well be true, but unless, in the planning stage, a proper assessment of the labour force needed has been made, site management cannot judge the efficiency of the subcontractors when the job starts. Records in the shape of labour histograms from previous contracts can be very useful in this respect. Where the labour force is directly employed, competent assessment of gang sizes is crucial in cost effective working in a highly competitive world.

Cost of labour

Whether sub-letting work or not, those who have a part in deciding construction methods and plant requirements must be able to build up what is usually known as 'the true cost of labour'. Unless they can, a true comparison of alternative methods with differing labour contents cannot be made.

True cost of labour

It is quite inadequate to assume that 'take home pay' is the true cost of labour. The employer is legally liable for a number of extras that do not appear in the operative's pay packet. These payments have to be made whether the employee does any useful work or not as long as he/she is on the employer's books. On this basis, the true cost of labour can be defined as: **the cost to the employer of having an operative on the books, having paid all legally laid down contributions for which the employee has no responsibility.** The build up of 'true cost' of labour is best considered in two parts: Take home pay and employers contributions over and above take home pay.

Components of take home pay
These are:

$$\left.\begin{array}{l}\text{Basic rate} \times \text{standard hours} \\ \text{Overtime rate} \times \text{extra hours} \\ \text{Bonus earned} \\ \text{Plus rates which qualify} \\ \quad \text{(as laid down in the} \\ \quad \text{Working Rule Agreement)} \\ \text{Tool money (in the case of} \\ \quad \text{tradesmen)} \\ \text{Travel time or subsistance} \\ \quad \text{payments, if appropriate}\end{array}\right\}$$ Take home pay: £(A)

Employer's additional costs
These are:

$$\left.\begin{array}{l}\text{National Health Insurance} \\ \quad \text{contribution} \\ \text{Holidays with pay contribution} \\ \text{Public holidays} \\ \text{Training levy} \\ \text{Insurances} \\ \text{Pension fund} \\ \text{Redundancy fund} \\ \text{Provision of tools and tackle} \\ \quad \text{(not covered by tool money} \\ \quad \text{allowance)} \\ \text{Allowance for inclement} \\ \quad \text{weather.}\end{array}\right\}$$ Employer's additional costs of employment: £ (B)

The total cost (true cost) of employing an operative is therefore £(A + B)

In developing the true cost, any contractor will have his own way of putting figures to the areas tabled above. Thus no standard tabulation exists. To provide an example, however, Table 6.1 shows a method used by a contractor to reach a true cost figure. In practice, many people are put off by the need to find all the cost figures for the calculation. Usually the true cost of labour is kept up to date on a running basis by the company estimating department, who are sometimes reluctant to divulge such figures. If so, a simple approximate value can be quickly obtained by the method shown at the bottom of Table 6.1.

In the planning function, this knowledge becomes very important when comparing alternative methods. An operative saved or increased must be costed at the true rate if the comparison is to be truly accurate.

Labour/plant relationships

In a world that is constantly changing, it is surprising how ill informed many site managers and planners are in respect of the comparative cost of plant versus labour. Labour often tends

Table 6.1   Components of true cost of labour

		Craftsmen		Labourer
Basic Wage (A)		£93.02		79.37
Bonus @ 60% (B)		55.81		47.62
National Insurance	10.45%	15.55	9%	11.43
	A + B		A + B	
Annual holidays with Pay	(Average)	12.15		10.36
Public Holidays with Pay		3.53		3.01
General Insurance 2½% A + B		3.51		3.00
Importation say – average		15.00		15.00
Inclement Weather 3% A + B		4.46		3.81
Tools and Tackle 2.5% A + B		3.72		3.17
Tool Money		0.72		–
Cost per Week to Employ (39 hrs)		£207.47		£176.77
or per hour		£ 5.32		£ 4.53

Notes:

1. Any calculation of this type will be particular to a given firm. The above example is a guide only. It is based on rates agreed at the July 1985 settlement for the Building Industry.

2. As working out such a tabulation is apt to put off many people, a close approximation can be achieved as follows:

Take the Basic Weekly Wage and convert to an hourly rate.

Take the Guaranteed Minimum Bonus and convert to an hourly rate

Thence:

Basic hourly rate	2.385	(for tradesmen)
G.M.B.	0.380	(for tradesmen)
	£2.765	

Multiply by 2 Thus 2.765 × 2 = £5.53

Comparing this value with the table above, the answer is within 4%

to be accepted as cheap compared to plant. While at one time this view was undoubtedly true, a second look at the true hourly rate established in Table 6.1 raises doubts about this at the present time. By carrying out an accurate comparison with plant operating costs, it becomes only too clear that, today, plant is relatively cheap compared to labour. Indeed, the old adage 'never keep plant standing' is no longer true. The opposite will now frequently give the most economical result, where plant allowed to stand can save expensive labour.

Method of comparison

Any particular item of plant can be costed on the basis of operating cost per week. The operating cost will include: hire rate, fuel, lubrication, servicing, maintenance charges and, in the case of plant using heavy duty tyres, an allowance for tyre wear and replacement. (This last item can be very expensive on heavy duty muck shifting vehicles). On the basis of a standard working week of 39 hrs, the hourly operating cost is given by:

113

$$\frac{\text{Operating cost/week}}{39 \, \text{hrs}} = \text{rate/hour}$$

If, now, the plant rate per hour is divided by the true labour cost per hour (usually by using the true cost of a craftsman), an equivalent number of tradesmen to the plant operating cost is obtained.

$$\frac{\text{Plant rate/hour}}{\text{True cost of craftsman}} = \text{equivalent labour}$$

At this point it must be recognized that the comparison does not include the costs involved in bringing plant to the site, setting it up or removal on completion. The components of the fixed charges involved are variable, depending on haulage distances, where appropriate, foundation costs relative to varying ground conditions, installation of electricity take-off points and erection charges. Dismantling may well be more expensive than erection if the plant is awkwardly placed after the structure is completed. As such, a figure can only be put to fixed charges when one has a specific site and the conditions are known.

In spite of the above limitations, it is useful to have some yardstick in respect of the influence of fixed charges when considering one item of plant against another as to its use potential in any situation. This can be achieved in general terms, by taking a generalized look at the fixed cost components. For example, in the case of a tower crane these are:

- Haulage to and fro from site. Many pieces, haulage costs relatively high. Increased a good deal if crane has to come from a long distance.
- Erect and dismantle costs are high compared to most plant involved in structural work.
- Service provision costs will vary, depending on how near the plant will be to a three phase electricity supply.
- Foundations are expensive, whether static base, need for track or as a climbing crane.

Even in such generalized terms, it will be clear that the fixed charges in relation to the use of tower cranes are very high. By comparison, the operating costs are surprisingly low for what the equipment can do. To be cost effective, therefore, the use of a tower crane depends on a long term use to amortize the high fixed charges.

Similar comparisons between plant and labour can be carried out for any item of plant. If the results are presented in a simple visual form, it is possible to build up a series of sheets as shown in Figures 6.1, 6.2 and 6.3, for quick ready reference when the method planning for a particular contract is started.

NOTE if the labour needed for the plant operation is also included with the plant silhouettes, the same number must be added to the equivalent labour, for the equation to remain equal.

With such diagrams, top management, too, can be given an immediate feel for the current comparative situation – often much to their surprise.

Updating from time to time is required to ensure that equivalent labour numbers have not significantly changed with plant or labour increases.

Labour gang sizes

With the increasing use of subcontractors, the knowledge of correct gang sizes for particular operations is declining in site managers and planners. This situation is to be regretted, as the ability to check on the efficiency of subcontractors is lost. With directly employed labour lack of knowledge means that the planning and management of labour strengths is not as accurate as it should be – both at the tender and contract stages.

When most contractors directly employed trades labour, site managers and planning staff built up a considerable knowledge of labour needs for all relative operations, including the most efficient gang sizes for individual operations. The building up of such records was, in

Plant operating cost 39 hrs Weekly hire rate, maintenance, fuel, lube, (excl. driver, banksman)	Equivalent in tradesmen (True cost to employer)	Haulage and erection cost	Economic use condition	
Richier G.T. 222B 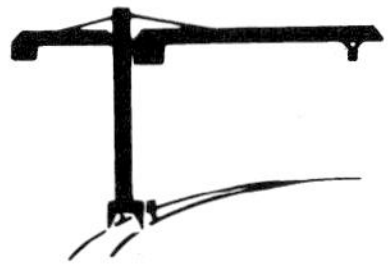		High	Long term or where access problems and reach critical factor	
22 R.B. (I.C.D.) 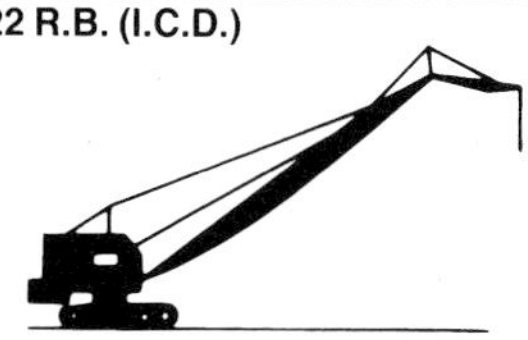		Small/low	Rough terrain general purpose Crane age short medium or long term	
J.C.B. 520 		Low	Low rise packaged materials short, medium or long term	

Figure 6.1 Plant/labour comparisons (sheet 1 of 3).

Plant operating cost 39 hrs Weekly hire rate, maintenance, fuel, lube, (excl. driver, banksman)	Equivalent in tradesmen (True cost to employer)	Haulage and erection cost	Economic use condition	
Sanderson S.B. 70 forklift 		Low	Short, medium long term Rough terrain Packaged materials	
J.L.G. 35S scissor platform 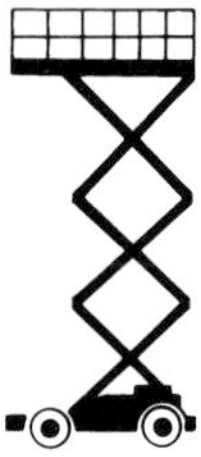		Very low	Short, medium, and long term Carries materials and erectors Industrial sheeting and services installation	

Figure 6.2 Plant/labour comparisons (sheet 2 of 3).

many ways, a personal thing. Much of such information has been lost with the change to subcontracting. In consequence, newcomers to the industry are ill equipped for the role of plan-

ning construction methods. In order to redress this situation three main principles have to be understood before any build up of labour gang sizes can be attempted.

Plant operating cost 39 hrs Weekly hire rate, maintenance, fuel, lube, (excl. driver, banksman)	Equivalent in tradesmen (True cost to employer)	Haulage and erection cost	Economic use condition
Schwing B.P.L 601 DG mobile concrete pump 		Low	High output Short, medium or long term or where access problems and reach critical factor
Warren 6 m belt conveyor 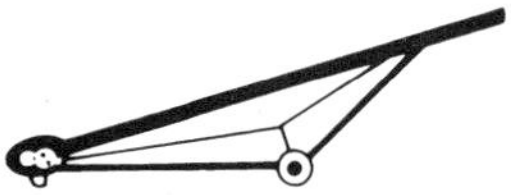		Small	Short, medium and long term
Kubota KH 8 mini-excavator 		Low	Short, medium or long term

Figure 6.3 Plant/labour comparisons (sheet 3 of 3).

1. Optimum gang sizes

In many trades, there are optimum gang sizes for particular activities. Such optimum figures are best determined from work study over the operation in question covering an adequate sample. From such studies, performance per operative will also emerge. The correct man hours for a given operation is a combination of these two items, that is: Optimum gang size × unit performance/operative.

While the use of work study will provide the most accurate answers in relation to labour performance, adequate recording of labour numbers involved in an activity together with performance achieved (derived from bonus measurement) is another approach that can be adopted, provided that the figures are properly analysed and interpreted, in particular, to establish the number of operatives who achieve maximum output.

By way of example, the most efficient gang for erecting the Metraform soffit formwork has been shown to be three men for the basic erection. Two further men are needed to achieve the final line and level. While the three-man erection gang is the optimum size, the performance

achieved will be variable to some extent – due to such things as variation in labour quality and the height of support to be dealt with. On the other hand, if a three-man gang is the optimum, any further increase in output needed must be achieved in multiples of three. Failure to do this merely increases cost as any other numbers added into the gang cannot work to the same efficiency as the optimum number.

2. Degrees of difficulty

While many items of work have established gang sizes (optimum) for a given method of working, there are other operations where the optimum gang size is variable – depending on the degree of difficulty of the work in question.

For example, consider a wet end concrete gang. Its size cannot be a constant as it is dependent on output, placing method and, most particularly, the degree of difficulty in getting the concrete into its final location, vibrating it and where appropriate, its final surface finishing, in relation to a given output from the handling plant. Table 6.2 illustrates this point. Three con-

Table 6.2 Degrees of Difficulty illustrated by three different concrete pours. All were carried out by pumping and were the subject of a work study

Pour	Pump output (m³/hr) while pumping	Wet end gang-men
1. Trough floor	19.5	8
2. Motorway viaduct (Flat slab deck 0.76 m thick)	37.0	5
3. House oversite rafts (200 mm thick)	19.11	4

crete placing situations, in relation to slabs, all having concrete handled by a pump were work studied with the results shown.

Study of items (1) and (3) in Table 6.2 shows that the pump output was almost identical in each case. Yet the wet end gang in one case was double that in the other. The only explanation of this difference can be in the degree of difficulty in placing the concrete in each situation. In the case of the oversite raft, placing and finishing the concrete is very straightforward. As the concrete comes out of the pump pipeline it almost places itself. Vibration is straightforward and the amount of surface finishing per m³ is 5 m².

In the case of the trough floor (Figure 6.4), concrete has to be placed and vibrated with care into the ribs first, while the topping concrete is placed and vibrated later. The surface finishing per m³ of concrete placed is greater than the raft, in this case 6.51 m² per m³ of concrete placed.

Finally, it should be noted that, on the motorway viaduct, 0.76 m thick, the concrete is to a high degree self placing. At the same time, the area of surface finishing per m³ of concrete placed is only 1.32 m². As a result an output per man of 7.4 m³ per hour was achieved. This is markedly better than the house rafts (4.7 m³/man hour) and even more significant in the trough floor where output per man has dropped to 2.44 m³ per hour.

This question of the degree of difficulty not only affects the manning level for a given output, but the unit cost also. It is also appropriate to any type of labour intensive activity.

3. Correct manning

The example immediately above, relating to the degree of difficulty and its influence on labour strength, leads naturally into the subject of the correct manning for any item of plant envisaged in the planning analysis. However the most cost effective plant is chosen, full advantage will not be gained if it is incorrectly manned with attendant labour.

Although the wet end gangs quoted above are an obvious case in point, the planner needs to consider all areas of plant operation. For example:

- In tower crane operation how many banksmen will be needed? Is it the standard one man or, due to visual difficulties, will a second one be required? If the latter is the case, would a closed circuit TV system be a cheaper alternative?

- Where trench excavation is involved, how many men are needed for bottoming up and placing the support material? In this case, the numbers will relate closely to the support system used.

- Standard bricklaying gangs are traditionally 2 bricklayers to 1 labourer. With packaged materials, handled by a fork lift onto a loading-out platform, studies have shown that 3 bricklayers to 1 labourer gang becomes

Figure 6.4 Illustration of Cordek trough (Cordek Ltd) floor forms in use.

feasible, without overstretching the labourer. Indeed, studies have shown that increases in production can also be achieved per bricklayer.

Summary

From the foregoing, in relation to the labour element in the construction process, it should be clear that labour is a significant proportion in cost terms of any method that may be selected, unless a fully mechanized solution happens to be possible. The planner and site manager need to be fully aware of this situation, particularly the relationship between plant and labour costs. Even if the work is subcontracted, such knowl-

edge is important if the efficiency of the subcontractors is to be evaluated and adequately controlled. Too often site management is inadequately experienced to provide counter arguments to subcontractor claims. Knowledge of labour performances is a key ingredient in this sphere.

The cost of construction plant

Cost elements

The cost of all handling plant is compounded from two different elements – weekly running costs and fixed charges. The first element relates directly to output per week, while the second, in its contribution to unit cost, depends on the total production of the item in question.

Weekly running costs

The components of weekly running costs for any item of plant are:

1. weekly hire rate for the item of plant;
2. maintenance allowances – including tyres for heavy plant;
3. fuel and lubrication costs (including electricity in appropriate cases);
4. true cost (to contractor) of all labour involved.

Where the plant is contractor owned, the relevant figures will be put together separately for a total weekly running charge. With plant hired-in from a hire company, an all-in figure may be given, inclusive of driver, or merely for the plant item with the contractor to provide the driver.

With a stated hours per week operating time, the cost per hour is easily derived. Alternatively, if the plant use relates to productive output, the unit cost of the handling plant per productive unit per week can be derived.

Fixed charges

In this case the components are:

1. haulage charges for the plant involved to and from site;
2. erect/dismantle costs – some plant may involve very high erect/dismantle charges;
3. cost of all temporary works associated with the plant in question, including such things as access roads, foundations, drainage etc.
4. cost of supplying and installing all necessary services needed – power, water, drainage, etc.
5. all labour charges associated with the above at a true cost to the contractor.

Fixed charges, when compiled, of necessity have to be spread over the whole time that the item of plant will need to be on site. Where the plant is

used on productive work, its contribution to unit cost levels will be:

$$\frac{\text{Total fixed charges}}{\text{Total production}} =$$

unit cost component addition

If the plant does not relate to measurable production, an hourly rate is determined which can be apportioned as needed to specific operation costs, thus:

$$\frac{\text{Total fixed charges}}{\text{Total hours on site}} =$$

cost per hour for fixed charges

Graphical presentation

Where plant is engaged on productive output – for example, handling concrete – the costing of its use can be taken a stage further. Let weekly running costs be A, fixed charges B and the total output required C. All these items can be considered as constants. (This is not exactly true as weekly output will vary to some degree, while the total output will not be exactly known until the final measurement has been completed). For the purpose of this exercise, the accuracy will be near enough. With these constants two variables arise. These are: output/week and unit cost. Let these variables be X and Y.

Linking equation

All the above variables and constants have a linking relationship. Expressed as an algebraic equation we have:

$$Y = \frac{A}{X} + \frac{B}{C}$$

This equation, in fact, is a particular form of an equation for a hyperbola. All graphs plotted in this context will have the same basic form.

By varying X over the probable output range required during the contract, a series of values for Y will be obtained from which a cost output

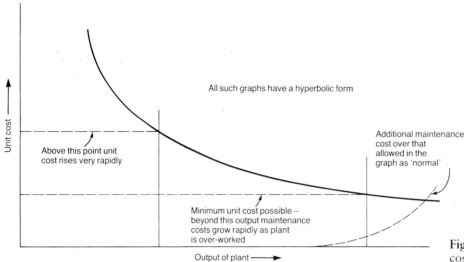

Figure 6.5 Plant – unit cost/output graph.

graph for the plant in question can be obtained (Figure 6.5). From this graph certain key issues can be established.

1. At the low output end, unit costs rise very rapidly.
2. At the high output end, beyond a certain point, increased output makes no real difference to unit cost.
3. Attempts to increase output beyond a certain limit will, in fact, increase cost as maintenance charges will increase dramatically.

From the above it follows that for a given item of plant on productive work (measurable item), there will be an output range between the two vertical lines within which the best value for money will arise.

This method of graphical presentation is particularly useful if graphs are produced in advance for a variety of construction plant. Alternatively, the planner can keep records of graphs produced from previous contracts. In either case, when the detailed planning has established the output required, a direct read-off from the graphs previously available will show which handling approach is the cheapest.

These same graphs also become a management tool during the contract. If output per week is less than planned an immediate idea of losses arising can be read off. To enable this to be reasonably accurate, the choice of vertical scale wants to be the largest possible in relation to the paper used.

The production graphs of this type lend themselves to computer programs for easy production, for recording past feedback and for matching records with new work.

Chapter seven

Selection of plant and methods

The construction industry could be said to be very fortunate in having such a wide choice of construction plant. To those studying for the construction industry, or those who have as yet little experience, this wide choice is a hindrance rather than a help. The choice seems vast, so how are decisions made?

Construction plant categories

A useful beginning can be made by recognizing that all construction plant can be located in one of three fundamental categories (Figure 7.1). From this diagram, the three basic divisions can be seen as: linear, two-dimensional and three-dimensional. Looking at each category in turn we see that: **Linear plant** only operates in the vertical plane and, as such, encompasses all the varieties of construction hoists currently on the market. **Two-dimensional plant** covers the bulk of wheeled vehicles – fork lift trucks and telescopic handlers excepted. In this category are the means of distribution on essentially two-dimensional surfaces. While it could be argued that a vehicle can travel up hills and down dales and therefore is three-dimensional, it is more realistic to consider that a vehicle travels on a plane where the plane is not necessarily level. **Three-dimensional types** are those with a true three-dimensional capability on the construction site, such a capability being only limited by the geometry of the plant in question. The tower crane, for example, typifies the three-dimensional concept. It can lift to considerable heights, has a relatively large coverage especially if mounted on rail tracks, and can move objects from close to the tower to the farthest point outwards allowed by the jib length. By comparison, on a much smaller scale, fork lift trucks and telescopic handlers each have limited three-dimensional functions, but appropriate to the handling they are asked to carry out.

The purist can argue that there are other items which do not fit into any of the three category groups. Welding sets, lighting sets and compressors, for example do not, in themselves, work as handling equipment. Nevertheless, they are adjuncts to the handling and assembly process, in much the same way as are hand tools.

Clearly, the first requirement for the construction method planner is an adequate knowledge of what construction plant in the three categories looks like and what their capabilities are – both in terms of range and capacity and the requirements for safe operation on site. Such knowledge takes time to acquire and is one of the reasons why those wanting to become construction planners are usually advised to get good site experience for some years before specializing in method planning. However, a useful introduction, in this respect, is given in a soft cover booklet, *Site Handling Equipment*, published in

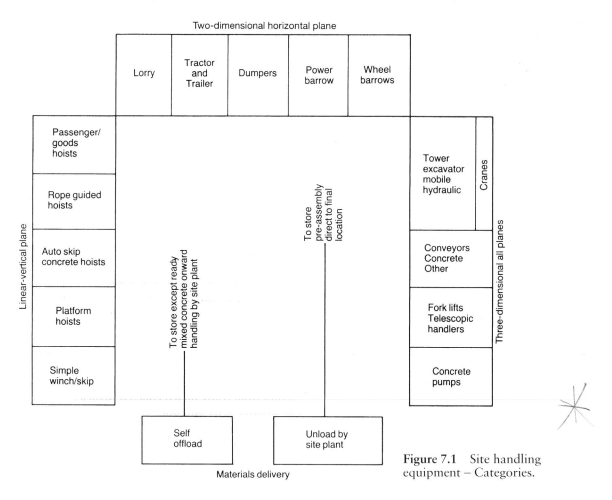

Figure 7.1 Site handling equipment – Categories.

the Institution of Civil Engineers Works Construction Guide series[1]. In this volume, examples of the three plant categories are given and illustrated, together with some information on selection. No attempt is made to tabulate performance of specific models as one sometimes sees in books on construction plant. The reason, of course, is that unless the tables are specifically related to a specific item of plant, they are not only useless for planning purposes, but downright dangerous to use.

While the above publication is a useful introduction, it is not entirely complete as it does not deal with excavation machinery on the building site. To fill this gap and to provide a more planning oriented introduction to construction plant

Tables 7.1 to 7.7 have been prepared. Tables 7.1 to 7.6 cover plant shown in Figure 7.1, while Table 7.7 deals with excavation machinery normally met with on the building site.

While the Tables and the booklet referred to provide an introduction, one needs always to build up knowledge from experience, which takes time. Even so, reference will need to be made to the manufacturer's literature for detailed points in relation to an item of plant use.

Availability of plant and equipment

In the past, main contractors carried out all the craft trades with their own labour and supplied

Table 7.1 Linear plant: hoists

Passenger/Goods hoists
Special safety devices have to be built-in for carrying people.
Special bases needed.
Anchorages to building required.
Such hoists essential for the handling of the workforce on tall structures.
Hoist speeds can be altered to suit situation.

See Figure 7.2

Rope guided hoists
Use situation primarily related to tall towers for TV or telecommunication purposes and in deep shafts.
Erection inside enclosed structure removes the effect of wind.
Rope guided systems automatically pay out and maintain tension of rope guides as structure rises/deepens. Ideal for combination with slip-form construction. No delays changing headgear to new levels.
Can be goods or passenger.

See Figure 7.3

Automatic trip skip concrete hoists
Specially designed for handling concrete.
Can be directly filled at bottom from ready mixed concrete vehicle.
Discharges automatically at working level into floor hopper.
Onward handling by crane skip or barrows.
High speed options available to speed up concrete delivery.
Often associated with tower crane for vertical lifting.

Platform hoists
Common user hoist for materials only.
Needs enclosing scaffold tower when working at heights.
Carries loaded barrows or packaged materials.
Winch and tower need safety enclosure at base.
Anchorage to building above low rise construction.
Interlocking gates at all floors necessary.

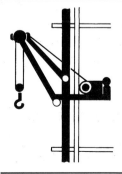

Simple winch and skip
Usually operates over pulley cantilevered from scaffolding.
Needs proper assessment of weights to be lifted and scaffold stability checked by a competent person.
Winch unit requires safety protection.

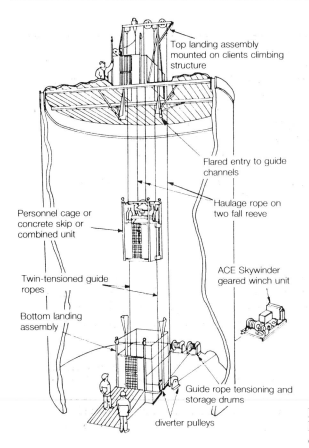

Top landing assembly mounted on clients climbing structure

Flared entry to guide channels

Personnel cage or concrete skip or combined unit

Haulage rope on two fall reeve

ACE Skywinder geared winch unit

Twin-tensioned guide ropes

Bottom landing assembly

Guide rope tensioning and storage drums

diverter pulleys

Figure 7.2 Schematic arrangement of rope guided hoists. (ACE Machinery Ltd).

Figure 7.3 Headgear and discharge level for twin high speed concrete hoists. One skip can be seen discharging into the right hand floor hopper.

all the major plant needed. Today, most of the craft trades are sub-let with the subcontractors expected to provide their own plant, with the exception of major common-user handling equipment and scaffolding. With specialist service subcontractors, the extent to which the main contractor is required to provide handling plant and scaffolding, in particular, is normally specified in the contract documents.

In today's world, with almost everything subcontracted, the main contractor no longer carries a lot of construction plant. In the United Kingdom at least, the availability of plant on hire, well spread all over the country, now makes significant plant holdings by the main contractor unnecessary. Hiring-in plant can also provide significant financial advantage.

If not in use plant can be taken off hire on demand (with whatever notice the contract of hire demands). With company plant, plant managers were not unknown to go on charging whether the item was working or not, as long as it was on site.

With contractor's own plant, it is necessary, in the tender stage, to make every effort to use what is available so that an income is coming in rather than having plant standing. In so doing, the most efficient and cost effective method may not be possible. By contrast, with plant hire organizations, one can shop around for the ideal equipment. That is, supply of plant by specification from the method planner. With this approach, knowledge of plant does not have to be so exacting. All that is needed is to determine the characteristics required and ask the plant hirer to provide a quotation.

124

Table 7.2 Two-dimensional plant: wheeled vehicles

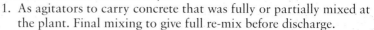

Ready-mixed concrete trucks
Can be used in one of three ways:

1. As agitators to carry concrete that was fully or partially mixed at the plant. Final mixing to give full re-mix before discharge.
2. Filled with dry batched concrete materials at the depot, water added and full mixing completed before leaving.
3. Filled with dry batched materials at the depot, travel to the site and water added on arrival. Mixing then until complete.

Two useful Reports on ready-mixed concrete, (one associated with concrete pumping) are available [2] [3].

General purpose transport
General purpose construction transport covers a wide range. At the heavy end are lorries and tractors and trailers capable of handling heavy loads of great variety: pre-assembled reinforcement, packaged timber and bundled reinforcement, formwork and falsework materials, contractors plant and so on.

Today, all the lorry type vehicles are usually fitted with hydraulic cranes or handlers of one type or another to allow loading and unloading with their own equipment. In the case of the heaviest vehicles (low loaders), the design allows for heavy construction plant to load and offload by driving the plant in question up heavy ramps onto the carrying body.

See also Figure 7.4

Handling mechanisms include grippers (for off-loading packs of bricks for example), pipe clamps, grabs for loose materials and clamping equipment to allow loads to be turned in position.

At the lower end of the scale are dumpers of varying capacity, and small barrows, either powered or moved by hand.

Apart from the wheel barrow, hand operated, all are provided with mechanical power:

1. Powered barrows – common in the United States of America, never really catching on in Europe. Operator walks behind in some versions. In others sits on to drive.
2. Dumpers are probably the most used (and abused) item of mechanical plant in the European scene. In the United Kingdom, they represent the main cause of accidents with mechanical plant.

See also Figure 7.5

Note: where dumpers are used for handling concrete, purpose built varieties must be used. They are built with hydraulic tipping to control discharge, either ahead or sideways.

Figure 7.4 Lorry loading surplus material with hydraulic grab mounted on vehicle. Note also, mini backhoe machine doing the excavation for a new driveway. The ability to hire all the equipment in the illustration allows a three man team to become fully mechanized.

Figure 7.5 Mini backhoe loading spoil into a dumper. To comply with safety legislation, the driver should not be sitting on the machine while it is being loaded.

Table 7.3 Three-dimensional plant: tower cranes

Rail mounted
Horizontal or luffing jib facilities.
Can operate at considerable heights, but hoist speeds slow. 360 deg slewing. With track, operating radius greatly increased.
If height requires anchorage to building, anchorage points have to be located at fixed operating locations along track.
Design of track must relate to ground conditions – weekly maintenance to keep track level.
Overswing of boundary requires permission of adjoining owners.

See also Figure 7.6

Static base
Horizontal or luffing options.
360 deg slewing, coverage limited by maximum jib reach.
Safety depends on base design. Needs competent person for design after study of soil report.
Anchorage to building no problem as always static position.
Some modern versions can operate at considerable heights without the need for anchorage to building.
Probably the most used version on central development sites.

See also Figure 7.7

Climbing
Horizontal and luffing versions.
360 deg slewing.
Initially erected on static base. When building reaches an appropriate height, crane uses its own winches to lift off the base onto climbing collars supported by the building structure.
Use only possible with the structural engineer's approval. Reach outside the structure more limited than the other two.

See also Figure 7.8

General notes
The use of tower cranes makes mandatory the solution of its dismantling method at the tender stage. Erection may be easy but once the structure is in place, dismantling becomes a different proposition altogether and money has to be assessed for inclusion in the tender sum.
Overswing of adjoining property is a trespass unless agreement reached with adjoining owners. See section on boundary conditions.
An item of dominant plant and efficient use requires careful planning and communication to all those using the crane.
From previous Chapter, tower cranes are long term use plant for best economy.

Figure 7.6   Rail mounted tower crane. Due to proximity to operating railway lines it has to work with a shortened jib (Chapter 2). As a result it needs a long length of expensive track to cover the building.

Figure 7.7   Static base crane. Enables the building to be erected around it. The most common situation on restricted sites.

Table 7.4 Three-dimensional plant: cranes

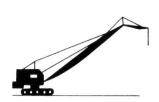

Excavator conversions

As the name implies, cranes converted from the old style rope operated machines. Provides a new lease of life as hydraulics have taken over for excavation.

Today, such cranes also associate with piling equipment. For auger bored piles provide the base machine in many cases on which the kelly bar auger system is mounted.

Advantages are:

1. Rough terrain stability as on tracks designed for digging loads.
2. Usually excess power in crane role.
3. Relatively cheap to hire as maid of all work.
4. Fast hoist and slew actions.

See also Figure 7.9

Disadvantages:

1. Slow travel speed.
2. Unless rebuilt with crane clutch, excavator clutch will snatch loads.

Lorry mounted strut jib cranes

Lower capacities have now been superseded by modern hydraulic lorry mounted cranes (see below).

Main role today in heavy lifting or with towers and luffing jibs for short term high reach handling:

1. With short jibs heavy lift capability on process plant equipment handling.
2. With tower and luffing jib for dismantling tower cranes or locating pre-assembled equipment on the roofs of buildings etc.

See also Figure 7.10

Hire rate expensive hence short term use on specific lifting operations.

Hydraulic, telescopic jib cranes

Lorry mounted in the main. Tracked versions have been made. Revolutionized the lorry mounted crane.

Advantages are:

1. Ability to go to work immediately on arrival. No assembly needed.
2. Range of use greatly extended from other mobile cranes.
   Can work in locations impossible for other cranes.
3. Basically pay for working time and travel time only.
4. New versions appear with bigger capacities daily.

More expensive than strut jibs. Short term hire plant.

Figure 7.8 Climbing crane located on a 35 storey office building.

Figure 7.9 Excavator conversion crane. The illustration shows the rough terrain capability of such machines, as well as the common user role due to having a cheap hire rate.

Figure 7.10   The lorry mounted strut jib crane. Here in the pre-cast concrete structural erection role. Now largely overtaken by the telescopic boom, hydraulically operated types.

Figure 7.12   Concrete pump – lorry mounted with hydraulic boom. High mobility has created the ability to pour large quantities of concrete in a short time. As such has revolutionized concrete handling and its relation to the overall planning of concrete operations.

Figure 7.11   Concrete pump – trailer version.

131

Table 7.5 Three-dimensional plant: concrete pumps and conveyors

	Concrete pumps Designed specifically for the transport of concrete – use calls for concrete to be designed to be pumpable.

Trailer pump
See also Figure 7.11
	Modern pumps available as trailer mounted or lorry mounted with hydraulically operated folding booms.
	
	Pump handling is very expensive for small quantities, but highly cost effective at high output. See Figure 6.5 in Chapter 6. Use on small quantities only tenable when no other solution is possible in the situation to be dealt with.
	
	With suitable aggregates the range of delivery is spectacular – world records currently claimed are:
	
	    Horizontally 1058 m
	    Vertically 432 m
	
	Note however, that when use has horizontal and vertical movement combined the above ranges are greatly reduced.

 Lorry mounted boom pump See also Figure 7.12	Proper cost evaluation is crucial as direct comparison only can provide misleading answers. In planning terms, allows the pump to be hired in and used with ready mixed concrete only when needed. This can be a big advantage when the concrete quantities are small in work load compared to that for the erection of falsework, formwork and the fixing of reinforcement. Greater freedom is possible in the cyclic planning of a structure.
 Single conveyor unit See also Figure 7.13 for conveyors used in a chain.	*Concrete conveyors* While much used in the United States of America, conveyors have never really caught on in Europe. Only limited use conveyors have found favour. For one example see Figure 7.13. Unlike the concrete pump can handle any type of concrete – dry mixes in particular. Must be designed to fulfil the concrete handling role – due to abrasive nature of the material to be handled. Conveyors need a good deal of movement and support when covering large area pouring, hence additional cost. American built conveyors forming chains can reach outputs of over 100 m³ per hour. Conveyor output is a function of a narrow belt width, running fast.

Figure 7.13 With the increasing use of the concrete pump, concrete conveyors have never really taken on in Europe. Yet they continue to be used in the United States alongside the use of pumps. (Morgan Manufacturing Co., Yankton, South Dakota, USA).

Figure 7.14 Variable geometry fork lift loading boxed roof tiles onto a loading out-platform (see also Chapter 4 concerning palletized loads on scaffolding).

Table 7.6 Three-dimensional plant: materials handlers

Variable geometry fork lift

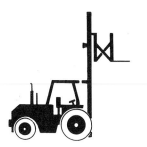

Scissor forward reach

Telescope handler

See Figures 7.14, 7.15, 7.16

Fork Lift Trucks
The concept of delivering materials in a packaged form opened the door to the use of Fork Lift Trucks for the unloading and distribution about the site of such packages. For such use, Fork Lift Trucks need to be capable of travelling safely over the normal rough terrain of the construction site.

In recent times, many materials suppliers have provided their own offloading equipment built as an integral part of the delivery vehicle–largely to obtain a quicker turn round than by waiting for the contractor's fork lift truck to appear.

Types of Fork Lift Truck
Three basic types are available:

1. Forks that move up and down at most.
2. Forks mounted on forward-extendable scissors.
3. Variable geometry versions.

In types 2 and 3 it must be noted that the further a forward reach is possible, so does the load stability decrease.

Safety in use
The safe use of fork lift trucks on construction sites is of crucial importance. An authoritative report on this aspect exists, published by the Building Employers Confederation [5].

Limitations
The three dimensional capability of Fork Lift Trucks is, of course, limited – usually to a maximum height of three floors in load bearing construction.

Telescopic jib materials handlers
The telescopic jib materials handler operates differently to the fork lift truck. Forks are provided at the end of a jib capable of extension in the same way as a hydraulic crane. As such, the motions that can be performed are increased in comparison to a fork lift truck:

1. It is possible to lower a pack over a guard rail and onto the scaffolding. With a fork lift, the guard rail has to be removed.
2. Within a structural frame, items can be lifted when the handler is within the structure, only extending its jib when in the final position.

Safety in use As with fork lifts.
Limitations As with fork lifts.

Figure 7.15 Scissor forward reach fork lift. (Sanderson (Fork Lifts) Ltd).

Figure 7.16 Telescopic handler. Note the ability to place loads on scaffold without the need to remove guard rails.

Table 7.7 Three-dimensional plant: excavation plant

See Figures 7.17 and 7.18

Those concerned with the Building Process generally become involved with excavation as related to foundations, trenches, basements or limited excavation on site to provide level work platforms for buildings. As such, the plant involved will generally be limited to hydraulic backhoe excavators of varying capacities, together with tracked excavators, usually designed to perform a multipurpose role – dozer, bucket excavator, grading and loading surplus material.

It is not the intention of this volume to deal with the Civil Engineering side of bulk excavation and the wide range of specialized plant involved.

Hydraulic backhoe excavators
In all its varieties, the hydraulic backhoe excavator is probably the most common sight on a construction site at the ground works stage. Such machines have a versatility that can be quite remarkable:

1. Being hydraulic in all actions, every motion is fast. Dig, slew, raise and alter bucket attitude. Because of this, output is high in relation to bucket capacity.

2. Can move up steep slopes using the backhoe arm as a stabilizer.

3. With the capability to alter the bucket angle to the arm, can dig almost rectangular holes without the need for a lot of hand trimming.

4. Can accept many attachments: grading buckets, grabs, concrete breaking equipment and varieties of gripping mechanisms.

5. Smaller versions mount a bucket at the front end for surface digging, loading and grading

Versions exist from the 'toy' size to those with large bucket capacities and considerable power.

Tracked bucket excavators
Consisting of a 'dozer type crawler chassis, this equipment has a front mounted bucket. The bucket is split into two sections, pivoted at either side. With this type of geometry, several actions become possible:

1. With the bucket closed loading from stockpiles and elevating for loading vehicles.

2. With the bucket open it can act as a grab for picking small quantities of material.

3. In the same configuration, it can be used for grading.

4. With the bucket closed, excavation of a face is possible.

Figure 7.17 The 'JCB'. Probably the best known name anywhere where small excavators and loaders are used. The 4CX shown is a well known model – four wheel drive, four wheel steer, giving both good rough terrain use and the ability to operate in 'tight' locations (J. C. Bamford Ltd).

Figure 7.18 Building ground works and key plant use. Heavy backhoe for levelling site and drain excavation; mini backhoe for excavating foundations, with dumper to remove surplus material. In addition, a compressor is present to break up old foundation concrete.

Selection of plant

When selecting plant and methods for a particular contract, a number of factors have to be taken into account.

1. The activities to be performed and outputs required;
2. Workloads for each discrete item of handling;
3. What access exists between the point of unloading and the point of fixing? (This will play a big part in assessing the handling chain of events);
4. Weights of all items to be handled and the range of weights. (Ideally, the most economic solution will be achieved if handling units can be kept as similar as possible – by design or by packaging).

These will enable an assessment of plant type and capacity to be carried out and a list of possible solutions to be listed.

Before any decisions can be made, however, further consideration needs to be given to the influence of:

1. the character of the site and its boundaries – do they affect choice of method?
2. the effect of obstructions below, on or above the site;
3. the ability to provide access (and the cost) to all plant envisaged and its removal at the end of the contract;
4. the influence of the Control of Pollution Act[4].

Do any of the above factors deny the use of any of the first thoughts of method and plant use?

With the above factors in mind, it is now possible to examine the selection procedure in more detail.

At this point it should be recognized that much of the information necessary for the correct answers to be achieved will come from the site inspection form. The need to carry out site inspections in a thorough manner will be only too clear.

Activities to be performed and outputs needed

To complete this section, a very careful study is needed of the work to be done.

1. The sequence of events plays a significant role at this stage. A hand produced critical path diagram is the desirable first stage in any planning assessment. In its preparation, the method planner obtains a detailed appreciation of what is involved and the sequence of events that will have to take place.
2. Once drawn, a much clearer picture of plant needs will arise, together with the degree of overlap that may be possible with an item of plant handling more than one activity.
3. At this point, calculation of the work loads involved must begin, both in relation to both plant and labour. See Chapter 11 for their preparation.
4. To achieve realistic output figures in assessing workloads, the planner, over time, has to build up a library of performance data or refer to books that tabulate such information for planners or estimators. The biggest problem here is that many data books inadequately describe the conditions under which the performance was achieved. Over a period of time, good planners build up their own files of performance data from past contracts' achievements. In building up such information it is important to use a structured method, so that all the factors involved are recorded. A typical record form of this nature is shown in Example 7.1. Its preparation can be done by a work study engineer or by the method planner asking the site staff to complete such a form to details given.
5. Decisions on plant type can only be made when the total workload has been established and related to the desired contract time.

Once the method planner has achieved a real working knowledge of the work to be done and the likely outputs needed, a list of potential

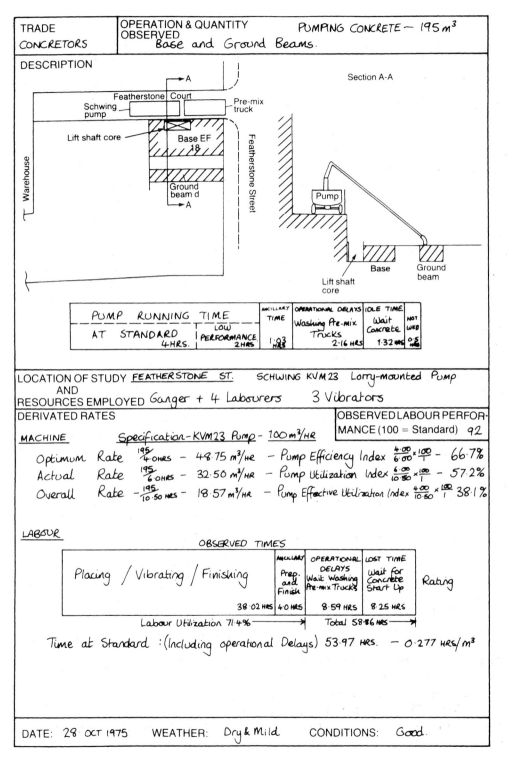

| TRADE

CONCRETORS | OPERATION & QUANTITY
OBSERVED | PUMPING CONCRETE — 195 m³ |
| | | Base and Ground Beams. |

DESCRIPTION

PUMP RUNNING TIME | AT STANDARD 4 HRS. | LOW PERFORMANCE 2 HRS | ANCILLARY TIME 1·03 HRS | OPERATIONAL DELAYS Washing Pre-mix Trucks 2·16 HRS | IDLE TIME Wait Concrete 1·32 HRS | NOT WKD 0·5 HRS

LOCATION OF STUDY FEATHERSTONE ST.   SCHWING KVM 23 Lorry-mounted Pump

AND

RESOURCES EMPLOYED Ganger + 4 Labourers   3 Vibrators

DERIVATED RATES

OBSERVED LABOUR PERFOR-MANCE (100 = Standard) 92

MACHINE   Specification – KVM23 Pump – 100 m³/HR

Optimum Rate $\frac{195}{4·0 HRS}$ – 48·75 m³/HR – Pump Efficiency Index $\frac{4·00}{6·00} \times \frac{100}{1}$ – 66·7%

Actual Rate $\frac{195}{6·0 HRS}$ – 32·50 m³/HR – Pump Utilization Index $\frac{6·00}{10·50} \times \frac{100}{1}$ – 57·2%

Overall Rate – $\frac{195}{10·50 HRS}$ – 18·57 m³/HR – Pump Effective Utilization Index $\frac{4·00}{10·50} \times \frac{100}{1}$ 38·1%

LABOUR

OBSERVED TIMES

Placing / Vibrating / Finishing | ANCILLARY Prep. and Finish 4·0 HRS | OPERATIONAL DELAYS Wait Washing Pre-mix Trucks 8·59 HRS | LOST TIME Wait for Concrete Start Up 8·25 HRS | Rating

38·02 HRS

Labour Utilization 71·4%   Total 58·86 HRS

Time at Standard : (Including operational Delays) 53·97 HRS. — 0·277 HRS/m³

DATE: 28 OCT 1975   WEATHER: Dry & Mild   CONDITIONS: Good.

Example 7.1   Plant performance record form

plant and method solutions can be tabulated (Table 7.8). Against these possible methods must be tested the other items previously listed in relation to the site conditions.

The character of the site and its boundaries

Five questions must be asked here.
1. Does the site geography influence plant choice? Is the site reasonably level? Very sloping?
2. What is access to the site like? Can heavy plant get to the site without damage to local roads? Are bridges strong enough?
3. Once on site, is there a need for expensive temporary roads, both for plant movement and materials delivery to the plant in question, e.g. to feed hoists with materials? Can the base course of permanent roads be used instead, to save money?
4. Does the soil report suggest any dubious areas on site for movement or storage?
5. Does the new work interfere with the natural site drainage?

All the above questions should be readily answered from the site inspection notes if they have been thoroughly documented.

Table 7.8 Concrete mix and place – initial listing of potential alternatives

Mixing	Transportation	Placing	Advantages	Disadvantages
Ready-mixed concrete	Hired pumps	Pump	Flexible Direct placing Small time loading on tower cranes	Cost
	Tower cranes	Skip	Relatively cheap	Large time loading on tower cranes
	Static pumps	Pump	Flexible Direct placing Small time loading on tower cranes	Not cheapest method
Site-mixed concrete	Dumpers and tower cranes	Skip	Well tried Relatively cheap	Busy streets Excessive time absorption for tower cranes
	Hired mobile pumps	Pump	Direct placing Small time loading on tower cranes	Cost Location of mixer set up
	Various combinations with dumpers, concrete hoists and tower cranes	Skip Barrows		Geometry of structure Cost Time absorption of tower cranes
	Static pumps	Pump	Direct placing Small time loading on tower cranes Flexibility Ease of positioning mixer	

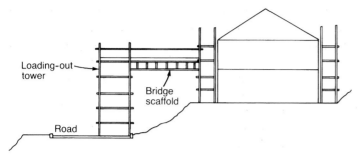

Loading tower sited on road bridge access to scaffold

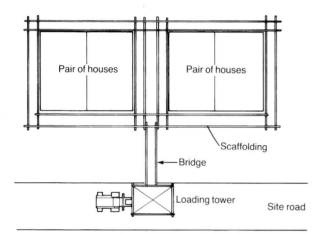

Figure 7.19 Economy in use of loading-out towers on sloping sites.

Example

A housing estate is on a sloping site. Houses are sited on level platforms cut into the hillside. Roads generally run parallel to the dwellings. All materials are delivered packaged for onward handling by fork lift trucks onto the building scaffolds. Figure 7.19 shows the solution adopted on the site, which minimizes the temporary works cost resulting from the site conditions.

The site boundaries may also have a major influence in the choice of plant and method.

1. If there is a river nearby, is there a risk of (a) construction work or temporary drainage polluting the river? (b) Do the local ground conditions allow river water to percolate into any excavations on site? How can this be dealt with?
2. If a railway is adjacent to the site, have you obtained a copy of the railway company's regulations regarding work adjacent to their running lines? Two matters will arise. First, no overswing of railway property adjacent to running lines is usually mandatory, in respect of cranes or activities involving piling equipment. Second, increased stability requirements over those used in the design are often demanded for cranes. Railtrack have a publication on this subject[6], while London Underground have a set of guidance notes[7]. Other railway companies will have similar conditions and should always be contacted by those involved in the planning of construction methods. Solutions to stability problems are given in a publication dealing with temporary works[8].

Figure 7.20 illustrates the way in which no overswing can affect the construction

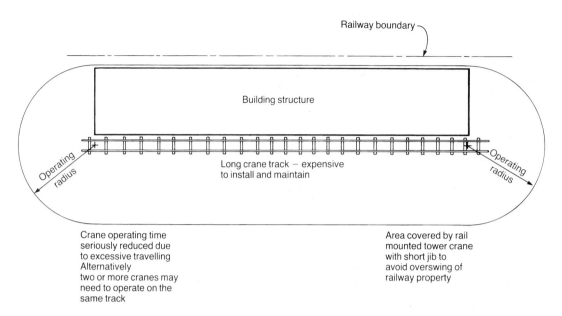

Railway boundary

Building structure

Operating radius

Operating radius

Long crane track – expensive
to install and maintain

Crane operating time
seriously reduced due
to excessive travelling
Alternatively
two or more cranes may
need to operate on the
same track

Area covered by rail
mounted tower crane
with short jib to
avoid overswing of
railway property

Figure 7.20 Influence of adjacent railway line to operation of a tower crane, see also Figure 7.6.

planning. Either the tower crane, with a shortened jib, has to travel much more than with a full length jib, losing paid work time, or an additional crane may be necessary to cope with lost operating time in order to maintain the programme required.

3. Have the heights of adjacent buildings been checked where there is a likelihood of crane overswinging such buildings? Has the attitude of adjoining owners to overswing been checked? It is an offence of trespass to overswing adjoining property without permission. This point has been tested in the courts.[9]. Remember, it covers roads and streets as well as buildings. In Scotland, the Highways (Scotland) Act 1970[10] requires more onerous action than in England and Wales.

4. Are there any other forms of obstruction close to the site boundary? Trees, radio aerials, overhead cables and so on? Will their height affect any overswinging plant?

5. The relationship of what has to be built to the site boundary can have a major impact on the choice of plant to be used. Figure 7.21 illustrates a case in point. A four-storey student accommodation complex at first sight could have all materials handled by fork lift trucks or telescopic handlers. Close inspection of the building's relationship to the specified site boundaries showed that, with scaffolding in place, there would be no access for the handling equipment to many parts of the buildings. The only solution was to abandon the materials handlers and substitute a tower crane capable of sweeping out the whole site. A much more expensive solution than was imagined.

Obstructions below, on or above the site itself

Reference back to the site inspection report should tell the method planner whether any problems are likely to arise from these particular sources. If such obstructions exist, the need for suitable temporary works must be considered, or allowance for relocation of the obstruction.

1. Are any temporary bridges required?

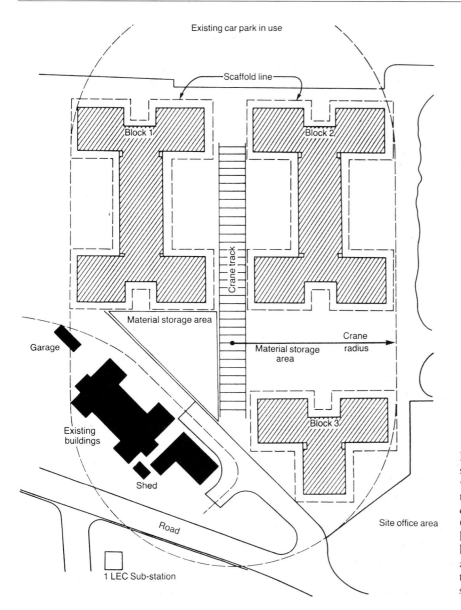

Figure 7.21 Four-storey student accommodation where a tower crane had to be used to obtain coverage of the site. Originally envisaged fork lift vehicles. Could not load out in certain areas as the site boundary was too close when the scaffolding was in place.

2. Special protection to stop plant touching overhead high tension lines?
3. Are any old basements safe or properly filled in and compacted?
4. If old basement slabs have not been removed, can piling take place without having to break through the slabs prior to piling commencing?

When any obstructions are identified, assessment of the effect on plant use is required and whether there is a need for temporary works and what the cost will be.

The effect of what has to be built

Plant is chosen to enable a particular handling problem to be solved. In other words, any consideration of plant use must take into account not just what has to be built, but also the effect of the completed building when the plant comes to be removed on completion.

It is usually quite straightforward to erect plant on an open site. In planning the location of any plant, however, it is much more important to consider the situation when the building has been completed – how can the plant be removed off site? It is this latter aspect that assumes crucial importance in the method assessment. Clearly, it must be resolved in the tender stage, and due allowance for the costs involved made and allowed for in the tender sum.

Noise limitations

With the passing of the Control of Pollution Act 1974[4], excessive noise being emitted from a construction site can become an offence liable to prosecution. Local Authorities, under the Act, have the power to set noise levels for any site as part of the planning consent. The particular sections concerning construction are 60 and 61. BS 5228: Parts 1–3: 1997 and Part 4: 1992[11] is the approved code for the carrying out of works to which sections 60 and 61 of the Act apply.

Those responsible for the tender planning of construction methods need to find out what noise limits have been laid down for the job in question. Once established, any method involving noisy plant will need to be checked against noise limits allowed. If it is too noisy, it will have to be abandoned. Better thus, at the tender stage, than to have to remove plant and equipment already installed on site. A change over to another method can have expensive repercussions, see check-list 7.1.

Comparative costing of alternative methods

When all the above questions have been satisfactorily answered, non-complying methods can be eliminated and the remaining options estab-

---

> **Check-list 7.1**   Control of Pollution Act 1974, Sections 60 and 61
>
> **At planning stage**
> 1. Check if noise limits imposed or have to be requested
> 2. Siting of plant and relationship to site boundary
> 3. Find out noise levels of plant envisaged for use
> 4. Establish need for screening and/or silencing
>
> **Helpful literature** – See Appendix A.

---

lished. As in any other commercial enterprise, the final basis of choice must be which of the alternatives will produce a least cost result. To assess the correct final choice it follows that adequate costing and comparison for alternatives must be carried out if the final decision is to be the correct one.

In the construction industry, comparative costing is often superficial and consciously or unconsciously drifted to suit the intuitive opinions of those involved. As a result, method and plant choice may not be the one which gives the true least cost situation.

If comparative costing is to provide the correct answers, it is essential that the basic ingredients of such comparisons are properly understood and a positive and systematic approach followed.

Three stage approach

To be complete, the cost analysis needs to be carried out as a three stage operation.

1. Direct comparison of alternatives.
2. Assessment of intangible factors.
3. The cost implications of method approach to related items and to the contract as a whole.

Only when all these matters have adequately been considered will a true cost comparison result.

Direct comparison of alternatives

The direct comparison between alternatives makes use of the individual plant cost assessment method given in the previous chapter. Where the method relates to output, the hyperbolic graphs are produced for each method and superimposed on each other (Figure 7.22). From the combined graph, several items of useful information result.

- The range over which method is the cheapest can be read off.
- If the planned output is not reached in any one week, a rough appreciation of the loss per unit item can be read off.
- An indication of when the method will need alteration when the programmed output has to drop towards the end of the operations involved.

Thus what started out as a planning tool is found to be a management tool as well.

From the method planning point of view, the production of graphs such as these readily translates into a computer program. Method planners can thus build up a library of plant cost data for frequent use plant. Comparison with alternatives becomes less time consuming as a result.

To provide a structured format for all the items that must be included in a valid cost comparison, the items to which cost must be assessed in any computer program need to be formalized. They are best illustrated by a typical standard form for manual comparison. See Example 16.1 in Chapter 16 and Example 4.2 in Chapter 4.

Note, finally, that every care has to be taken to see that comparisons are on an equal basis. For example, if the supply and handling of concrete is the item being evaluated, all methods must include the cost of ready mixed concrete supply or site mixing, if both are part of the alternative methods.

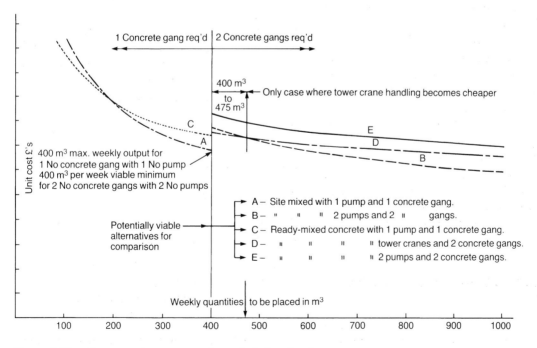

Figure 7.22 Comparative cost graphs and their value in management.

When a direct comparison has been completed it is not the end of the story.

Intangible factors

Once the direct comparison has been completed satisfactorily, attention is now necessary to the second item in the full comparison method.

By intangible factors is meant those matters to which money cannot readily be put at the time of the comparative exercise. They represent items which cannot be foreseen in terms of money at the planning stage. In other words, they cover matters which may or may not arise in the contract duration. Examples are:

1. vulnerability of method to delay by lack of labour availability, that is, a labour intensive solution in an area of labour shortage;
2. vulnerability of method to inclement weather;
3. hold-ups due to plant breakdown in plant intensive solution;
4. possibility of the client changing his mind during the contract, and impact on plant standing and who will pay standing time?

With such items, the assessment of risk and the attendant cost will be, to some extent a gamble, or if a sum of money is added in for such contingencies, an insurance premium. The action to be taken at the tender stage becomes largely a matter of experience related to the circumstances of a particular contract. In particular, the likely extent of the competition for the work.

Intangible factors can be very real and if two alternative methods show near enough the same direct costs, decisions may depend entirely on an adequate assessment of intangible factors.

Effect on related items and the contract as a whole

The final aspect of a complete cost analysis is the consideration of the effect of the method solution on related items and the contract as a whole.

Related items

Any method decision will usually have repercussions on related items. For example, a method of concrete supply may well affect the steel fixing and the formwork and falsework, in terms of unit cost for these items. A fast method of concrete placing may require a greater quantity of formwork and falsework than a slower system. Thus, for the comparison to be complete, the changes in formwork utilization need to be costed against an alternative solution. Any increase/decrease can then be set against savings/increases from the direct comparison. It is by no means certain that the cheapest direct comparison will end up the cheapest method overall. On the other hand, it may be found that all three components add up to savings in each case.

Contract as a whole

Method choice can also affect the contract as a whole. Preliminaries and overheads now represent a very significant part of the contract sum. Any method, even if not the cheapest, which effects a considerable saving in contract time can produce financial savings to the contract as a whole – which could well be the deciding factor in chosing the method to be used.

Conclusion

It should be clear, therefore, that comparative costing of alternatives requires a careful examination and, often, the apparently obvious solution will not turn out to be the right one. This can particularly be true when labour costs are escalating faster than plant charges. In such times, plant intensive solutions are likely to be a better bet than those which are labour intensive. The reverse could be true when plant cost grows faster than that for labour.

Case study

The philosophy put forward above is best illustrated by an actual case study. This demonstrates a situation where, in the event, both time and money were saved by introducing additional plant to that originally envisaged.

This example relates to the extension of a public building in the south of England. It consisted of adding two wings, at right angles to each other, which together with an existing side of the original building were designed to create a ceremonial square, with one open side. Both new wings were seven storeys in height. The general arrangement is indicated in Figure 7.23.

The structural form was in reinforced concrete, using plate floors with columns of identical cross-section in all locations, both longitudinally and vertically. Concrete walls were provided in stair wells and lift shafts to provide lateral stability by acting as shear walls.

Major quantities were: floors 300 mm thick giving a quantity of concrete per floor of 790 m³. Floor width in each wing was 18 m and the specification called for concrete to be poured in bays full width, but only 36 m³ in volume. Thus the bay length was 6.67 m.

To meet the construction programme, it was only necessary to average 50 m³ per week of concrete poured in floors.

The method priced for in the tender is shown in Figure 7.23. Two tower cranes were employed to give full coverage of the structure – one static

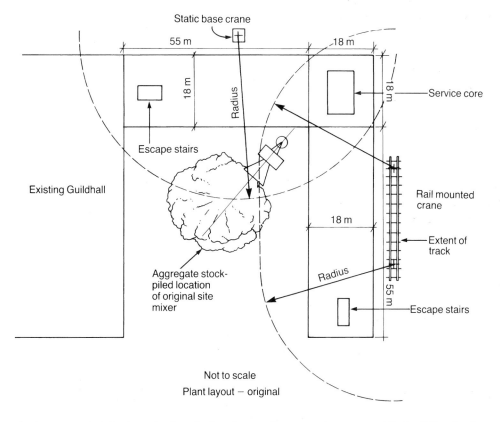

Figure 7.23 says (in figure labels): Static base crane, 55 m, 18 m, 18 m, Radius, 18 m, Service core, Escape stairs, Existing Guildhall, Rail mounted crane, 18 m, Extent of track, Aggregate stockpiled location of original site mixer, Radius, 55 m, Escape stairs, Not to scale, Plant layout – original

Figure 7.23 Original method and plant layout for extension to a public building in the south of England. (See also the extension to this case study used as the illustration for construction cycle preparation in Chapter 11).

based, while the other was on track to master the longer wing. Because the concrete quantities were very low, a small site mixer was provided in the location shown. On this basis the contract was started.

After some weeks into the contract it was realized that the contract programme was not being met. Analysis of the situation showed that the work load on the tower cranes had been underestimated, to the extent that one day per week was being lost on programme. From the study, it was clear that the floor concrete was occupying each crane for nearly a day – the tower cranes being very slow, handling only on average 6 m³ per hour – thus one bay was taking 6 hours for the pour of 36 m³, without allowing for start up and clean up times.

Because of this situation, it was decided to take the floor concrete out of the crane work-load altogether and use a hired-in concrete pump. For this to operately efficiently, it would need to be fed with ready mixed concrete. The site mixer was removed and even the columns and walls (still poured by crane) used ready-mixed concrete.

To confirm the wisdom or otherwise of this decision, a detailed costing of the new method was compared with the actual costs being obtained from the original method. To the surprise of all concerned, not only was the time being retrieved, but the overall cost of the concrete operation would be reduced. While at the start of the new method, the cost benefit was only an estimate, the following weekly costs proved that the estimate was indeed accurate. In Example 7.2 are shown the actual costs achieved for both methods derived from the weekly cost sheets. This tabulation is, to a degree, unique. When alternatives are compared to decide the most cost effective one, alternatives are not normally tried out in practice. Here both methods were used and real costs obtained for both.

As further proof of the new method's validity, the weekly work loads for the two cranes were recorded and are shown in Example 7.3. As will

be seen, even without the floor concrete, the full working week of 48 hours (including overtime) was needed to keep on schedule.

To ensure that the pump could operate as efficiently as possible, permission was sought to make the bay poured much larger – to approximately 113 m³, thereby reducing the number of bays per floor from twenty-two to seven. How this reduced the cost of the provision of stop ends is demonstrated in Example 7.4. This is an excellent example of the best influence of a change in direct method on a surrounding item. It will also be apparent that larger pours mean less frequent pours. In this case, what will happen to the cost of falsework and formwork? It was found that no increase arose because the method used was a proprietary one which was hired in at a figure of so much per m² per month. In such cases, the quantity used has no impact on unit cost provided that the turn round per m² is not affected.

Plant chains

While many situations in the handling process may be solved by the use of one item of plant, there will be other situations where two or more items of plant may be required to work in sequence. For example, in excavation the sequence of events is likely to be: dig – load vehicle – travel – tip. To achieve this sequence economically, the vehicle needs to be of a suitable capacity in relation to the excavator output (vehicles must not be filled so quickly that many vehicles are needed with attendant drivers); the return time to tip and time to fill one vehicle will give the number of lorries – to keep the excavator digging as much as possible, and avoid vehicles standing. At the tip a bulldozer or other plant will be needed to spread tipped material and compact and clean up. This again needs to be of a size that can adequately deal with the incoming material.

To summarize, any chain of plant operating together must match in capacity to maintain all

Example 7.2 Cost comparison between two methods of concreting used on site

Comparative costs – Pump vs Crane

Original method. Site mixer and tower crane

per m³ of concrete

Concrete materials	26.45	
Mixing plant	1.35	
Mixing labour	3.35	31.15
Crane cost (incl. driver + banksman)		7.80
Labour place concrete		8.35
Placing plant (vibrator + standby)		Common to both
Cost per m³		£47.30

New method. Ready mixed concrete and pump

Supply ready-mixed concrete	32.70
Pump hire (incl. 2 men)	4.75
Labour place concrete	2.65
Placing plant (as above)	Common to both
Cost per m³	£42.10

Data Base: Labour cost at the time (full cost to employer) £7.50 per hour.

Example 7.3 Crane workloads after method change on site

	Crane 1	Crane 2
Workloads – hours per week		
Handling formwork	17.30	17.80
″ reinforcement	4.90	6.10
″ concrete (cols/walls)	6.20	7.90
″ other materials	9.30	8.40
Standing time (aggregate/week)	6.30	4.40
	44.00	44.60

Official working week at the time 44 hours

Concrete output per week to meet programme av. 50 m³
Crane time to pour 50 m³ was 8 hours.

Example 7.4 Saving in stop end costs due to method change

Comparative costs – large bays vs small bays

Original method – Crane pouring   22 No. pours needing 21 No. stop ends
New method – Pouring with pump   7 No. pours needing 6 No. stop ends
In each case: Slab thickness   300 mm Each bay stop end is 18 m long
                Slab concrete per floor 790 m^3

Stop end costs
Labour fix and strike   8.70
Material   1.50
Scabble joint   0.50
                            _____
                            £10.70 per m

Stop end cost per floor
With crane pouring
    10.70 × 21 × 18 = £4044.60 per floor or £5.120 per m^3 of concrete placed
With pump pouring
    10.70 × 6 × 18 = £1155.60 per floor or £1.462 per m^3 of concrete placed
    Saving in stop end cost per floor is   4044.60
                                           −1155.60
                                           _____
                                           £2889.00

If the method had been used over the whole contract an overall saving of £2899 × 7 = £20 223 would have resulted. Alternatively, looking at the E O cost per m^3 of stop ends, a saving of £3.658 per m^3 would have been achieved.

---

items working as efficiently as possible. Failure to do so means that one item of plant could prevent others from achieving maximum output.

As an alternative, the addition of extra plant to a chain of events may allow overall economy by improving the output of the original plant concept. The case study given earlier in this chapter is a good example.

Conditions for efficient use

Once a decision has been made on the plant and method to be used, the moment has arrived to consider whether anything needs to be done to boost the efficiency of the plant in question.

Example 1

A concrete structure has a tower crane handling formwork, reinforcement and the concrete. Concrete is delivered by a three cubic metre capacity tipping lorry from a site batching plant. The crane capacity allows the use of one cubic metre skips at maximum radius. The distance to the batching plant is short and one concrete carrier could easily cope with the output if it could fully discharge its load in one operation. To allow this to happen, the contractor designed a turntable on which three one cubic metre skips could stand. By means of an electric motor with suitable cut outs, the turntable could be operated by a push button, to rotate from one skip to another automatically. When the concrete vehicle arrived, the first skip was filled, the turntable rotated to the next skip, filled and so on to the last skip. The delivery vehicle is then freed for the return journey to the mixing plant, while the crane has three full skips to handle before a new load returns. Figure 7.24 shows this 'aid to efficiency' in action.

Example 2

Of all the motions of a tower crane, the hoisting speed is the slowest. On a tall structure, hoisting from the ground can take up a great deal of the crane's time in the course of the working day. The introduction of a high speed hoist to deal

Figure 7.24 Improving the crane handling of concrete as well as reducing transport needs. A method seen in Norway. Use of three 1 m³ skips on an electrically operated turntable, see full description in the text.

with the vertical movement of materials can allow the crane to operate in a distribution role at the working level. (This concept, of course, presupposes that the material in question can be carried by the hoist. Concrete is an ideal medium in this respect). Figure 7.25 shows a high speed concrete hoist fulfilling this role on a 35-storey structure, by feeding two climbing cranes.

The above example is a very common solution to improving a tower crane's output. A hoist is relatively cheap to operate and usually pays for itself by increasing the crane's output on other handling items in a given time.

Example 3

With modern hydraulic excavators, there is a wide range of bucket sizes available, both in relation to width and type. For greatest efficiency, therefore, use of the correct type and size of bucket is crucial if unnecessary digging is to be avoided or, alternatively, maximum output is to be achieved in bulk digging. As almost a bonus, hydraulic machines can be also used to push trench sheeting into the ground when installation is required for safe support of excavations as required by the Construction (Health, Safety and Welfare) Regulations 1996[12]. In other words, without the need for anyone to be in the trench itself. For greater detail see Chapter 4 which describes methods and gives illustrations of methods available.

Dominant plant

Of all the available categories and types of construction plant, it is possible to re-divide the list into two groups of special consideration: substitute plant and dominant plant.

Substitute plant

This group encompasses the bulk of construction plant in use. Substitute plant can be defined

Figure 7.25 Use of high speed hoist to feed two tower cranes with concrete. More crane time becomes available for other materials handling.

as items of plant that can be used to replace other items of plant without altering the planned sequence of events. Such items may provide the whole or part of the handling chain. The particular chain of events chosen would normally be those giving the most cost effective result.

Dominant plant

Some types, small in number compared to substitute plant, exert an influence on the planning which is quite different. Their adoption controls the sequence of events and the planning of the method has to revolve around their efficient operation – i.e. they dominate the situation. Examples in building construction are tower cranes, concrete pumps and fork lift trucks.

How this concept works is best demonstrated by the tower crane. When a crane of this type is seen as the main item of plant to solve the handling problems of a particular site, it becomes immediately obvious that a number of factors in its use stand out.

- It will be the major item of plant on site (other than any item of bulk excavation).

- If required to handle a variety of materials, conflict will arise between different gangs as to who has priority.
- It will stand out like a sore thumb if it is not working.

As an item of plant that is relatively expensive to operate and considerably more expensive to establish on site and remove at a later stage (as seen in the previous Chapter) management will speedily recognize that the crane's use will require careful planning and control. Indeed, a cycle of events will be needed to which all parties will be required to adhere. In achieving this, all other items of plant and the planning and organization must fit into this philosophy.

The use of dominant plant creates a number of planning and management advantages.

- Management requires no pushing into adequate planning. It is self apparent.
- Better planning creates improved management.
- Dominant plant that is idle is only too apparent. Action to remedy the situation is more immediate. Better management again.
- Consequential benefits also often arise, in unexpected ways (see case study earlier in this Chapter).

Note, in the case study referred to above, that more than one item of dominant plant can be used on any given site. In so doing, the cyclic solution becomes more complicated. As a result, more care is needed in its preparation.

To conclude, therefore, dominant plant has a greater influence on contract efficiency than just the handling activity.

Keeping up to date

Those involved in the construction process, and especially site managers and staff specializing in construction method planning, need to keep up to date with new developments in plant and methods. Not all new developments are necessarily improvements on the previous ideas, but the planning activity must be able to recognize a cost saving idea when it appears. To do so, calls for the ability to see the implications beyond the new concept itself. The following two examples make the point quite clearly.

Ready-mixed concrete vehicle with integral conveyor

Certain companies now offer ready-mixed concrete deliveries by a vehicle which has a conveyor mounted upon it so that the range of direct discharge is greatly enhanced. The advantages of this approach become very clear when one considers the problem of concreting oversites on a housing estate. The normal routine would be: discharge part load into wheel barrows – wheel to oversite – tip – return to ready mixed concrete vehicle – start cycle again. Such a cycle involves at least four men with barrows, apart from the laying and compaction gang. It is also time consuming in emptying the ready mixed truck, which will probably have a capacity of $6\,m^3$ at the least.

By the introduction of the conveyor system (Figure 7.26), no men with barrows are needed (provided that the vehicle can get reasonably close to the foundations) and the productivity of the laying gang will be greatly increased as the conveyor output will be faster by far than with barrows.

To evaluate real savings properly it will be necessary to price the old method with the new. In particular, the concrete cost will be greater in the new method. The reason, of course, is that with the weight of the conveyor now within the allowable axle load, less concrete can be carried, and a return is necessary on the conveyor cost. Nevertheless, what can be left out in cost terms in the new method will often be found to make the method profitable compared with the old. At this point reference needs to be made to the requirements for a complete cost analysis given earlier in this chapter.

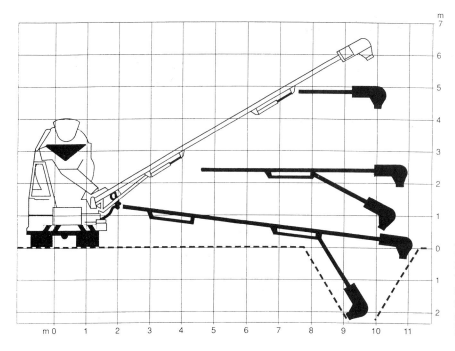

Figure 7.26 Ready-mixed concrete delivery vehicle with in-situ conveyor discharge. Graph showing range of use. (Tarmac Topmix Ltd).

New floor construction methods

In the last decade, the construction of industrial floors has undergone a revolution in method terms. In the past, specifications demanded that floor bays had to cast in areas not exceeding 12 ft × 12 ft (as it then was) and in a chequer board pattern, to allow shrinkage in the cast bay before another bay was cast against it.

More recently, the casting of floors was revolutionized by the use of the 'long bay' concept. That is, floors cast in one continuous strip down the long dimension of the floor – the whole area being sub-divided into a series of such strips. Before casting, crack inducers are positioned below the reinforcement at specified centres. After casting, when the concrete has reached a suitable strength, the bay is saw cut across the width, over the location of the crack inducers. The immediate advantage of the method is that saw cut joints have truly level areas on both sides. This compares with bay-to-bay construction which, for some reason, always seems to cause the concrete to lift up at the stop end (Fig-

ure 7.27). When trucks or fork lifts pass over such joints there is a heavy dynamic load on the joint which, in time, usually cracks the edge of the bay.

There are a number of equally if not more important features of the continuous bay approach.

1. Continuous bays make the pouring sequence much easier and faster with good access. Chequer board bays make access very difficult.
2. If the bays are poured along the outside walls first, cladding and service installation can begin much earlier on a good hard level floor foundation.
3. In the winter period, production can be markedly improved by the use of 'vacuum mats'[16]. Instead of having to stop concreting early, to allow the concrete to go off enough for finishing, the vacuum mat method allows concrete to be laid up to three-quarters of an hour before finishing time. After the concrete has been laid and levelled, a special type of mat is laid over the

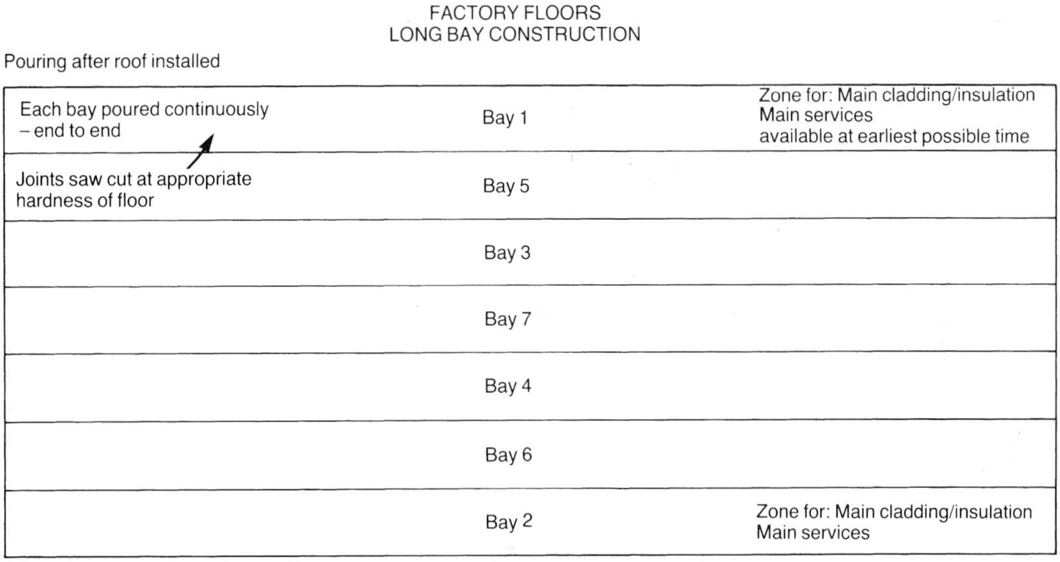

FACTORY FLOORS
LONG BAY CONSTRUCTION

Pouring after roof installed

Each bay poured continuously – end to end	Bay 1	Zone for: Main cladding/insulation Main services available at earliest possible time
Joints saw cut at appropriate hardness of floor	Bay 5	
	Bay 3	
	Bay 7	
	Bay 4	
	Bay 6	
	Bay 2	Zone for: Main cladding/insulation Main services

Figure 7.27 Principles and advantages of continuous bay construction. For specification, methods and operational details see references[13][14][15].

wet concrete. The immediate surface in contact with the concrete is full of fine holes, such that water can be sucked out but not the cement particles. Above this layer is a porous layer connected to a vacuum pump by a pipeline. Suction is applied to the concrete for about twenty minutes, when the mat is removed. At this stage the concrete is hard enough to walk on and finishing by power float can be started.

Item three above is a classic example of the principle of spending money to make money. The cost of the vacuum equipment in the colder months of the year more than pays for itself in increased production from the floor laying team. In one case known to the author, the vacuum equipment added ten percent to the cost per cubic metre of concrete but, in the process, increased output from the work force by twenty-five percent. In consequence, profitability also increased.

More recently, further and more spectacular developments have taken place. Many clients

have demanded higher specification floors. The main result has been the provision of jointless slabs, creating lower maintenance costs. One company[17] can now achieve 3000 m² of jointless construction in one working day, including finishing. This approach meets design requirements for a high standard of surface tolerance and where reduced maintenance is as important as maximum flexibility for the prospective user.

Foundation design is critical in ensuring that the slab can meet its life expectation. Floor flatness achieved will normally be in Category 2, as defined in the supplement to Concrete Society Technical Report 34[13]. In defined areas, flatness to Category 1 can be delivered. The supplement to the above Report[14] gives additional details. A good practice guide is also available[15].

Where ground conditions are unsuitable for a ground bearing slab, Stuarts provide a suspended slab design where reinforced ground beams and piles support a suspended slab, steel fibre reinforced, and can achieve 12 000 m² in two working weeks.

Demand for jointless floors has grown

rapidly with an increase of 15% in the past year for Stuarts alone. The company produce upwards of 15 000 m² every day of the week. To achieve this figure, automatic, laser controlled levelling equipment is used. Figure 7.28 gives a general impression of a floor being laid, with all the equipment in view. Figure 7.29 gives a closer view of the automatic laser levelling machine.

In addition to the references given, a new book has been published (March 1999) on the whole subject of jointless floors. It is detailed in Further Reading section at the end of this chapter.

While the above examples mainly relate to concrete technology, they provide clear examples of how new methods can have repercussions beyond the basic reason for use.

Safe use of construction plant and methods

From the contractor's point of view, plant and method operational safety depends on fulfilling certain readily defined principles.

1. Knowledge of, and compliance with, manufacturers' erection, operating, dismantling and maintenance instructions.
2. Adequate training of erectors, drivers and banksmen.
3. Knowledge and practical implementation of safe operating conditions and procedures, which are the responsibility of the user.

In all these, it must be understood that compliance with all relevant statutory requirements and codes of practice is deemed to be included. Here, the Health and Safety at Work etc. Act 1974[18] is the overriding legislation. See also [12][19][20].

More recently this has been added to by three sets of Regulations:

1. The Management of Health and Safety at Work Regulations 1992[19];
2. The Construction (Design and Management) Regulations 1994 [20];
3. The Construction (Health, Safety and Welfare) Regulations 1996 [12].

Figure 7.28 Large area pouring of ground slabs. (Stuarts Industrial Flooring Ltd).

Figure 7.29 Large area pouring of ground slabs. Close-up view of automatically controlled, laser guided levelling machine. (Stuarts Industrial Flooring Ltd.)

All of these affect the use of plant. Their contents are outlined in some detail in Appendix A.

It follows that, for the safe use of construction plant and methods to be effective, construction site managers and those responsible for the method planning have an essential contribution to make. Safety has to be seen as an essential part of the method planning and operational process, not an after thought when the decisions have been made.

The implication of the three principles are now considered in more detail.

Manufacturers instructions

Manufacturers are required by the Health and Safety at Work etc. Act 1974, Section 6 (1) (c): 'To take such steps as are necessary to secure that there will be available in connection with the use of the article at work adequate informa-

tion about the use for which it is designed and has been tested, and about any conditions necessary to ensure that when put to that use, it will be safe and without risks to health'.

In addition, Section 6 (3) states: 'It shall be the duty of any person who erects or installs any article for use at work in any premises where that article is to be used by persons at work to ensure, so far as is reasonably practicable, that nothing about the way in which it is erected or installed makes it unsafe or a risk to health when properly used.'

In law, therefore, the manufacturer is required to provide all the information necessary to use his product safely, both in respect of erection/installation and operation where these items will be carried out by the purchaser/hirer. In fact, this is not strictly true. The wording of Section 6 (1) (c) reveals what is often a popular misconception. The manufacturer or supplier's legal duty is not to **supply** adequate information, but '. . . to secure that there will be **available** . . . adequate information . . .'. It is important that contractors recognize this fact and ensure that all relevant information is in their possession and that they act on the requirements laid down.

It is to be regretted that the law is worded in this way. Responsible manufacturers will supply a comprehensive use and erection manual to the purchaser without the need to ask for them. It would appear, however, that the purchaser/user has the ultimate responsibility, as far as is reasonably practicable, of making sure that his information is, in fact, complete. By oversight or error it may not be so.

Within this question of manufacturer's instructions, the user needs to be sure that (a) general compliance with erection/dismantling and operating instructions takes place and (b) under no circumstances are modifications and alterations made to items of plant or equipment without the approval of the maker.

Safe use of cranes

Of all the items of plant likely to be in use on a construction site, the use of cranes is arguably the most likely source of accidents due to improper use. The construction planner, particularly at the tender stage, needs to be well briefed in the necessary requirements for safe installation and operation, so that adequate monies can be allowed in the tender.

Strict compliance with the requirements of the maker/statutory regulations is necessary in respect of:

1. use of outriggers;
2. static bases for tower cranes or track provision;
3. anchorage needs for tower cranes;
4. inspection and renewal of ropes;

5. maintenance generally;
6. statutory inspections;
7. sequence and specific rules relating to erecting and dismantling;
8. calibration and regular checking of safe load indicators, audible warning devices and electrical cutouts.

Where the plant is contractor owned, inspections, testing, erection and dismantling and renewal of ropes etc. will be carried out by the company plant department. If the plant is hired in, the plant hire firm will be responsible.

Whichever the method of supply, it must be understood that the remaining items (1) to (3) fall into the temporary works category and the method planner will need to obtain expert advice as to what will be needed and ensure that the estimator allows an adequate sum of money for their execution in the tender.

The whole subject of crane safety is best summarized, so far as the local conditions of use are concerned, by the production of the above tabulation of the key issues which can be copied at will and circulated to all relevant levels of supervision. It also provides an *aide memoire* for those doing the method planning of things that need to be remembered and allowed for. The first part ('Section 2') covers on-site use; the second ('Section 3') provides a check-list for cases where cranes have to swing outside the site boundary.

From these it will be clear that the main contractor has a major contribution to make to the safe use of cranes. For this to be fully effective, knowledge of the crane's characteristics and adequate training of operating personnel is not enough. Site planning and organization must recognize and practice the needs for safe operating conditions.

Knowledge and practical implementation of safe operating conditions and procedures

In this table, the concern is not with the crane itself, but with the local conditions in which it is operating. Sections 2, 3, 6 (3), 7 and 8 are particular relevant portions of the Health and Safety at Work etc. Act 1974. Site management and planning are very much involved, however well the crane may be designed, however well trained the drivers and banksmen, the site organization has a direct responsibility under Section 2 and Section 3.

Section 2

In respect of Section 2 adequate consideration of the choice of crane type best suited to the site conditions and the work to be performed is of prime importance and relates directly to the planning concept. Safety in use should always be an integral part of planning – not an afterthought.

Under this section the requirements are tabulated as follows:
(i)   Realistic assessment of load weights to be lifted.
(ii) Provision of the correct type of slings, lifting bridles and brothers – together with safety cages, nets etc. where appropriate.
(iii) Site conditions. Are they suitable for the safe operation of the crane in question?
    (a) Local ground conditions capable of supporting the loads imposed by outriggers – if not, need for special precautions.
    (b) Impact of backfilled trenches etc. on the stability of mobile cranes or excavator conversions.
    (c) Provision of firm level areas for the crane when lifting.
    (d) Impact of obstructions – particularly HT wires – and the measures needed to prevent accidents.
(iv) With tower cranes:
    (a) Bases or track sufficient for the purpose. Proper consideration of ground conditions and the loads to be carried.
    (b) Base or track construction strictly to specification.
    (c) Anchorages to building to specification.
    (d) If using climbing crane, has the Engineer agreed that the structure can carry the loads involved?
(v) Proper attention to driver/banksman communication. Over significant heights use of radio or closed circuit television. Often cheaper and more effective than using an additional man. Or where driver is blind at lower levels.
(vi) Strict supervision against the temptation to 'overreach a bit and ignore the audible warning'.
(vii) Constant vigilance for any unforeseen conditions arising.

Section 3

Section 3 of the Health and Safety at Work etc. Act 1974 states, *inter alia*, that:

> It shall be the duty of every employer to conduct his undertakings in such a way as to ensure, so far as is reasonably practicable, that persons not in his employment who may be affected thereby are not thereby exposed to risks in their health or safety.

Thus, in crane operation planning, due care must be taken to safeguard the safety of the public. In order to achieve this, the following aspects need to be considered by the contractor.

1.   If possible avoid crane jibs overswinging the public highway or adjacent property.
2.   Where railways are adjacent it is a statutory requirement that jibs must not overswing railway property. The railway concerned must be consulted when working near and may impose additional requirements to improve crane or piling plant stability against collapse or overturning. See references [6] [7].
3.   Notwithstanding (1) above, overswing may be the only practical and economic solution. If so, the agreement of all adjoining owners is necessary as overswing constitutes a trespass in law [9]. If agreement is reached, the following precautions should be taken:
     (a) try to limit overswing to non load carrying movements, or
     (b) by the use of luffing or trollying movements keep the load inside the site boundary to the greatest possible extent.
4.   Where there is high risk property adjacent – e.g. a school or playing field – method planning should find an alternative solution to overswing regardless of any increased cost.
5.   Where offloading vehicles on the public highway, allow for restraining the public whilst the load passes over the pavement.

Technical aspects apart, site management on site must ensure that **all** employees, whether working for the main contractor or not, are aware of the dangers of crane operations. In particular:
(i)   If not working with the crane – keep well clear;
(ii) If working with the crane:
     (a) Watch tail radius movement – especially if limited clearance is present.
     (b) Never stand under the load.
     (c) Exercise vigilance when loads are being lowered into position or lifted from vehicles or

References

1. Illingworth, J.R. (1982) *Site handling equipment.* Institution of Civil Engineers Works Construction Guide. ICE, London.
2. Anson, M. *et al.* (1986) *The pumping of concrete – a comparison between the UK and West Germany.* Research report funded by the Science and Engineering Research Council and the British Concrete Pumping Association. University of Lancaster.
3. Anson, M. and Cooke, T.H. (1988) *The operation and capacity of the ready mixed concrete industry – a regional study.* University of Lancaster.
4. *Control of Pollution Act 1974.* HMSO, London.
5. Building Employers Confederation (1976). *Safe use of rough terrain fork lifts.* Report of a working party. Building Advisory Service.
6. Railtrack plc (1978) *Notes for the guidance of developers*

and others responsible for construction work adjacent to the Board's operational railway. Civil Engineering Department Handbook 36. Railtrack plc, London.

7. London Underground. Department of Civil Engineering. (1986) *Notes and special conditions for work affecting London Underground.* Originally issued by the then London Regional Transport, London.

8. Illingworth, J.R. (1978) *Temporary works – their role in construction.* Thomas Telford, London.

9. *Woolerton & Wilson vs Richard Costain (Midlands) Ltd* 1970 1 All England Law Reports p 483.

10. *Highways (Scotland) Act 1970.* HMSO, London.

11. British Standards Institution. *Noise control on construction and open sites.* BS 5228: Parts 1–3: 1997 and Part 4: 1992. BSI, London.

12. *Construction (Health, Safety and Welfare) Regulations 1996.* HMSO, London.

13. Concrete Society. *Concrete industrial ground floors.* Technical Report No 34. Second Edition. 1994. Concrete Society, Slough.

14. Concrete Society. *Concrete industrial ground floors.* Supplement to Technical Report No 34. 1997. Concrete Society, Slough.

15. Concrete Society. *Concrete industrial ground floors. Good practice guide No 1.* Concrete Society, Slough.

16. The principle of extracting surplus water from concrete floors when freshly laid by covering with a porus mat, to which is applied a vacuum. Patented method marketed by Alimack Ltd.

17. Stuarts Industrial Floorway Ltd. Birmingham, Borough Bridge and Loanhead.

18. *Health and Safety at Work etc. Act 1974.* HMSO, London.

19. *The Management of Health and Safety at Work Regulations 1992.* HMSO, London.

20. *The Construction (Design and Management) Regulations 1994.* HMSO, London.

Further reading

Knapton, J. (1999). *Single Pour Industrial Floor Slabs.* Thomas Telford. London.

PART TWO

Establishing Methods and their Planning Control

Introduction

In this part, each element of the sequence of construction is examined and the principles established in Part One applied to arrive at and determine cost effective plant, method and labour elements, together with time scales, when related to specific situations.

Before doing so, it is important to recognize that, within the sequence of operations necessary to complete a particular structure, some are what may be called critical elements. That is to say, activities which inevitably control the attainable rate of progress to completion.

Critical items

The critical items can be listed as: foundations, deep basements, superstructure, cladding and services and, in the case of air conditioning, final balancing after the building has been occupied. Each will interrelate with the others in the form indicated in Figure 1. In a world where clients increasingly want fast construction performance, it will be clear that designers need to ensure that the critical items are capable of the best possible construction time, by designs tailored to this objective.

Each of the critical items will be considered later in detail, but the following brief notes are useful at this stage.

Foundations range from straightforward simplicity to great complexity. Assessing an accurate time scale with an effective method is an essential feature to get the contract off to a good start.

Deep basements incorporated into the foundation situation can take a long period of time to construct. The ability to overlap basement construction with other operations is often very limited. Planning is therefore often unable to telescope one item with another to shorten time.

Superstructures. The structural form of a building is clearly a critical path item. This will be true whether the structure is low or high rise. Until the roof is in place, final waterproofing cannot be completed. Thus the speed at which the superstructure can be built determines the speed possible for all following operations – or those already in progress – which carry on after the superstructure.

Cladding is not always a critical item – in the sense that most carcass work does not depend on a weathertight building. On the other hand, in many cases, service carcass cannot be completed until the cladding is in position. For example: perimeter pipe work supported by the cladding support; cill-line heaters and other items which may depend on support from the cladding support structure. In so-called fast track construction, the development of a proprietary 'dry envelope' method was designed to overcome the difficulty of cladding programmes being unable to keep up with fast service installation. In planning terms, it is always desirable to keep cladding as tight up with the structure as possible.

Services today, in commercial and industrial premises, are highly sophisticated, especially where air conditioning and computer installation are involved. The main carcass needs completing at the earliest possible moment. All carcass work including testing of services needs to be complete before final finishes can take

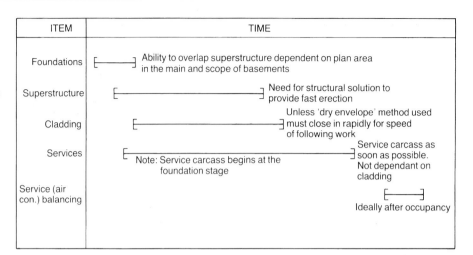

All finishes tend to be quantity related, so present no significant problem

Foundation situation a major factor in time scale.
If basements with diaphragm or piled external walls, together with ground anchors, time scale is very long, because:
– regardless of site size, each specialist sub-contractor needs all, or most, of the site to operate.
– overlapping of specialists not readily possible to any large extent.

Figure 1 Significant elements of construction.

place. Failure here may involve taking down previously erected units and ruining completed decoration. Service installation planning – the sequence of operations, supervision and integration with other activities – is a key factor in achieving the maximum overall time reduction. **Service balancing.** Many clients often fail to realize that final balancing of an air conditioning system cannot take place until the building is occupied – and that such balancing may take as long as three months.

Time/cost solutions

Construction planning should begin at the tender stage. An essential part of the planning role,

here, is to look for potential savings, both in time and money, originating from the plant and method solutions. Such savings may well be instrumental in obtaining a contract against the competition. In this aspect of planning, much will depend on the technical knowledge and experience of those doing the planning. In each of the following sections, examples are given to illustrate the way one should be looking to create savings of this kind. Note: such savings have to be proven by means of true cost comparisons (Part One, Chapter 7).

With the above observations in mind, a detailed examination of the key elements of construction and how their components are dealt with in the planning process can begin.

Chapter eight

Non-piled foundations

The foundations to a structure can range from simplicity to cases of great complexity. Some will be very shallow, where light loading and good ground come together. At the other end of the scale, heavy loads to be supported in bad ground conditions will need the use of piling to considerable depths to support the loads involved. Foundations can become further complicated by a structure having a deep basement, which also provides lateral support to the surrounding ground and adjacent buildings and streets.

For the sake of clarity, this chapter deals with non-piled foundations and Chapter 9 with piled foundations. The specialist nature of deep basement construction is dealt with in Chapter 10.

Non-piled foundations divide into three specific types: strip footings, pad foundations and rafts. In each case, the choice is related to ground conditions and the loading conditions that have to be supported.

Strip footings

As the name implies, strip footings are long continuous strips, usually of concrete, created in the ground at a suitable depth to provide adequate support for the loads brought upon them. Such foundations may or may not be reinforced, depending on the ground conditions and the loading involved.

The loading on this type of foundation is of a linear pattern. It will arise from load bearing construction involving brick, block, masonry or other material in low rise housing, utility buildings or low rise industrial structures (Figure 8.1). A further application arises where a structure may be carried on individual foundations but the cladding is supported on strip footings which span between the individual column foundations. In this case such footings will need to be reinforced and act as a beam.

Planning considerations

A number of matters need examination at the pre-tender planning stage.

1. If the excavation for the footing requires a trench over 1 m deep, the Standard Method of Measurement requires a minimum width of trench of 0.75 m to ensure sufficient working space for the bricklayer. This minimum width also applies to the footing width[1].
2. If the excavated depth of the trench is greater than 1.2 m, the Construction (Health, Safety and Welfare) Regulations 1996[2] require that the trench shall be adequately supported. This, in turn, may require the excavation to be wider still to accommodate the support system while maintaining the working space for the bricklayer.
3. In addition to the points above, it is necessary to examine the service drawings to see what entries pass through or under the footings. One can then consider what effect these may have on the foundation sequence of construction and whether delays are likely if Public Utility companies have to install.

From the answers to the above, the actual trench width will be established. This, in turn, determines (a) the bucket width for excavating the trench (b) the width of the foundation concrete required and (c) the extent of temporary works in trench support to be priced – if any. (Chapter 4, Support of excavations.)

At this point an activity sequence can be established of all the possible items that may arise in this type of foundation. From this it will be obvious those items that occur in a particular situation (Table 8.1).

Those items relevant to a given situation can now be assessed in terms of plant and labour requirements. By applying appropriate con-

stants for performance (Chapters 6 and 7), a time scale for each item can be found. Such information will form the basis of a method statement for the estimator to price (Chapter 17) and also provides the starting point information for programming the foundations (Chapter 18).

The trench fill method

The traditional strip footing method already described has a number of disadvantages. As many as ten separate operations can be involved, each of which has to follow the previous one. As a result the time scale becomes elongated in relation to the amount of work to be executed.

The alternative is a method known as the trench fill system (Figure 8.2). Where conditions are suitable, the method involves only four operations to bring the foundation structure up to ground formation level. These are: (a) excavate over site to reduce level; (b) excavate trench to required depth; (c) concrete to trench; (d) remove excavated material off site.

The method, to be effective, needs the trench concrete to follow the excavation as quickly as possible. Even trenches which will not stand

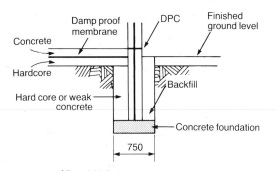

Cross section   Housing typical strip foundation

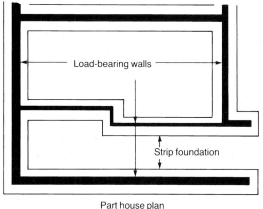

Part house plan

Figure 8.1   Principles of strip foundations.

Table 8.1   Activity sequence for strip footings

Item	Description
1.	Remove topsoil and store on site, followed by reduce levels
2.	Excavate trenches
3.	Support excavated trenches – if more than 1.2 m deep
4.	Prepare formation of trench – including make up levels if needed
5.	Lay blinding, if footing reinforced
6.	Fix reinforcement if specified
7.	Concrete to footing
8.	Brickwork or alternative to DPC level
9.	Backfill to GL or formation level of floors as specified
10.	Remove surplus excavation off site

open for long without support can be accommodated by rapid filling tight behind the excavation. Brickwork or other material is avoided altogether below ground level (Figure 8.3).

The method is a good example of the philosophy expounded in the first part of this volume

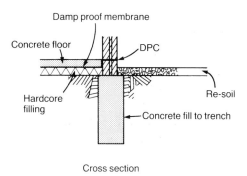

Figure 8.2 Trench fill method for strip foundations.

Figure 8.3 Direct concreting of trench with ready mixed concrete. (British Cement Association).

– savings in materials do not necessarily produce savings overall. A detailed cost comparison was carried out by quantity surveyors S Lazarus and Partners[3] in order to establish the relative costs between strip foundations and the trench fill method. While the results varied for different sizes of the house types chosen, savings using the trench fill system amounted to between 11% and 36%.

A British Cement Association publication[4] provides practical details for applying the method.

Planning considerations

Table 8.2 presents the sequence of activity that results from this method.

If the ground conditions are very bad, a trench support system may be needed. In most cases, the time lapse between dig and fill is so small that no support will be required (Figure 8.3).

While the cost comparison referred to shows savings over the conventional strip footing, further savings arise from the faster construction time through reduced overheads.

Key issues here are: balancing the rate of excavation with the rate of ready mixed concrete delivery, together with the labour needed to maintain the tamping and levelling of the top surface at the same rate; can ready mixed concrete vehicles get to the trenches without bogging down? or are some form of temporary roads and their cost going to be involved?

Table 8.2   Activity sequence for trench fill

Item	Description
1.	Excavate to reduce level after removal of topsoil to store
2.	Excavate trenches and position ducts for service entries
3.	Fill concrete in trenches
4.	Remove excavated material off site

Pad foundations

While strip footings are designed to support linear loads of low intensity, pad foundations are designed to support high loads over a limited area. Such foundations are common where a structural form brings loads to the ground by way of columns. As such, they are applicable to reinforced concrete, pre-cast concrete and structural steel design solutions. Depending on ground conditions, pads will be found in a wide variety of structures – warehousing, low-rise industrial plants requiring large clear areas and high rise office and domestic accommodation with favourable ground below.

Pad types

Depending on the nature of the structural solution, the type of pad foundation will vary. The three main variants are illustrated in Figure 8.4. All need to be in reinforced concrete.

Foundations of this type normally have a rectilinear shape in plan and can be of some depth where a lot of poor soil overlays, for example, stable rock. In this case, extensive support to the excavation may be necessary for the safety of operatives who have to work in the excavation to fix reinforcement and place concrete, fix templates for holding down bolts etc.

Planning considerations

There are a number of items that have to be considered in the planning of their construction, before the most economical solutions are achieved. Temporary works decisions, in particular, are important.

1. In reinforced concrete construction, it is not uncommon for the designer to detail a pad foundation as shown in Figure 8.5. The idea, of course, is to spread the column loads to the much larger area of the pad. By tapering the base up to the cross-section of the column, the designer is trying to save concrete.

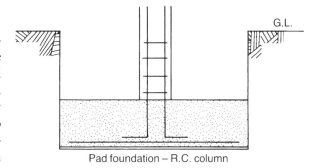
Pad foundation – R.C. column

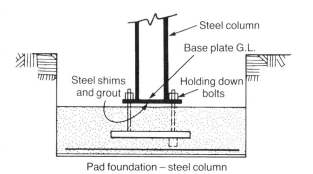
Pad foundation – steel column

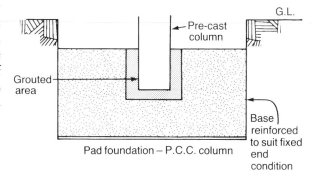
Pad foundation – P.C.C. column

Figure 8.4 Pad foundation types for concrete frame, pre-cast concrete frame and steel frame.

Unfortunately, saving concrete often increases costs elsewhere, to a greater extent than the saving in concrete cost. In this case, the excess labour element is in tamping and trowelling the sloping surfaces. All experience shows that it is both cheaper and quicker to give away concrete to the shape indicated by the hatched area.

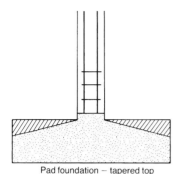

Pad foundation – tapered top

Better to give away concrete as
shaded area – easier and quicker
to do – easier to support column kicker
former, especially if external formwork
is needed.

Figure 8.5 Tapered top bases. Uneconomical economy!

2. Some form of suspended former will be needed to create a 'kicker' for starting off the column, unless the kickerless method is used.
3. Columns in steel framed structures normally have to be held down on the foundation by bolts cast into the base. Where the base is shallow, the temporary system for holding such bolts in the right location is relatively simple (Figure 8.6). More careful examination is needed in deep locations.
4. In the case of precast concrete structures, end fixity is usually called for with the base. This is usually achieved by forming a suitable pocket in the base to receive the bottom end of the column (Figure 8.6(c)). See Chapter 12 on PCC Structures for the installation method for the columns.
5. Where bases have to be large, it is often found, in practice, that the distance between individual bases excavations is quite small. In such situations, it will pay to examine whether it would be cheaper and quicker to take out a series of bases as a trench (Figure 8.7).
6. With deep pad foundations, the type of support that will be necessary can be both time consuming and costly. Adequate assessment

and costing of alternative solutions at the tender stage is essential.

7. All pad types will need additional temporary works, in addition to any for the support of the excavation, for forming kickers, supporting holding down bolts in position while pouring concrete and recess formers for precast columns etc. They all cost money and some solutions are given in Figure 8.6.

As the varieties of pad foundations are numerous, it is better to create a checklist approach to the activity sequence. In planning, one only needs to tick off those items that apply in specific circumstances (Table 8.3).

The good planner will consider the effect of all the listed matters given above and the sequence of events, for the case in point, before deciding the plant, labour and time scale required.

Raft foundations

The name gives the lead to this type of foundation. In broad terms, the raft foundation is designed to 'float' on poor ground while distributing local heavy loads which come down upon it, to an acceptable final ground pressure.

Table 8.3 Possible activity sequence for pad foundations

Item	Description
1.	Excavate pad foundation.
2.	Support to excavation – if more than 1.2 m deep.
3.	Bottom up and inspect.
4.	Place blinding.
5.	Fix reinforcement.
6.	Edge formwork, if needed.
7.	Support to: column upstand formwork and sloping sides of base formwork; template for holding down bolts; suspended formwork for hole to receive PC column.
8.	Pour concrete to base.
9.	Remove all formwork and support material.
10.	Backfill, compact and dispose of surplus.

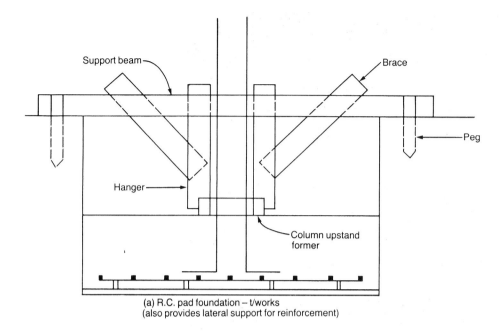

(a) R.C. pad foundation – t/works
(also provides lateral support for reinforcement)

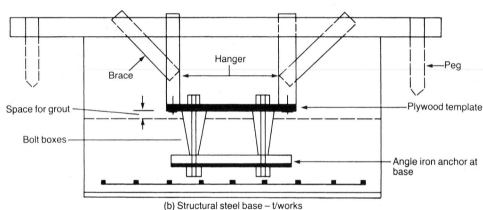

(b) Structural steel base – t/works

In its simplest form, a raft slab may be only 150 mm thick to carry the weight of the walls of a garage to a house, together with the weight of the car using it. In addition, a similar but thicker raft is often used for house foundations in mining settlement areas to limit damage due to uneven movement of the ground.

At the other end of the scale, rafts can be massive structures of considerable depth (2 m or more) used to link together high concentrations of piles and supporting thousands of tonnes of high rise structure above.

In between, many varieties are possible. Rafts may form a linkage between more widely spaced piles while forming the bottom slab to a basement. Alternatively, the raft may be relatively shallow and support the columns to a structure above. This will often be the case where no basement is involved.

Planning considerations

Each type of raft will need varying approaches to the planning.

(c)

(d)

Figure 8.6 Temporary works needs for pad foundations, in addition to any excavation support (a) in-situ concrete (b) structural steel. (c) R.C. pad foundation for pre-cast concrete. (d) Pre-cast concrete foundation – t/works. Note: if oversize excavation and external base formwork is needed, this will provide support for temporary works, (c) above, more easily.

House rafts

As will be seen from Figure 8.8, this type of raft is really a series of ground beams linked together by the floor slab. As such, the edge-beam sections have to project above ground and also become retaining walls for the fill material under the floor slab. This is especially important in mining areas, where compacted sand is used as the filling material, to give greater flexibility to ride any mining 'wave'.

The sequence of construction tends to be dif- ferent from that expected. For maximum econ- omy in construction, once all topsoil has been removed, hardcore, or other approved filling is laid and compacted to underside of floor con- crete. The edge beam is next excavated to the profile required (Figure 8.8). From this illustra- tion it will be seen that formwork will be neces- sary around the perimeter to provide the external beam edge as well as the rebate to receive the external brickwork.

At this stage, all service pipework, ducts,

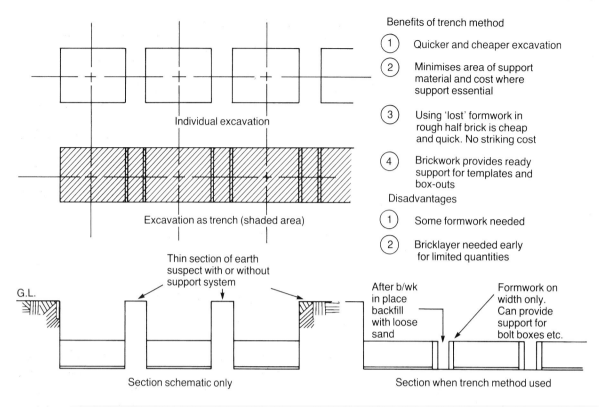

Individual excavation

Excavation as trench (shaded area)

Benefits of trench method

(1) Quicker and cheaper excavation

(2) Minimises area of support material and cost where support essential

(3) Using 'lost' formwork in rough half brick is cheap and quick. No striking cost

(4) Brickwork provides ready support for templates and box-outs

Disadvantages

(1) Some formwork needed

(2) Bricklayer needed early for limited quantities

Thin section of earth suspect with or without support system

G.L.

Section schematic only

After b/wk in place backfill with loose sand

Formwork on width only. Can provide support for bolt boxes etc.

Section when trench method used

Figure 8.7 Trench excavation alternative for pad foundations.

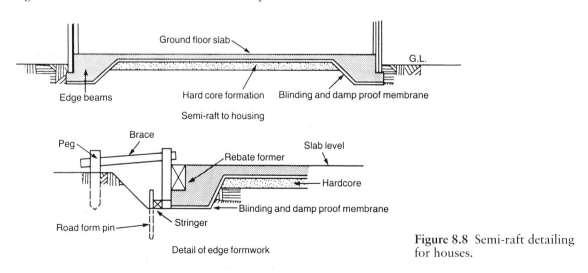

Ground floor slab

G.L.

Edge beams   Hard core formation   Blinding and damp proof membrane

Semi-raft to housing

Peg   Brace   Slab level

Rebate former

Hardcore

Blinding and damp proof membrane

Road form pin   Stringer

Detail of edge formwork

Figure 8.8 Semi-raft detailing for houses.

internal gullies, together with service entries through the edge beams must be installed and held rigidly to avoid displacement when placing the concrete. The installation of such items can take much longer than anticipated and time scales need to be assessed with the specialist subcontractors concerned.

Working in this sequence, it becomes possible

to concrete both the beams and the floor slab at the same time after the blinding and damp proof membrane have been positioned. This is much better than splitting the concrete item into two – ground beams first and floor after. It is both quicker and less expensive.

Activity sequence

The activity sequence with this approach is shown in Table 8.4. At this point, plant and labour needs can be established and this component of the method statement formulated for the estimator, as well as the data to assess the programme times for the raft as a whole.

Intermediate type rafts

A generalized raft of this type is illustrated in Figure 8.9. Such rafts are used in a number of situations: (a) where isolated pads would overlap with each other, and not provide the required support area, a continuous raft of suitable depth provides the answer; (b) where a single basement is specified and a raft can provide the structure foundations and that for the retaining wall to the basement; (c) to provide extra weight to a basement structure when the possibility of hydraulic uplift in wet ground may arise, before the weight of the structure above is applied.

Table 8.4 Activity sequence for rafts for housing

Item	Description
1.	Strip top soil.
2.	Excavate to reduce level.
3.	Hardcore bed and compact over whole floor area.
4.	Install service entries, ducts, pipework, gullies etc. under the floor area.
5.	Fix external raft formwork, install blinding and damp proof membrane.
6.	Fix reinforcement and starter bars, and link to pipework, gullies etc.
7.	Pour concrete.
8.	Strike edge formwork.
9.	Brickwork to DPC level.

Planning considerations

If the depth of excavation is greater than 1.2 m the Construction Regulations[2] require that the perimeter must be supported. How this will be best achieved will relate to the construction method as a whole. A number of alternatives exist. (a) If space permits, an economical answer is to excavate for the raft in open cut. Carried out to a batter suitable to the ground in question, no further support will be necessary. Once the retaining wall has been completed, backfilling is not an expensive item and is relatively quick. (b) Use of the H piling method, (Chapter

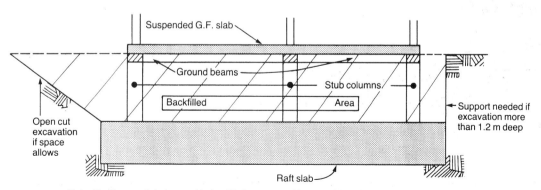

Note: Venting needs to be provided to filled area to avoid any accidental build up from leaking gas mains.

Figure 8.9 Intermediate raft types – general arrangement.

4), where the main support material is predriven before excavation takes place, or, in wet ground, the use of sheet piling are obvious answers where the excavation must support adjacent land or buildings. In such cases, initial excavation is carried out (Figure 8.10) and the available area of the raft completed. Raking shores can then be used to support the sheeting method, using the completed areas of raft as the kicking block. With the top shores installed, excavation can take place of the dumpling between the shores. At this stage, the rest of the raft can be completed, as well as the retaining wall sections between the shores. Once the wall is mature,

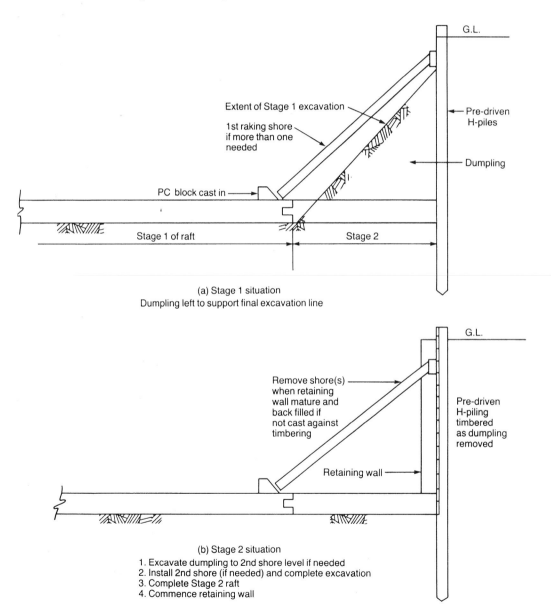

Figure 8.10   Intermediate raft types – construction sequence.

backfilling can be carried out and the shores removed, to allow the last parts of the wall to be constructed. (c) Techniques used in deep basement construction may also be applied (Chapter 10), although the economy of such methods will need examination for the single basement situation.

Where no basement is involved, storage for the excavated material will have to be found until required for backfilling. To minimize cost, it will need to be as near as possible. The cost of compaction must be assessed as well.

Where the raft is of considerable size, the method of handling reinforcement into place, the need or otherwise for external formwork and the concrete handling system all have to be carefully studied, to arrive at the most economical as well as quickest time solutions. In today's thinking, the use of concrete pumping will be a front runner because of the speed of placing possible.

Where external formwork is necessary, the

Table 8.5   Activity sequence for medium rafts for structural loading

Item	Description
1.	Strip topsoil and store on site.
2.	Excavate to formation of raft (if battered sides are possible).
3.	Make up levels where necessary in weak concrete.
4.	Provide support to sides of excavation if deeper than 1.2 m. If needed, this activity will come before excavation starts, in part, see planning considerations above and Chapter 4.
5.	Install blinding concrete and erect external formwork if needed.
6.	Install service entries under or within the raft slab coming to the top surface of the raft and accurately level and support.
7.	Fix reinforcement including any starter bars to columns and walls.
8.	If any recesses in raft (lift pits, drainage sumps etc.) fix box outs or other formwork.
9.	Pour concrete to raft slab.

Table 8.6   Additional activities where the raft is supporting columns in situ

Item	Description
1.	Fix steel to columns to GF level
2.	Erect formwork to columns to ground level
3.	Pour concrete to columns
4.	Strike formwork
5.	Backfill and compact to form reduced level for GF slab

most economical solution is likely to be the use of rough half brickwork left in place. To be economical such brickwork should be backfilled with uncompacted sand before concreting takes place.

At this point, the activity sequence can be tabulated (Table 8.5). As previously, with all the above information to hand this section of the estimator's method statement can be developed, as well as time scales for the construction programme.

Heavy rafts for major structures

Such rafts, often 2 m plus in thickness, are more likely to be associated with piled foundations. That is, forming one continuous pile cap over a large group of piles. The detailed planning approach is therefore more appropriately dealt with in the following chapter dealing with piled foundations.

References

1. *Standard Method of Measurement for Building Works.* SMM 7. Revised 1998. RICS, London.
2. *Construction (Health, Safety and Welfare) Regulations 1996,* HMSO, London.
3. Orchard, C.F.A. and Hill, P.H. (1976) Concrete trenchfill for house foundations. *Surveyor* 16 July 1976. Reprinted by Cement and Concrete Association (now British Cement Association), London.
4. Barnwood, G. (1993) Strip foundations for houses. Includes advice on trench fill method. *British Cement Association Guide 1993*, British Cement Association, London.

Chapter nine

Piled foundations

Where ground conditions are incapable of supporting structural loads by the methods already described at an economic depth, it becomes necessary to use piling. Piling can be defined as the provision of structural columns below ground level to transfer the structural loads down to a strata capable of accepting them. The majority of such piles are likely to be in some form of concrete, but H-section steel, steel tubes and solid timber sections are also used in suitable situations.

Types of piling

BS 8004:1986[1] divides piles into three main groups, depending on their effect on the surrounding soil.

1. **Large displacement piles.** These include all types of solid pile, including timber and precast concrete and steel or concrete tubes closed at the lower end by a shoe or plug, which may either be left in place or extruded to form an enlarged foot.

2. **Small displacement piles.** These include rolled steel sections, such as H-piles, open ended tubes and hollow sections if the ground enters freely during driving. However, it should be recognized that open-ended tubes and hollow sections frequently plug and become displacement piles particularly in cohesive soils. H-piles may behave similarly.

3. **Replacement piles.** These are formed by boring or other methods of excavation; the borehole may be lined with a casing or tube that is either left in place or extracted as the hole is filled.

These groupings can be split into varieties of pile type. For those who wish to go into the science of piling and pile design in detail, a definitive book by M.J. Tomlinson[2] should be consulted.

Piling systems

While the number of piling systems available is quite large, the most common general methods cover pre-cast concrete solid piles, solid timber piles, steel H-piling, steel tubing, together with methods using continuous flight augers and large diameter bored piles using augers of short length.

The following examples illustrate examples of large displacement, small displacement and replacement piling.

Solid pre-cast concrete piles

Such piles are manufactured off-site under factory conditions using high strength concrete and fully reinforced with high yield steel to resist handling and driving stresses. One company[3] produces a pile that can be varied in length to suit individual site conditions. A patented integral steel mechanical joint is used to connect units together when required.

Installation

The company mentioned above installs its precast piles as shown in Figures 9.1 and 9.2. Note

179

the ability to join piles together to avoid excessive handling lengths by the patented jointing method, or to deal with variable length needs of particular site conditions.

Rolled steel H-piling

Rolled steel H-piling has already been mentioned in Chapter 4, in connection with the support of excavations. It can equally be used as a bearing pile in suitable ground conditions. H-piles are universal column sections with equal dimension X and Y axes.

Installation

In the temporary works role, H-piles are normally installed by means of hanging leaders on an excavator crane and a drop hammer, rope operated. For permanent structures, however, penetration will need to be much deeper into solid strata. For this reason, power assisted hammers are usually needed, diesel, compressed air or hydraulically operated. They can be mounted on hanging leaders as with drop hammers or on purpose built piling frames. Figure 9.3 shows a diesel hammer on a travelling piling frame installing H-piles to a shipbuilding slipway.

Continuous flight auger piling

Also known as auger-injected piles, continuous flight auger (CFA) piles are constructed by drilling a long auger with a central hollow stem

Pile Installation

Piles are driven using crawler-mounted rigs, normally with conventional rope-operated drop hammers. Diesel or hydraulic drop hammers can also be used. Most Hardrive piling rigs operated by Westpile are capable of installing piles at a rake of up to 1:3 away from the machine and 1:3½ towards the machine.

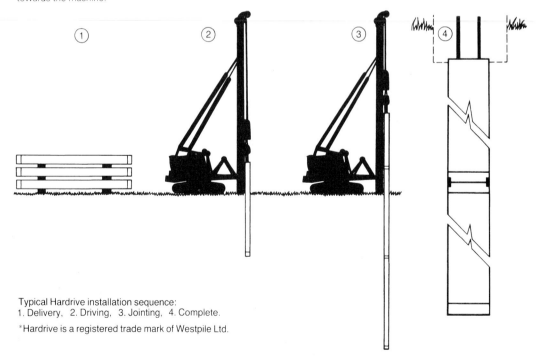

Typical Hardrive installation sequence:
1. Delivery, 2. Driving, 3. Jointing, 4. Complete.

*Hardrive is a registered trade mark of Westpile Ltd.

Figure 9.1   Typical Hardrive sequence. (Westpile Ltd).

Figure 9.2 Hardrive rig in foreground with rotary auger rig behind

into the ground to form the pile bore. After reaching the required depth, the auger is slowly withdrawn at a controlled rate while a highly workable concrete is pumped through the hollow stem to fill the pile bore below the rising auger. The process, using cementitious grouts, was developed by Intrusion Prepakt in the United States in the 1950s. It was introduced into the United Kingdom in 1966 by Dowsett Prepakt Ltd, now part of Westpile.

The use of concrete rather than cementitious grouts was developed in the United Kingdom and Europe in the 1970s and since that time CFA piling has secured an increasingly important place in foundation engineering techniques.

Installation

The piling rigs for this type of pile may be either crane mounted or purpose built crawler mounted rigs which are self erecting. The general principles of how such piles are installed are illustrated in Figure 9.4.

As a method, an important feature is that installation is virtually vibrationless and can be used in delicate locations. Figure 9.5 shows the crawler type rig completing a pile.

Figure 9.3 Diesel hammer and leaders mounted on piling frame – capable of longitudinal and lateral movement. In use for driving H-piles for foundation of a ship building slipway. Rig assembly raised above ground level to give uninterrupted driving at the seaward end where the ground was submerged at high tide.

Rotary augered piling

In this type of pile, the boring method is based on the use of limited length augers, as distinct from the continuous flight system already described. Figures 9.7 and 9.8 show such augers. As the depth bored is limited by the depth of the auger, at any one time, power becomes available for the auger to be used for much larger diameters than with the continuous flight method. Piles of 1.8 m in diameter can readily be achieved.

The large diameter pile is a much favoured method for many structural applications. The ability to carry even more enhanced loads is given by the use of under-reaming equipment. The under-reaming attachment is shown in Figure 9.9.

Installation

The installation procedure normally starts with the sinking into the ground of a steel tube the same diameter as the pile required. Its purpose is to provide a guide to the boring equipment in the initial stages as well as to support the initial soil conditions against falling in to the pile bore. The length of such liners is determined by the need to reach a stable strata beyond which the bore will be self-supporting during the pile making period.

The liners are usually vibrated into the ground by special equipment attached to the top of the lining tube. As the vibration can be felt for some distance away, consideration is always necessary as to whether the effects of the vibration might be a source of damage elsewhere. Vibration may cause compaction of loose material behind earthworks support systems, for

Pile Installation

Construction sequence for a typical Prepakt CFA pile:
1. Whilst the pile bore is being drilled the central stem is plugged with a temporary steel cap. Relatively little spoil is removed during this operation.
2. and 3. The central hollow stem of the auger is connected by a swivel assembly to flexible steel hoses leading to a high pressure concrete pump. Once the required drilling depth has been reached, a high slump concrete is pumped through the swivel assembly, down the stem of the auger, and, after blowing the temporary cap into the base of the pile. The auger is steadily extracted at a controlled and predetermined rate, maintaining a slow rotation in the drill direction, whilst concrete is continuously pumped through the stem to form the pile. Spoil from the bore is simultaneously brought to the surface by the auger and the rising column of concrete.
4. and 5. When the bore has been completely filled with concrete the top of the pile is cleaned off and a reinforcement cage with suitable spacers inserted into the fluid concrete. This reinforcement is usually only installed in the upper portion of the pile.

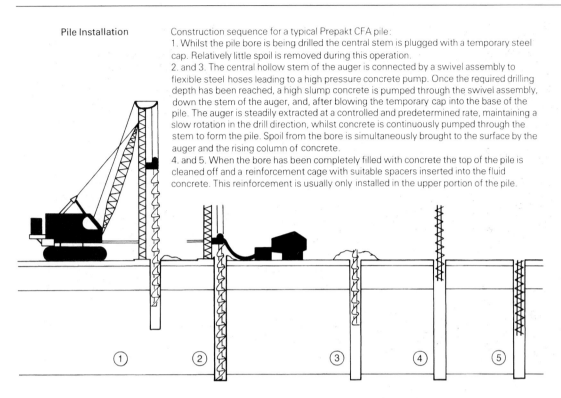

Figure 9.4 Construction sequence for Prepakt continuous flight auger pile. (Westpile Ltd).

example, or become a pollutant under the Control of Pollution Act 1974[4]. In this case, vibration can cause unpleasant effects on human beings such as sickness. In Figure 9.10 a trial is being carried out adjacent to an electric signal box to see if the electronic switches are likely to be affected. Clearly, it would be extremely dangerous if the vibration could cause signals to change colour.

Once the lining tube is in place, boring can commence. The auger is lowered to the liner and bores into it until the whole auger is full of excavated material. It is then raised to ground level and by reversing the turning motion the material is spun off the auger. This sequence is then repeated until the specified depth is reached. If at this stage under-reaming is called for, the auger is removed and replaced by the under-reaming equipment.

With under-reaming, it is usual to inspect the pile to see if the under-reaming is clean of excavated material. A special safety cage is lowered into the pile bore, with an engineer inside to do the inspection. Telecommunication is provided and if previous tests have shown any suspect air in the bore, a blown air supply needs to be provided. With the development of miniature television cameras an alternative, and less claustrophobic, inspection is now available.

The final stage is the insertion of pre-assembled reinforcement and the concreting of the pile by means of a tremie pipe.

The complete sequence of events is shown in diagrammatic form in Figure 9.6.

General comment

While the foregoing piling methods give examples of the main pile classifications, they are by no means a complete summary of the methods

183

Figure 9.5 Close up view of CFA rig in action. (Westpile Ltd).

of piling available. Special methods exist for installing piles in limited headroom situations, and individual firms may claim special features to their systems which give advantages over

those of competitors. Piling systems also have a major role, today, in relation to their ability to provide the support to deep excavations, in the temporary sense, while eventually becoming part of the permanent works. Their use in this respect is examined in detail in the next chapter.

For those who wish to study the subject of piling in more detail, BS 8004 : 1986[1] and a book by M.J. Tomlinson[2] are a mine of information.

Cutting down to level

With the exception of continuous flight auger piles, all piles have to be finished to a level above that specified for cut-off (the final level above which the pile cap will be cast). The reason for such action is a dual one: (a) to ensure that the point of cut off will be in solid material – in displacement piles below any area damaged by driving or any concrete at the top of the pile that has not been fully compacted – and (b) to allow for the projecting reinforcement, after cutting down, to have the desired bond length within the pile cap. Piling specifications usually give the minimum finishing level of the pile as well as the cut off level. Figure 9.11 makes the point clear.

In the case of the continuous flight auger pile,

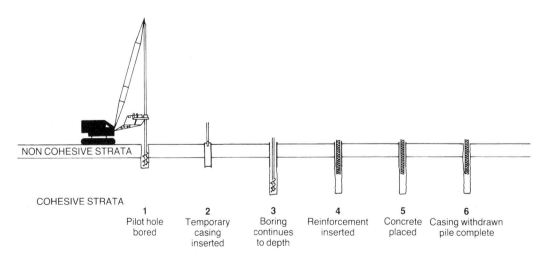

NON COHESIVE STRATA

COHESIVE STRATA

1	2	3	4	5	6
Pilot hole bored	Temporary casing inserted	Boring continues to depth	Reinforcement inserted	Concrete placed	Casing withdrawn pile complete

Figure 9.6 Sequence of operations for constructing a rotary augered pile. (Westpile Ltd).

Figure 9.7   Rotary auger rig in action. (Westpile Ltd.)

Methods of cutting down

Following the completion of an area of piling, the trimming of the piles to cut-off level has to follow. This operation will normally be the responsibility of the ground works contractor, who may be the main contractor or a subcontractor. Four main methods for cutting down can be adopted.

1. Using a compressed air jack hammer to break out the concrete down to the cut off level.
2. Fitting a chisel bit to a hydraulic excavator to break out the first half of the excess pile, followed by the use of hand held hammers.
3. Using a proprietary cutter on the pile (rather like a pipe cutter to a much bigger and stronger scale[5].
4. The Elliot method[6]. (Patent Application Number 9606079.3). This recent new approach is a clever and rapid way of cutting piles down to the length needed. It is stated that the cost of using the method is comparable to traditional methods in the middle range of pile sizes, but is 300% quicker. Considerable savings in contract time result, with indirect cost savings.

The method used is as follows. Before the reinforcing cage is placed in position, each bar is covered in a sheath of polyethylene down to the required cut-off level. With the pile poured and excavation to cut-off level completed, a saw cut is made round the pile at cut-off level. A hole is then drilled at cut-off level and a hydraulic splitter inserted and used to sever the top part of the pile from the body. The disposable part of the pile is then lifted off. The steel is left clean and undamaged. Figure 9.12 illustrates the method in use on an actual site.

Whatever method is used, with the exception of method 4 above, the reinforcement has to be freed from adhering concrete and the final cutting to level, making sure that there is no damage to the pile below cut off level. In method 4,

a different situation arises. The technique of installation is such that CFA piles can only finish at ground level. While a cut off level can be specified, those planning the cutting down of such piles must recognize that the amount to be cut off may well be considerable if the cut off level is well below ground level. Such extra work needs to be established at the tender stage, if possible, so that adequate allowance can be made in the tender figure for the costs involved.

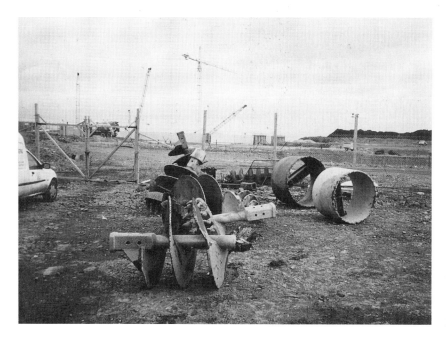

Figure 9.8 Close up of augers used in rotary augered piles. (Westpile Ltd).

Figure 9.9 Under-reaming equipment for rotary augered piling. As the kelly bar rotates the under-ream cutters spin outwards.

Figure 9.10 Test installation of pile bore liner with vibration equipment adjacent to London Underground signal box. The test was to determine if the vibration could safely be used and not trip electrical relays causing signals to change colour.

the polyethylene sheaths ensure that the reinforcement in the cut-off section does not adhere to the concrete. Hence the ability of the excess concrete to be lifted clear.

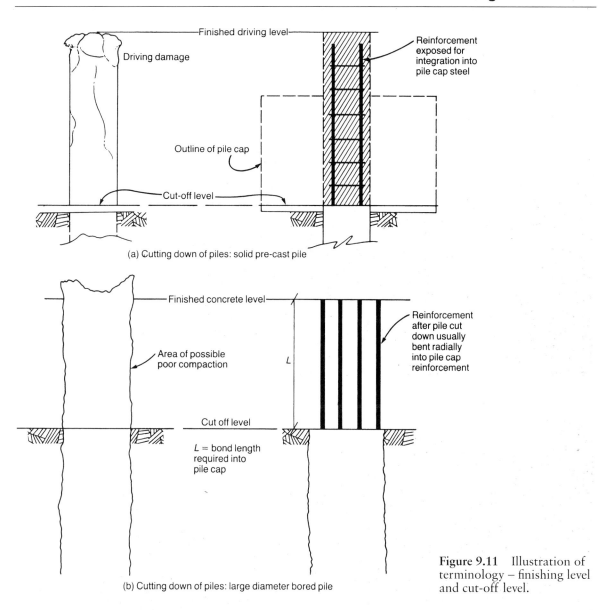

(a) Cutting down of piles: solid pre-cast pile

(b) Cutting down of piles: large diameter bored pile

Figure 9.11 Illustration of terminology – finishing level and cut-off level.

The cost of the cutting down is expensive and the cost comparison between two methods in Example 9.1 makes this clear. Those responsible for the estimating and planning at the tender stage need to consider adequately all the factors that may affect them, so that, as far as possible, adequate monies have been included by whoever is to carry out this operation. Note that ear protectors are essential for the operatives involved and jack hammers need to be silenced as effectively as possible.

Calculation of quantities involved

If the levels for pile finishing and cut off are specified, the theoretical cost for cutting down the difference can readily be established from previous performance figures. See Example 9.1.

Example 9.1 Comparison of costs and performances for two different methods of cutting down pileheads

The following information was obtained from work studies on a site in docklands, London, in 1980. The man-hour figures should still be reasonably accurate.

METHOD 1

Amount of concrete to be cut off per pile = <u>4.6 m³</u>

Method Hymac 580c with hydraulic breaker used to remove the top 2 m³. The remaining 2.6 m³ removed by hand using pneumatic hammers, to avoid damage to pile concrete at cut off level.

Performances achieved
Hymac 580c with hydraulic breaker – machine hours to remove 2 m³
$$= \underline{3.68\,hrs}$$
$$\text{Performance rate} = \underline{1.84\,hrs/m^3}$$

Labour with pneumatic hammers – Man-hours to remove 2.6 m³
$$= \underline{14.88\,hrs}$$
$$\text{Man-hours performance} = \underline{5.72\,man\text{-}hrs/m^3}$$

Cost make up
All in labour rate (True cost of labour) at the time £2.50/hr

Labour element 14.88 hrs @ £2.50 = £37.20
Unit cost will be $\underline{\frac{37.20}{2.6}} = £14.31/m^3$

Hymac 580c with hydraulic breaker
All in hourly hire rate (including driver) 14.36
In use for 3.68 hrs (used elsewhere when not on pile cutting) Cost = £14.36 × 3.68 = £52.84
or   Unit cost will be $\underline{\frac{52.84}{2}} = £26.42/m^3$

Compressor and tools for hand breaking
All in hire rate for compressor and tools £1.39
Hours worked to cut down 2.6 m³   8.72
Total cost is 8.72 × 1.39 =   £12.12
Unit cost is $\underline{\frac{£12.12}{2.6}} =$ £4.66/m³

Note. The discrepancy between the plant hours and labour man-hours is due to labour time clearing away cut-off concrete.

To summarize

We have Hymac 580c all-in cost	£52.84
Compressor and tools all-in cost	12.12
Labour (True cost)	37.20
Overall cost	£102.16

£102.16 is the cost to cut down 4.6 m³ of concrete in pile.

Unit rate becomes $\frac{102.16}{4.6} = 22.21/m^3$

METHOD 2

In view of the cost of the Hymac machine, a comparative cost was carried out doing the whole cutting down with labour and pneumatic hammers.

Performance for cutting down 5.72 man-hrs/m³ as before.
Quantities to be removed 4.6 m³
Total time taken 4.6 × 5.72 = 26.31 man hours

Cost factors Using the rates for labour and pneumatic hammers as in the first method we have:

   Cost/hour

Compressor and tools hire	1.39
True cost of labour	2.50
	£3.89

Total cost for the operation = 26.31 hrs × £3.89 = £102.34

This figure is almost identical, in cost terms, with that recorded for part cutting down with the Hymac.

$$\frac{£102.34}{4.6} = £22.25/m^3$$

From the results above it was decided to keep to the method using the Hymac for the first 2 m³ cut down. The reason was that method 1 reduced the number of operatives needed. There is little difference in overall time if one takes a three man gang to do the breaking out.

Hymac time 3.68 hrs + gang time $\frac{14.88}{3}$ = 3.68 + 4.96 = 8.64 hours

All hand operatives and compressor $\frac{26.31}{3}$ = 8.77 hrs

been filled with concrete and the lining tube withdrawn, a significant quantity of concrete has to be added to top up the pile to ground or other specified level. The reason for this is often that the pile at this level is in made ground and, as the liner is withdrawn, some concrete fills voids in the fill material. It follows that at the cut off level the pile can be larger in diameter than expected or specified with a greater cost in cutting down because of greater quantities being involved. Table 9.1 shows the results of a work study on a London site near to the River Thames. Note that although the piles were 42 m long in many cases, the excess concrete was always in the top 11.5 m – the length of the lining tube used.

Those involved in the planning and site managers need to keep a close watch on situations as described above, so that proper records are kept and agreed with the Resident Engineer or Clerk of Works, as the work proceeds.

Attendances on piling contractors

The main contractor will always be required to provide services to the specialist piling contractor. As such, the construction method planner needs to be well versed in what such attendances are likely to be required and check any BOQ requirements so that adequate allowance, in cost terms, can be made at the tender stage. The following items are the most likely to be involved. Nevertheless, every effort needs to be made to establish with the piling contractor the precise needs when piling quotations are being sought.

Figure 9.12   Method of cutting down to level as described in the text. A big difference to cutting down by compressed air tools.

If, however, the finishing level is greater than specified, extra concrete will have to be cut away which is not accounted for in the bill of quantities. A direct loss will result. In consequence, whoever will be paid for the cutting down should endeavour to record the actual finished levels of all piles and agree them with the Resident Engineer or Clerk of Works, as the case may be.

Further financial loss on cutting down can arise when a pile is bigger in diameter than the specified size. While the top part of a pile would be expected to be accurate as it is made within a liner, in most cases, there is a good deal of evidence from work studies that, after a pile has

Preparation of site before piling starts
1. Soil strip and take to store for future use.
2. Temporary access.
   (a) To area allocated for piling contractor's yard, providing an all weather surface.
   (b) Access to piling areas and hard surface for the rigs to work on. Suitable for ready mixed concrete delivery and muck away as well.

Table 9.1 Volume of concrete wastage in piles of various diameters recorded from work studies on rotary bored piles

Pile diameter and length	Theoretical volume	Actual concrete used	Difference
750 mm			
18 m long	8 m³	12 m³	4 m³
	Average over 4 No. piles		
1050 mm			
43 m long	37	44	7
42 m long	36.5	43.5	7
	36.5	48	11.5
		Average =	8.5 m³
1200 mm dia			
42 m long	47.50	56.0	8.0 m³
	Average over 3 No. piles		
1350 mm dia			
42 m long	60.0	70.4	10.4 m³
	Average over 9 No. piles		
1500 mm dia			
42 m long	74.0	87.0	13.0 m³
	Average over 5 No. piles		
1800 mm dia			
42 m long	107.0	129.5	22.5 m³
	Average over 15 No. piles		

As mentioned in the text, the main wastage occurred in the top 11.5 m of the pile when the liner of this length was withdrawn and topping up was needed. Piling firms are well aware of this wastage and make allowance for it in pricing.

3. Temporary services. Provision of services to subcontractor: water, telephone, electricity. If offices are the responsibility of the subcontractor, is heating also his responsibility?
4. Wash down facilities. Piling is a very early operation and the main contractor will have few facilities or labour on site. Local authority requirements with regard to keeping the roads adjacent to the site clean are usually stringent. A decision will have to be made as to who will be responsible for the provision of vehicle wash facilities before they leave the site.

During operations

All excavation involved in the piling process (replacement piles) has to be removed from site, unless the cut off level is well below ground. A degree of backfill on a temporary basis may be required for safety reasons. Who is contractually responsible?

Where displacement piles are specified, ground heave will take place. The degree to which this takes place will depend on the nature of the ground and the piling used. At some stage, on completion of the piling, additional excavation will be necessary. In this case, levels need taking before and after the piling to determine the additional excavation involved. Who is contractually responsible for its removal?

Pile caps and ground beams

The building's structural loads are transferred to a piled foundation by means of a pile cap. To provide stability, groups of piles providing the

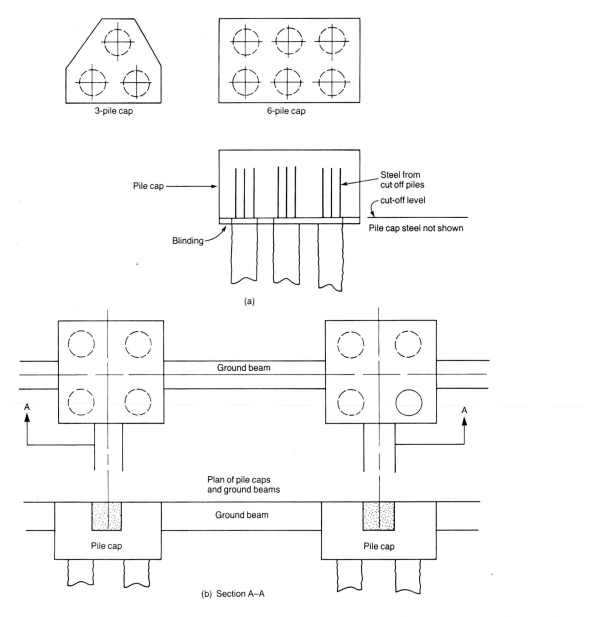

Figure 9.13 (a) Simple pile caps, linking two or more piles in a group. (b) Ground beams linking pile caps for improved stability and to carry suspended ground slabs.

foundation for one structural column usually have a linking pile cap as shown in Figure 9.13(a). The cap shape is determined by the number of piles to be linked. Where additional stability is called for, individual pile groups and their caps may be interconnected by means of 'ground beams', Figure 9.13(b).

In situations where the pile layout calls for close spacing due to high loads above, the designer will usually work on the principle of using an overall raft slab above the piles. It is easier to construct and only requires edge forms instead of too many individual caps with little room between. Better load distribution results. Such a situation is illustrated in Figure 9.17. Note the use of steel H-piles to provide the temporary support to the raft excavation. The details of the method used are fully described in another book by the author [7].

Planning considerations related to pile caps

It is generally the case that pile cap construction will be the responsibility of the main contractor – whether by direct action or sub-letting the work. Which ever is the case, the main contractor will need to have made adequate allowance for the construction of pile caps at the tender stage. To do so in a proper manner, the construction planners will need to have built up

adequate information so that costs can be established for work outside the piling specialist's brief. The main items will be:

- pile finishing level and amount of cut-off;
- shape and depth of pile caps, hence excavation required, disposal of surplus;
- assessment of temporary works needs viz, formwork requirements, reinforcement needs, provision for hanging bolt templates and so on;
- reinforcement – prefabricate or fix in place;
- access for handling reinforcement;
- method of supply and placement of concrete.

Planning considerations necessary for ground beams

Where both pile caps and ground beams arise it will be necessary to consider the two items as an entity. Pile caps can be required on their own,

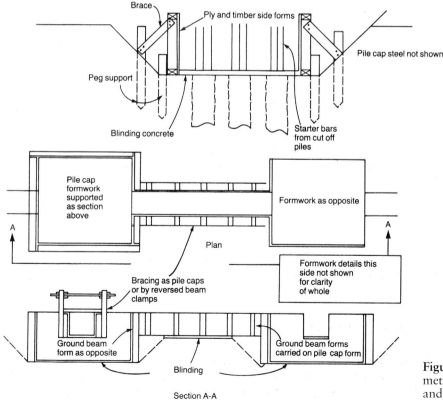

Figure 9.14 Traditional method for forming pile caps and ground beams.

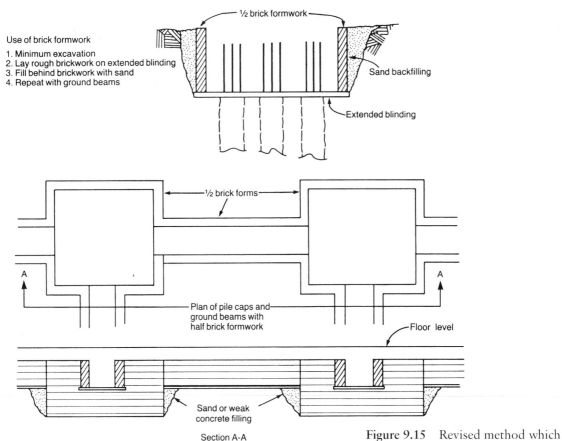

Use of brick formwork

1. Minimum excavation
2. Lay rough brickwork on extended blinding
3. Fill behind brickwork with sand
4. Repeat with ground beams

½ brick formwork

Sand backfilling

Extended blinding

½ brick forms

A

A

Plan of pile caps and ground beams with half brick formwork

Floor level

Sand or weak concrete filling

Section A-A

5. Once all brickwork complete backfill to make up levels for floor
6. Place blinding to floor
7. Fix reinforcement to caps, beams and floor
8. Concrete whole in one pour

Figure 9.15 Revised method which speeds up construction by using lost formwork which, in turn, allows pile caps, ground beams and floor slabs to be poured in one operation.

ground beams are always a linkage between pile caps. Therefore, in association with decisions regarding pile caps, the following matters in relation to the linking beams will also need resolution:

- shape, length and depth of ground beam and excavation necessary;
- requirements for temporary works – method for formwork;
- striking of formwork and backfilling and compaction before commencing floor slab works;
- reinforcement – blinding needs, fix in place or prefabricate reinforcement;

- concrete supply and method of placing; are access needs going to involve cost or can such items as cranes reach without trouble? would a different method of handling speed construction and reduce cost? (for example, to pump the concrete to allow the crane to concentrate on other items of workload);
- are any service entries required to pass through or under ground beams? are the necessary pipes or holes known as to position and size? have they been ordered?

When planning the methods to be used for the construction of pile caps and ground beams, it is necessary to be looking for the easiest methods to allow the fastest performance, as these items are on the critical path. In particular, the formwork needs merit careful examination. Figure 9.14 shows what might be called the traditional approach. Timber or plywood forms are assembled in the usual way and suitably strutted from the excavation. After the reinforcement is in place the concrete is poured. Once set, the forms are struck, the whole area backfilled and compacted and blinding laid for the floor slab.

In Figure 9.15 a different and more efficient method is illustrated. Instead of using traditional formwork, half brick or hollow blocks are laid in the usual manner to provide the formwork which will become 'lost' after the concrete has been poured. In other words, no striking is required.

The effect of this approach is to allow the floor to be poured at the same time as the pile caps and ground beams. No striking of formwork is needed, the backfilling under the floor is carried out in sand when the brick or blockwork has been completed, the floor damp proof membrane laid and the blinding concrete laid underneath. Finally, the floor reinforcement is positioned at the same time as that for the pile caps and ground beams and the concrete poured in one continuous item.

A system of plastic lost formwork[8] which can be supplied to specified lengths is also available. The method allows the capability to pour all items simultaneously as with the lost brickwork example. The installation is very fast (Figure 9.16).

Figure 9.16 A more up-to-date version of lost formwork. It can be supplied to any shape or size[8]. (Cordek Ltd)

Figure 9.17   2 m deep raft slab covering close-spaced piles to provide the foundation for a 29-storey office building. Note the steel H-pile support method to uphold the excavation to the raft slab.

The final link in the planning chain is to determine the anticipated performance of the piling contractor to meet the overall programme needs. From this information the number of rigs will be determined. This information will, in turn, provide the available time scale for the follow-up items: cutting-down piles, pile caps and ground beams and the floor progress possible.

References

1. British Standards Institution. *Code of practice for foundations*. BS 8004: 1986. BSI London.
2. Tomlinson M.J. (1993) *Pile design and construction practice*. Fourth edition. E. & FN Spon, London.
3. Westpile Ltd, Uxbridge.
4. *Control of Pollution Act 1974*. HMSO, London.
5. Euro Construction Equipment Ltd. Rochester, Kent.
6. Jim Elliot and Co. Liskeard, Cornwall.
7. Illingworth, J.R. (1987) *Temporary works – their role in construction*. Thomas Telford, London.
8. Cordek Ltd. Billingshurst.

Chapter ten

Deep basements

Most commercial developments in town centre areas involve the construction of deep basements, while centres of entertainment usually require basement construction, both for car parking facilities and production facilities below stage. Industrial construction, also, will have basement needs which are likely to vary to a greater degree than those under commercial developments.

Basements in commercial developments in town centre areas almost invariably occupy the full area of the site and will vary from a single basement to perhaps five levels if parking is the objective. It follows that their construction will involve major upholding problems, in relation to both adjoining streets and buildings. It is in these situations that the greatest care is needed in deciding upholding methods. Temporary works can be extensive or, with the co-operation of the structure designer, become a joint operation designed to minimize construction cost.

Temporary works

Any temporary works method proposed requires design by designers experienced in this field. Only two methods are suitable structurally in these situations: H-pile support where there is no free water, or steel sheet piling where water is likely to be a problem. Both these approaches have been examined in Chapter 4. Where an excavation occupies a large area, single face support is usually the only approach likely to be economically satisfactory. In either method, as has already been made plain, the sheeting support by internal raking shores has a disruptive effect on the new works within the excavated area (Figure 10.1). By the use of ground anchors, the excavated area can be left completely clear for new works construction (Figure 10.2).

Ground anchors

If the planning method indicates the use of ground anchors would be best, care needs to be taken to follow the legal requirements necessary for such action.

1. To install a ground anchor under a public highway or adjacent building is a trespass in law, unless the necessary permission has first been obtained. In the case of the highway, the local highway authority must be consulted for permission. With adjoining owners, every single one has to be consulted insofar as his own property may be affected.
2. In the case of highway authorities, there is usually understanding of the reasons for the request. What may happen, though, is that certain conditions may be imposed. For example, Figure 10.3 illustrates a requirement that no ground anchors are to pass nearer than 2 m to existing old brick sewers in the adjoining roads. A second requirement was that all ground anchor tendons had to

Figure 10.1 Raking shore support system, showing disruptive effect on new works.

Figure 10.2 Clear working space by use of H-pile support with ground anchors.

be distressed along one main road, when the permanent retaining wall had been supported by the new structure.

3. With adjoining owners, the problems can be more difficult. Some may see no difficulty after consulting with their building surveyor. Others may have old settlement problems and do not want any more. As a result, they may refuse point blank to allow any anchors below their property. Where this is the case, an alternative support method has to be used which does not infringe the next door boundary.

In all such cases, agreement, if given, needs to be in writing and Schedules of Conditions completed and agreed with the adjoining owners, whether in relation to roads and pavements or property. Written lists should always be backed up by good commercial photographs, of the dated variety. Such action is essential protection against adjoining owner claims at a later date.

Permanent works as temporary works

Where the contractor has to provide the entire upholding system to a basement excavation, its installation time and cost can be extremely

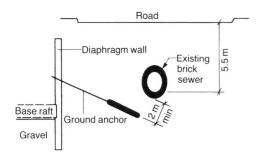

Notes: 1. Minimum distance from brick sewer may cause change from ideal angle for ground anchor.
2. Change in angle usually causes increase in anchor length. May increase cost if longer than allowed for.

Figure 10.3 Portion of a contract drawing showing restriction on distance between ground anchors and nearby brick sewer.

Figure 10.4 Use of secant piles and H-piling on running tunnel construction. Hannover Metro construction.

expensive and very little else can take place while the support is installed.

In recent times, the use of temporary support sheeting has become increasingly unnecessary. Modern techniques allow the installation of permanent wall construction before excavation commences and all the contractor has to do is provide a degree of temporary support until the permanent support works take over.

The methods allowing such action to take place are diaphragm walling, contiguous bored piling and secant piling (Figure 10.4). All are described in some detail in a book on temporary works [1] or in specialist firms' literature, to which the reader is referred. In this chapter, the methods in question will briefly be outlined and the main emphasis placed on the method planning.

Diaphragm walling

The technique of diaphragm walling allows the creation of a concrete wall within the ground before any other excavation takes place. The fundamental principle involves excavating a trench of the required width on the line of the required wall, and to the required depth. The

excavation is stabilized by keeping it full of a mixture of bentonite (Fuller's earth) and water. This mixture has the effect of forming a gel against the faces of the excavation which stops loosening taking place. In addition, the hydraulic head of the liquid provides a support system. An alternative to bentonite has been the use of a polymer slurry[2]. Once the correct depth has been reached, pre-fabricated reinforcement mats are lowered into the bentonite which does not stick to the reinforcement. Concrete is then placed through a Tremie tube, displacing the liquid. This is run off for filtering before re-use. The Tremie tube is held at the bottom of the excavation until buried in 3 m of concrete. From then on, the tube is lifted and sections removed, but still maintaining the bottom buried 3 metres at all times (Figure 10.5).

It will be clear that diaphragm walling is a specialist operation, requiring specialized equipment and highly trained operatives (Figure 10.6).

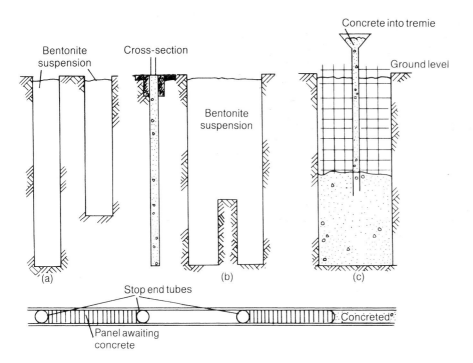

Figure 10.5 Schematic diagram of diaphragm wall construction. (a) Initial excavation on each side of a bay. Excavation kept full of bentonite suspension. Section shows guide wall. Today usually L-shaped to provide better stiffness to guide trench width. (b) Middle section of excavation removed. (c) Reinforcement lowered into bentonite suspension, followed by placing of concrete by use of tremie. Bottom of tube always kept buried one metre into rising concrete. Bentonite pumped back to storage for de-sanding and re-use at the same rate as the concrete is placed.

At the pre-tender planning stage, the main contractor will need the following information:

1. A guide trench is required to be constructed with concrete walls, some 0.7–1.5 metres deep to assist in maintaining plumb and accuracy of line. Their construction is the responsibility of the main contractor. Details will be needed from the subcontractor. Table 10.1 lists the activity sequence.
2. Agreement on rate of progress required to meet overall programme. Diaphragm walls are built in bays and progress should be expressed in bays per week.
3. How many rigs will be in operation and space required for bentonite mixing and storage together with needs for access – muck away and ready mixed concrete in?

4. Sequence of bays – hit and miss or other sequence.

It should be obvious the method creates a lot of mess – spillage of slurry, especially when loading muck away vehicles. The planning needs to allow the specialist contractor full use of the site and not attempt much other work until the walling has been completed. Such action allows having a contract clause making the contractor responsible for cleaning up the site on completion.

A useful Concrete Society Current Practice Sheet has recently been published on the requirements for concreting diaphragm walls[3].

While these activities complete the main contractor's responsibility for getting the diaphragm wall contractor started, the pre-tender planning

Figure 10.6 Grabbing rig. Assembly above grab helps to keep excavation plumb. (Kvaerner Cementation Foundations).

needs to allow for a final activity that is often overlooked. When excavation starts after the diaphragm wall is complete and matured, the internal half of the guide trench wall becomes redundant. For safety reasons it will have to be removed before too much excavation takes place. This situation becomes another example of the need for the temporary works designer to be cost conscious. Breaking out hard concrete is expensive. The experienced planner will look for a method which will save money and Figure 10.7 illustrates such a way. When constructing the inner part of the guide trench wall, instead of a continuous strip of concrete, it is split up into short sections, as shown, each divided from the next by Flexcell jointing material. At the time of casting, steel loops are cast into the top of the wall sections. When removal is required, a crane hook can be attached to the loop of an individual section and the whole lifted away and

deposited into a vehicle for removal to tip. No expensive breaking out is required.

To Table 10.1, therefore, it is necessary to add an eighth item – Remove internal section of guide trench wall. Figure 10.8 shows a typical section of diaphragm wall. The extent of the guide wall inside the excavation is clearly visible

Table 10.1 Activity sequence for guide wall construction

Item	Description
1.	Excavate to outlines shown in figure
2.	Provide temporary support as necessary
3.	Erect framework to internal faces of the guide
4.	Concrete to guide walls
5.	Strike formwork
6.	Backfill between walls and compact
7.	Hand over to specialist subcontractor

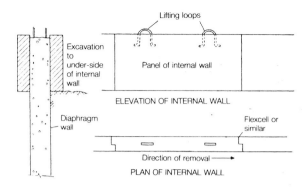

Figure 10.7 Method for economical removal of inner guide trench wall.

Figure 10.8 Diaphragm wall with ground anchor support.

by its smooth appearance, compared to the rougher texture below.

Contiguous bored piling

With this method, bored in-situ piles are installed in such a way that they touch as nearly as possible. Either continuous flight auger or rotary bored piles can be used in this context. See Chapter 9 for details of installation.

Due to the irregular nature of such piles, the resulting wall will not be watertight. Where this is essential, at least two firms can offer a grout pile in the join between the contiguous piles, which greatly reduces the ingress of any water between the piles[4][5] (Figure 10.9).

Stent Wall secant piling

A more positive pile system which prevents water penetration, but which is claimed to compete with standard contiguous pile walls, has been developed by Stent Foundations Ltd. Known as the Stent Wall system, the basic principle is a pile wall consisting of alternate bentonite-cement-PFA piles and reinforced concrete piles which interlock to form a continuous wall (Figure 10.10). The interlocking is achieved by driving the concrete piles partially into the bentonite-cement-PFA piles. This secant cut out ensures a watertight join. The method of installation is described in full detail in the book on temporary works already mentioned, Chapter 8[1]. The use of bentonite, as in the case of diaphragm walls, creates a messy site and the comments made in this context for diaphragm walls apply equally well here. The main contractor does not have to provide any preliminary works.

Secant piling heavy section walling

Where there is a need for a heavy section, watertight wall or where obstructions exist below ground, a much heavier system is available which is capable of drilling through boulders or old foundations. The type of rig used is of a special design to deal with the heavy boring involved. Formally known as the Lilly Libore rig, Lilley Construction Ltd now are using rigs manufactured by Casagrande. The principle of this method is shown in Figure 10.11. As with the Stent method, the method of installation is described in detail in the book on temporary works referenced above.

Figure 10.9
Contiguous piles to
form basement walls.

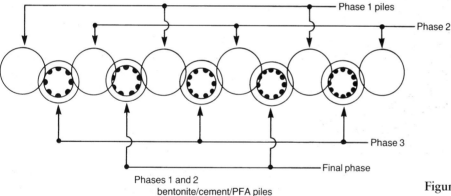

Phase 1 piles

Phase 2

Phase 3

Final phase

Phases 1 and 2
bentonite/cement/PFA piles
Phases 3 and 4
concrete piles reinforced

Figure 10.10   Stent Wall
secant piling. (Stent
Foundations Ltd).

Planning needs for permanent piled retaining walls

Whichever solution is appropriate to the circumstances in question, piled walls involve specialist installation as does the diaphragm wall
method.

At the pre-tender planning stage, the follow-

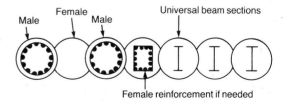

Male   Female   Male   Universal beam sections

Female reinforcement if needed

Figure 10.11   Heavy duty secant piling method
(Lilley Construction Ltd).

203

ing information will be needed from the specialist firm chosen:

- rate of progress and number of rigs on site to
  meet the desired progress;
- allocation of site space required for plant
  and equipment, including pre-fabrication of
  reinforcement;
- access needs for muck away and ready mixed
  concrete in.

Unlike the diaphragm wall approach, the main
contractor does not have to provide any temporary works before piling starts.

Temporary support of permanent works

Whichever of the walling systems used, they will
become part of the permanent works and measured as such. Some cosmetic overlay will usually be provided for in the design.

The contract, in such cases as these, will specify the extent of the contractor's liability to provide temporary support for these walls until the
permanent structure is advanced enough to take
over the support role. The loadings to which the
walls have been designed will be provided to
allow the temporary supports to be designed.
Most usually, the contractor will opt for a
ground anchor solution. In so doing, account
must be taken of all the actions necessary in this
respect already described in the early part of
this chapter. If the conditions cannot be met,
other solutions will have to be adopted.

Situations such as these involve divided
responsibilities. The structural engineer will be
responsible for the design of the walls, while the
contractor is responsible for the temporary
works in their support. At first sight this may
not seem to present any problems. Unfortunately, this is often not the case. It is in the contractor's best interest to work to certain rules.

1. A clear division of responsibility should be
   agreed with the structural engineer from the
   start.

2. The temporary works engineer should not
   just accept the loading to be supported as
   indicated on the drawings or stated in the
   specification. The engineer should be asked
   to provide a copy of his calculations for the
   wall design. If the wall has been designed by
   a specialist contractor, a copy of his calculations should also be examined. Only then
   will the temporary works designer have a
   complete picture of what he has to deal with.
   The relevance of this precaution is well illustrated in the case study that follows.

3. All agreements and doubts must be
   expressed in writing.

Case study

While at first sight it may seem presumptuous to
ask to see the engineer's calculations, it must be
understood that a temporary state may be different from the final one when the permanent
structure is providing the final support. There
will usually be a third party involved – the specialist contractor who will actually do the calculations for the wall design to specified loadings
given by the consulting engineer. If a mistake
arises in this chain of events, who will be
responsible in the event of failure? The following case study from an actual site makes the
point very forceably.

The situation is illustrated in Figure 10.12. In
checking the engineer's calculations, it was not
clear for what span the diaphragm wall had been
designed. It materialized that the contractor
designing the wall had been given the distance
between the top of the basement slab and the
point specified for temporary support by the
main contractor. What had been completely
overlooked was the fact that the diaphragm wall
would need to span a greater distance in the
temporary condition while the basement slab
was excavated to formation level. If this point
had not been picked up by the main contractor's
temporary works specialist, and excessive
deflection had taken place (estimated at 40 mm)
resulting in the wall cracking, three parties

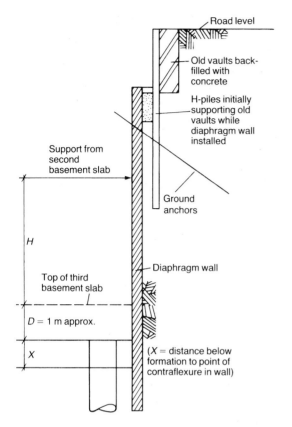

Road level

Old vaults back-filled with concrete

H-piles initially supporting old vaults while diaphragm wall installed

Support from second basement slab

Ground anchors

H

Diaphragm wall

Top of third basement slab

L

D = 1 m approx.

X

(X = distance below formation to point of contraflexure in wall)

Figure 10.12 Temporary support for permanent diaphragm wall. Accounting for temporary conditions during construction.

would have been in the subsequent argument as to who was responsible. For the smooth running of the contract, it is far preferable to analyse such matters and sort them out before work begins.

As a result of the error being discovered, an additional line of ground anchors was specified to break the span down to safe proportions.

Programming

In programming this type of work, there is a temptation to believe that other work can be started once the diaphragm walls or piled walls are under way. This, in fact, is far from the truth. Specialist contractors carrying out this type of

work require large areas of working space and access needs. Experience shows that it is better to let the wall support contractor have the site to himself, with a degree of follow up by the ground anchor installer, who in turn will require an appropriate degree of staged excavation to his installation operation levels. Figure 10.12 illustrates this argument quite clearly.

Only when the anchors are in place and excavation down to formation level has been completed in an area large enough to provide economic working should other work be programmed to start.

Permanent construction within secured boundaries

The installation of permanent works, with temporary support, to secure the site boundaries is a time consuming operation. Site managers are usually under pressure to get the follow-up work under way as fast as possible. Optimistic programmes get prepared which in the event fail to be met. What is often not understood is the amount of work that has to take place before the base slab can be concreted. It is wise to examine thoroughly all items involved before preparing basement programmes. The reasons are:

1. The frequent complexity of cast iron drainage to be positioned in or below the foundation slab.
2. Such complexity may be coupled with long delivery times of the materials.
3. If internal shoring has to be used, speed of construction will be limited by such shoring.
4. Any requirement for lift pits, drainage sumps and ducting for other services than drainage.

The impact of services in basement slabs is illustrated in Figure 10.13. Gullies have to be set to accurate levels before concreting takes place as well as any other items.

All these items can take much longer than anticipated. All subcontractors involved need to

Deep basements

Figure 10.13 Basement slab services.

be required to provide programmes for the work at basement level and, in the case of multiple basements, level by level. In upper basement levels, other items arise: cast-in fixings to support sprinkler systems, suspended drainage, cable trays and racks, together with fixings for heating and ventilation systems.

Table 10.2 Activity sequence for basement slab

Item	Description
1.	Excavation to cut-off of piles
2.	Concrete blinding to underside of basement slab and forming thickening over pile heads if required
3.	Trim piles down to cut-off level
4.	Construct lift pits, sumps. Additional excavation formwork and reinforcement and concrete. Allow for lift guide fixings in lift pits
5.	Install cast iron drainage, together with gulleys
6.	Fix reinforcement to basement slab
7.	Fix box-outs and any surface fixings specified together with stop-ends and screeds
8.	Pour concrete and trowel to specified finish
9.	Remove stop-ends and screeds

Table 10.3 Activity sequence for suspended basement slabs

Item	Description
1.	Fix reinforcement to columns – basement slab
2.	Erect formwork to columns as above
3.	Pour concrete to columns
4.	Strike column forms
5.	Erect falsework and soffit formwork to first suspended slabs
6.	Fix on soffit all built-in fixings for soffit services suspension, together with all electrical conduits cast within the slab
7.	Fix reinforcement to slab, together with specified box-outs or cast in anchorages
8.	Fix stop-ends and screeds
9.	Pour concrete to suspended slab
10.	Strike formwork and falsework

Activity sequence for basement elements

Within the perimeter structure, the sequence of activities that the construction planner will have to evaluate is shown in Table 10.2.

For upper levels of basement the sequence will be as in Table 10.3, bearing in mind that all upper levels will be suspended slabs.

References

1. Illingworth J.R. (1987) *Temporary Works – their role in construction*. Thomas Telford, London.
2. Cole, (1990) Sinking in slurry, *Construction News*, February 8, p 26.
3. Current Practice Sheet No. 115 Concreting Diaphragm Walls. *Concrete*, February 1999.
4. Lilley Construction Ltd. See Figure 8.4 in reference 1 above.
5. Westpile Ltd (1990). Hospital gets pile treatment, *Construction News*, Piling and Marine Engineering supplement. 13 December p 11.

Further reading

Institution of Structural Engineers (1975) 'Design and construction of deep basements.' ISE, London.
Frischmann, W.W. and Wilson, J. (1984) Top down construction of deep basements. *Concrete*, November, pp 7–10.

Chapter eleven

In-situ concrete structures

This and the following two chapters cover the method and planning aspects related to the structural elements likely to be used in relation to the following types of buildings:

- Industrial and commercial;
- Service industries – schools, hospitals, hotels and so on;
- Accommodation – apartments, hostels, but not including load bearing brick, block or masonry housing or similar.

Structural solutions for these will normally adopt one of three approaches: in-situ concrete framing, pre-cast concrete frames or structural steel frames. In the pre-cast concrete solution, the method may also include external structural elements which are also part of the building cladding.

In-situ concrete structures

Those involved in establishing construction methods need to understand fully the system they have to plan – the components involved, temporary works needs, handling methods that can be used, safety requirements [1][2][3][4][5] and where the money is likely to be, in terms of distribution through the components that make the whole. It is best, however, to start with the advantages and disadvantages of a structural system.

Advantages and disadvantages

Advantages of in-situ concrete

1. Amenable to almost any shape.
2. Connections are homogenous with the rest of the structure.
3. Structural concept resistant to disaster or can be made so – earthquakes, explosion or collision, for example.
4. Whole production activity on site – in broad terms.
5. Alterations can be made at the last minute.
6. Design can proceed as the structure is built.
7. Rates of production (given the right

Components of in-situ concrete construction

In-situ concrete construction, in relation to construction methods, has three discrete components: formwork and falsework, reinforcement and concrete. Such clear-cut divisions are a major advantage when considering the planning of concrete construction. However, the planner and site manager need to understand that the three components can combine in several different forms to achieve the overall structure. Basic forms are: columns, beams, walls and floors. In turn these forms combine in whole or in part to produce a structural entity.

Disadvantages of in-situ concrete

1. Subsequent alteration difficult if change of use needed.
2. Errors in setting out and in formwork and falsework make corrections, once the concrete has been cast, very difficult and expensive.
3. All activities involve high labour numbers and plant on site.
4. QA linked to materials used and/or supplied.
5. Reinforcement and formwork/falsework tend to be labour intensive.

- Structural form with all elements. Less amenable to labour saving methods, mechanized formwork and falsework, and with a greater degree of difficulty in placing concrete than the alternatives (Figure 11.1).
- Structural form with no beams. Removes the most labour intensive element – beams. Amenable to mechanization of formwork and falsework.
- Structural form with columns and slabs only. The simplest form of structure. Highly amenable to table forms, flying forms. Prefabrication of reinforcement in columns and slabs can reduce cycle time. Some walls will be needed to provide for lateral shear. In many overseas countries, brick or block infill between columns is used in lieu (Figure 11.2).
- Crosswall construction (often known as 'egg box'). Walls and slabs only. Longitudinal stiffness achieved by floor to wall connec-

Figure 11.1 Example of structural form containing all elements.

tions being made very stiff, or by end bays having walls at right angles to the cross-section. Main use in domestic high rise accommodation either for flats or hotel rooms. Provides good repetition and allows the use of sophisticated steel formwork (Figure 11.3).

While the comments on each structural configuration may not be comprehensive, they give the flavour of the differences between them.

Financial implications

As a beginning, in planning the methods for in-situ concrete construction, it is useful to have an appreciation, in percentage terms, of the breakdown of the cost of a concrete structure. It should be the aim of those involved in both planning and in site management to build up such knowledge for different types of in-situ concrete configurations.

Such information is best compiled from the priced bill of quantities for each contract dealt

Figure 11.2 Multi-storey building with columns and slabs only which is highly amenable to sophisticated forming systems. Structural stiffness is achieved by box formation walls in service cores.

Figure 11.3 Walls and floors only. Commonly known as 'egg box' construction. Longitudinal stiffness provided by occasional walls at right angles to cross walls. Highly repetitive formwork.

with. In this way the effect of various configurations can be seen. The method used is to take the bill rates for all the structural concrete elements involved, together with the total value of the concrete structure. With this information, it is now possible to express each structural element as a percentage of the total concrete cost. At the same time, the individual components of a structural element, formwork, reinforcement and concrete, can be shown as individual percentages of the element in question.

Table 11.1 shows how the final presentation will look; in this case for a reasonably typical office building – inclusive of the cost of materials. The figures in brackets show the cost percentage breakdown for the components of structural elements, while the non-bracketed figures relate to the overall percentage costs for each component. By studying the Table, a good idea of where the key cost elements are will become apparent.

In contrast Table 11.2 shows the percentage breakdown for an in-situ bridge structure.

The big difference in the figures is accounted for by two related factors. In Table 11.1, the formwork and falsework could have six uses on the contract, while the bridge was one-off with a very complex soffit shape. The second related factor was the height of the falsework needed – that for the bridge was 14 m high and supporting a very heavy concrete load, while the office building falsework was 3 m high supporting a 150 mm slab.

From these two Tables, it will be clear that repetition of both formwork and falsework, together with the height of the falsework play a significant part in the overall structural cost. With these matters in mind, detailed consideration can begin of the planning process for in-situ concrete.

Plant and method assessment

Applying the knowledge of plant and methods appropriate for in-situ concrete construction requires, in addition, an understanding of how to formulate work loads for both plant and labour. With such information to hand, all the operations have to be integrated together to achieve the least overall time while, at the same time, balancing the available resources to the best degree. The result of such activity is, of course, a programme. If the structure is repetitive, the programme becomes what is usually known as a construction cycle. If the structure is not repetitive, the programme defines the work sequence.

Plant workloads

An initial assessment of what will be the dominant item of plant on the site is the first step needed. All other items of plant will be subordinate to it. For example, a tower crane on a building would be the dominant item. At the same time, it must be remembered that more than one item of dominant plant can be involved – more

Table 11.1   Cost breakdown for the reinforced concrete frame of a fairly typical office building, six storeys in height; the figures include cost of materials and are expressed as a percentage of total cost.

	Columns		Beams		Slabs		Walls		Overall
Formwork	4.5 (24.4)	+	8.5 (33.0)	+	18.5 (47.7)	+	8 (53.1)	=	39.5
Reinforcement	12 (60.9)	+	14.5 (56.3)	+	3 (7.1)	+	3.5 (23.8)	=	33.0
Concrete	3 (14.7)	+	3 (10.7)	+	18 (45.2)	+	3.5 (23.1)	=	27.5
	(100.0)		(100.0)		(100.0)		(100.0)		(100.0)

Table 11.2 Cost breakdown for a heavy bridge structure with a complex soffit shape. One use only for formwork and falsework.

Item	% of cost
Falsework and Formwork	62%
Reinforcement	20%
Concrete	18%
	100%

than one crane, for example, or the decision to pour the concrete with a pump and leave the crane(s) to handle all other materials. In deciding such matters, the available time for completion, if specified by the client, or what seems to the contractor as a competitive time and price, will be the main decision factor.

While the dominant plant items will be on the critical path, either singly or integrating with each other, work loads need to be established for all plant seen as necessary.

The work load for any item of plant must be based on the time taken to carry out all operations required in the working day. For example, in the case of cranes, the work load for a day might be: slew to stockyard (or vehicle); fasten slings to load; hoist and slew to working level; lower to required position, release slings and return slings to ground level for further load. In a multi-storey structure, such a cycle of events needs to be based on the average height of the building. By way of example, assume one crane only is proposed. The number of loads, of various kinds, will be assessed as above in terms of time. The number of repeat loads must be established for each particular item (e.g. bundles of reinforcement) and the overall time for the particular item worked out. All such calculations are then tabulated in the form shown in Table 11.3, to provide the workload for the crane in question to cover the working week.

Subordinate plant is dealt with in a similar

Table 11.3 Make up of crane workload

Items to be handled	No. of loads	Time per loads	Crane hours
Assessment of crane workload per week			
1. Handling formwork. Allow 15 min per load	70 lifts	15 min	17.5
2. Handle reinforcement. Av. 0.25 tonnes per bundle. Allow 6 mins per bundle for 55 loads	55 lifts	6 min	5.5
3. Concrete to walls and columns. $14\,m^3$ per week with $\frac{1}{2}\,m^3$ skip = 28 loads	28 lifts	15 min	7.0
4. Hoist other building materials. Estimated at 87 loads	87 lifts	6 min	8.7
Net crane workload			38.7
Working hours per week = 44.0			
Established workload 38.7			
Crane idle 5.3			5.3

5.3 hrs idle time = 12% which is very good for a crane. Perhaps even a little optimistic.
Note: If desired, the loads to be lifted can be broken down into more detail.

Working week 44 hrs		Total hours 44.0

way. Once established it must be checked against the dominant plant workload to make sure that it will fit in.

In finalizing the dominant plant workload, an allowance needs to be made for unavoidable delays. To achieve 85% utilization of a crane is about the best that can be expected in the planning stage. Note, however, the 15% idle time would be expected to include inclement weather – rain, wind or fog – unless the contract was taking place in a very exposed location.

Labour workloads

Labour workloads are more involved than those for plant. All activities on a structure need delineation, gang sizes decided upon and performance data used to assess gang weeks. When this has been done for all activities, the various gangs have to be examined to see how they can be made to interact, overlap and give the shortest possible time – again with an allowance for potential delays. As with plant, all activity gang weeks should be tabulated.

Once decided upon, the minimum time can be compared with that for the plant proposed. If the plant time is longer, but not by enough to justify further plant, that will be the best time possible. If the labour time is longer by some margin, efficient utilization of the crane will not be possible. In this case, the gang sizes need reconsideration to reduce the labour workload time to fall in more with the crane overall time. In reassessing the labour it must be remembered the optimum gang size requirement discussed in Chapter 6.

The labour workload calculation is best understood from an actual example from a real contract. In Chapter 7, a comparative costing exercise relating to a real contract was used to illustrate how a change in method had resulted in a reduction in cost, after it had been found that the initial work loads for the two cranes planned were higher than had been anticipated and time was being lost.

As an extension to this study, the following worked example illustrates the way in which the labour work loads were re-assessed for the change in method in taking the concrete to floors away from the cranes. It provides an example of how the sequence of events is built up and, in this case, the development of a construction cycle for the whole building of seven floors (Figure 11.4).

Case study

Determination of labour work loads for a seven storey office block extension

A first requirement is to determine all the quantities involved, per floor, together with both plant and labour outputs anticipated. In addition, the weekly work loads for the plant to be used are required for comparison with the best time achievable when the labour cycle per floor is known from the workload calculation.

The summary of quantities, together with operative performance data is given in Table 11.4. Plant performance and work load per week is provided in Table 11.5. The make up of the crane work load has already been given in Table 11.3.

Determination of key work groupings

Consider what are the main structural elements involved.

1. Main building core structure Components are:
   (a) fix reinforcement
   (b) erect formwork
   (c) pour concrete
   (d) strike formwork
2. Escape stair area – (two number)
   (a) fix reinforcement
   (b) erect formwork
   (c) pour concrete
   (d) strike formwork

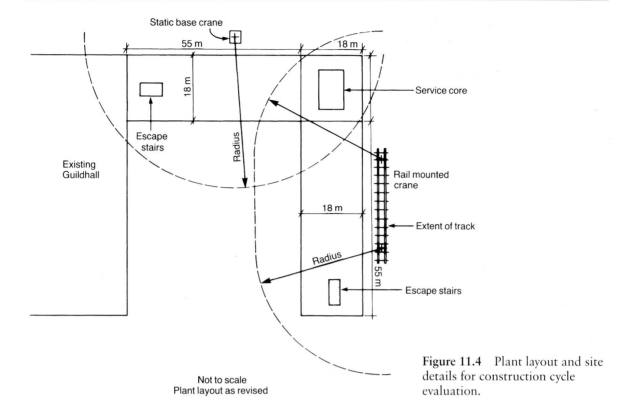

Figure 11.4 Plant layout and site details for construction cycle evaluation.

Not to scale
Plant layout as revised

3. Floors
   (a) erect formwork and falsework
   (b) fix reinforcement
   (c) pour concrete
   (d) strike formwork and falsework

When the workloads have been established for each group, the integration of each with the others will determine the construction sequence. Note: columns are always capable of integration to suit, as the workload is relatively small.

Determination of work loads in each group

From the quantities and performance data available, establish:

1. Manhours involved in item;
2. Manweeks involved in item;
3. Gang size needed (may require alteration later);

4. From (3) above calculate gang weeks.

For each group produce trial programme for group items only. Repeat for all groups.

Integration of groups

On completion of above items make trial integration of groups to assess least time sequence and cycle time per floor. In so doing, consider:

1. Degree of overlap possible
   (a) within group
   (b) between groups
2. Need for gang size changes to balance
   (a) resources
   (b) programme with crane workload.

Note: in any construction rhythm, there will be a period needed to get into the sequence (run in period) and at the end of the structure a period to get out of the sequence (run out period).

Table 11.4   Quantities and planning data

Summary of quantities per floor

Concrete
Floor Slab   790 m³
Columns   53 m³ for 82 No. cols
Walls   210 m³ for 1036 m² of wall, or 2072 m² of formwork. For planning purposes assumes that the two
              escape stairs contain 518 m² of formwork each, while the central core contains 1036 m² of
              formwork, with the concrete pro rata

Reinforcement
Floor Slab   105 tonnes
Columns   3 tonnes for 82 columns
Walls   12 tonnes for wall division as above

Formwork
Floor Slab   2440 m²
Walls   2072 m² Divided into three areas as above
Columns   402 m² for 82 columns

Notes
There are 4 No. columns across the width of the structure and longitudinally at approx. 6 m centres. There
will be some variation due to the incidence of walls to escape stairs and the central core.

Operative performance data

Formwork
Formwork to floor – Fix and strike 1 man-hr/m²
Formwork to columns Take two men strike and erect six columns per day
Formwork to walls (Fix and Strike) 2.5 man-hrs/m²

Reinforcement
To walls   0.045 tones per man-hr
To columns   0.045 tonnes per man-hr
To floors   0.060 tonnes per man-hr

Concrete
Pumped in floors 16 m³/hr av. over pour

Handled by crane in columns – allow pour rate of 3 m³ per hour
Handled by crane in walls – as for columns

Thus: the overall programme time is NOT the number of floors × cycle time. It is the summation of:

(Number of floors × cycle time)
+ Run in time + Run out time

In carrying out the above procedures, make-up sheets are prepared for each group to provide a permanent record of how the programme time and cycle were established. This information may provide valuable evidence, if alterations to work load are made by the designers as work

Table 11.5   Plant workloads

Concrete pump
There are 7 No. pours per floor for the pump each containing 113 m³ poured at a rate of 16 m³ per hour.

Two tower cranes
In order to simplify the exercise, the workload for each of the two tower cranes has been taken as equal. When comparing the labour workload with that for the tower cranes, take the tower crane workloads as given below:

Workload for each crane

Handling formwork per week (hours)	17.5
Handling reinforcement per week, hrs	5.5
Handling concrete to walls and cols per week	7.0
Handling other materials	8.7
Idle time per week	5.3
Total hrs	44.0

proceeds, in establishing claims for extensions of time and additional cost.

Core structures

In multi-storey structures, the vertical elements – core structures – control the rate of vertical rise. They have the biggest workload for a given area in plan. The planning must, therefore, achieve the best time possible for these activities.

The crane workload assessment given in Table 11.3 only allows 7 hrs per week per crane for pouring concrete in walls/columns. The pour time is likely to be the controlling factor in over-all time. Analyse this first.

Total concrete in cores and columns is 263 m³ the breakdown of which is as follows:

Main core	105.0 m³
Escape stairs(1)	52.5 m³
Escape stairs(2)	52.5 m³
Columns	53.0 m³
	263.0 m³

Taking the pour rate assessed per crane as 3 m³/hr/crane, we have:

$$7\,\text{hrs} \times 3\,\text{m}^3 = 21\,\text{m}^3/\text{week/crane}$$
$$\text{or two cranes}\ 42\,\text{m}^3/\text{week}$$

Thus, the best possible time for pouring walls and columns:

$$\frac{263.0\,\text{m}^3}{42\,\text{m}^3/\text{week}} = 6.25\ \text{weeks}$$

Consider now each core in turn for all items.

Main core: make up sheet

Reinforcement
$$6\ \text{tonnes}\ @\ 0.045$$
$$\text{tonnes per man-hour} = 133.33\ \text{man-hours}$$
$$44\,\text{hr week gives}\quad 3.03\ \text{man-weeks}$$
$$\text{Say 2 man gang} = 1.51\ \text{gang-weeks}$$

Formwork
$$1036\,\text{m}^2\ @\ 2.5$$
$$\text{man hours per m}^2 = 2590\ \text{man-hours}$$
$$44\,\text{hr week} = 58.86\ \text{man-weeks}$$
$$\text{Try 6 man gang} = 9.81\ \text{gang-weeks}$$

Important note
Formwork is different from reinforcement and concrete. It has to be fixed in place and, when the concrete has set, removed from the concrete (struck), cleaned, repaired and re-oiled ready for the next use. The rate for use will either show fix and strike as one value or as separate values. In this exercise an overall value is used. In such instances, it is usual to take fix = 2/3 value and strike at 1/3 value.

In the above calculation, therefore, we have:

$$\text{Formwork} = 9.81 \text{ gang-weeks}$$
$$\text{of which erect} = 6.54 \text{ gang-weeks}$$
$$\text{and strike} = 3.27 \text{ gang-weeks}$$

Concrete
As the main core has the highest workload, it needs to be planned to achieve as fast a time as possible. The concrete time tends to be the problem. Both cranes, however, can reach the whole of the main core. It is therefore possible to pour the walls at a rate of $2 \times 21 \text{ m}^3 = 42 \text{ m}^3$ per week (see above).

Best time for concreting main core can be

$$\frac{105 \text{ m}^3}{42 \text{ m}^3/\text{week}} = 2.5 \text{ weeks}$$

But as only one day's pour per crane per week, allow 3 weeks.

Summary
From the above, the programme times for the four elements in the main core can be summarized for the initial trial programme and cycle (see Cycle No. 1 below).

Escape stair core (2 No.)
Reinforcement (per core)
$$3 \text{ tonnes} @ 0.045 = 66.67 \text{ man-hours}$$
tonnes per man-hour
$$44 \text{ hr week} = 1.51 \text{ man-weeks}$$
$$\text{Allow 2 man gang} = 0.75 \text{ gang-weeks}$$

Formwork
$$518 \text{ m}^2 @ 2.5 \text{ man}$$
$$\text{hours/m}^2 = 1295 \text{ man-hours}$$

$$44 \text{ hour week} = 29.43 \text{ man-weeks}$$
$$\text{Try 6 man gang} = 4.90 \text{ gang-weeks}$$
$$2/3 \text{ erect forms} = 3.27 \text{ gang-weeks}$$
$$1/3 \text{ strike forms} = 1.63 \text{ gang-weeks}$$

Concrete
$52.5 \text{ m}^3 @ 3 \text{ m}^3$ per hour. Allow crane time 7 hrs per week:

$$\text{Pour time} = \frac{52.5}{3.0} =$$
$$17.5 \text{ hrs} @ 7 \text{ hrs/week} = 2.5, \text{ say 3 pours at}$$
$$1 \text{ No. per week}$$

Escape stair 1st programme assessment
From above timings, 1st time assessment for an escape stair can be established (see Cycle No. 2)

Make-up sheet – floors
Formwork and falsework to floors
$$2440 \text{ m}^2 @ 1 \text{ m}^2/ = 2440 \text{ man-hours}$$
man-hour
$$44 \text{ hour week} = 55.45 \text{ man-weeks}$$
Gang 6 men
$$\text{Erect formwork} = \tfrac{2}{3} = 6.16 \text{ gang-weeks}$$
$$\text{Strike formwork} = 3.08 \text{ gang-weeks}$$

Reinforcement to floors
$$105 \text{ tonnes} @ 0.060 = 1750.0 \text{ man-hours}$$
tonnes per man hour
$$44 \text{ hour week} = 40.0 \text{ man-weeks}$$
$$\text{Say 6 man gang} = 6.63 \text{ gang weeks}$$

From the figures obtained for formwork and falsework and reinforcement, adjust performance to make both items operate at the same pace.

Reinforcement
1.51 gang weeks

Formwork (erect)
6.54 gang weeks

Concrete 3 pours over
3 week period

Strike formwork and
falsework

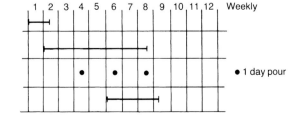

Cycle No. 1 Trial cycle for main core structure.

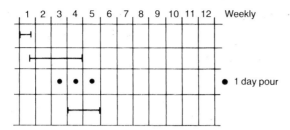

Cycle No. 2 Trial cycle for 2 No. escape stairs.

Allow 6.50 weeks in both cases. From the decisions taken to pump all concrete to floors (see Chapter 7), each floor is divided into 7 No. bays, 18 m wide by 21 m long (Figure 11.5). In assessing the amount of formwork and false-work needed, it is necessary to consider the length of reinforcing bars detailed. In this case all bars were straight and 12 m long. Minimum of 21 m of formwork needs to be erected before any concrete can be placed. In turn, reinforcement erection cannot start until at least 12 m of forms in place.

As proposed above, the rate of formwork progress is

$$\frac{147\,m}{6.50} = 22.61\,m/week$$

which ensures that a bay 18 m × 21 m will be available every week for the pumped pour.

Concrete
With one bay a week of formwork and reinforcement completed, the concrete can be programmed to be one pumped pour (one days work) every week. The actual quantity of formwork in place to achieve such progress is shown in Figure 11.5. The total amount is finally dependant on the specification in relation to striking time. If one can strike at ten days, at least 4 No. 21 m bays of formwork are required. For twenty one days the figure would be 5 No. bays and so on (see Cycle No. 3).

Columns make-up sheet
Reinforcement
    3 tonnes at 0.045 = 66.67 man-hours
tonnes per man-hour
            44 hour week = 1.51 man-weeks
                Allow, say = 2.00 man-week
    Allow 2-man gang = 1.00 gang-week

In fact, an achievement rate of 6 No. columns/day will not fit in with crane pour time or progress need. In this case it is more logical to include column erection, as needed, to be handled by core and escape stair gangs.

Concrete labour force

While the labour workloads have been based on performance figures for labour output, the time

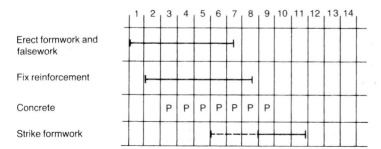

Cycle No. 3 First trial cycle for floors.

scales for concrete production have been established on plant output. This is normally the case, as both types of plant used, cranes and the concrete pump, are dominant items.

Once performance has been established, the labour needed can be apportioned. In this case, the concrete does not dictate the overall programme – it fits into the high workload, labour wise, in respect of formwork and falsework, and the fixing of reinforcement.

In establishing the wet end concrete gang, it has to be remembered that throughout the programme only one day's pour a week will be needed. In other words, a full time concrete gang is not needed. The labour used needs to be kept to a minimum and put to work elsewhere when not needed for concrete. Numbers for concrete work:

1 No. concrete ganger
1 No. vibrator operator
2 No. spread and finish
1 No. pump wet end or handle skip

5 No. for pumping or for walls

Formwork requirements

The final item to be established is the quantities of formwork that need to be provided to meet the programme.

1. Core structures. All core structures will need a full set of forms. This is because of the lim-

ited space for assembly, and the need to keep time down to a minimum.

2. In the case of the floor forms and attendant falsework, the labour work load and the amount of formwork will eventually depend on the striking time specified. If three weeks have to elapse after pouring, six bays will be needed (Figure 11.5). If the falsework used allows for removal of the soffit forms yet leaves falsework to provide dead shores in place, the time might be reduced. The problem with this method is that extra sets of falsework are required. In many cases, it can be easier to to provide a whole floor – both to counter possible delays and allow the formwork/falsework to move vertically only, when struck, which is more straightforward.

Method statement

All the above information is now ready for inclusion in the planning method statement which will go to the estimator at the tender stage. The compilation of these method statements is described in Chapter 17.

Construction cycle

The information derived from the planning analysis and individual group programmes can be put together to provide the overall pro-

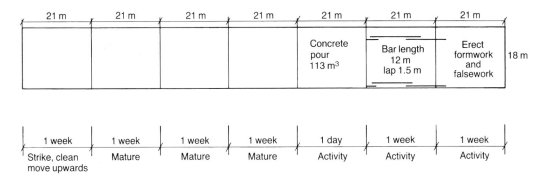

Figure 11.5 Assessment of soffit formwork and supporting falsework.

gramme for the structure as a whole. A section of such a programme is illustrated in Figure 11.6. From this can be derived the cycle time per floor, together with the run in and run out times at the start and finish.

The example in this case study is based on a real contract. It must be realized however, that simplifications have been made to make things easier to follow.

When construction cycles have been prepared, it is a useful management tool to produce a cycle time record programme to provide performance achieved for future use. An example of this type is shown in Figure 11.7. Taken from an actual contract, it provides reasons for cycles which were unusually long and vividly illustrates the sharp increase in time once the repetitive floors have been completed.

It is also good practice to maintain a record

of the methods proposed – plant locations and type, formwork and falsework arrangements etc. Again a useful record for future method statements and planning purposes (Figure 11.8).

Speeding up construction

Structural steel has proved a major competitor to reinforced concrete construction in recent years. As a result, the in-situ concrete industry has done a great deal of research on how to speed up in-situ concrete construction.

Those responsible for construction method planning need to keep well versed in ways of speeding up construction – if only to try to keep ahead of their competitors. At the same time, those who design such structures should also see the advantages to their clients.

Much has been made of the word 'Buildability'

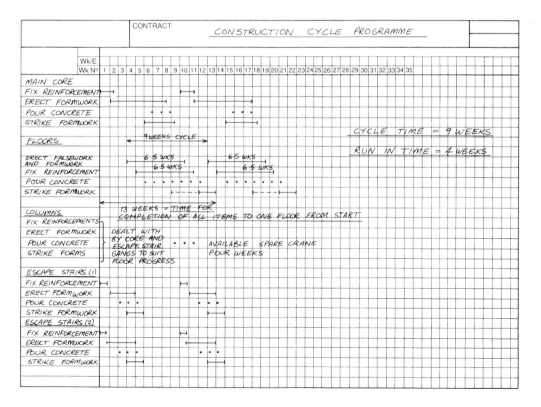

Figure 11.6 Construction cycle development from programming case study.

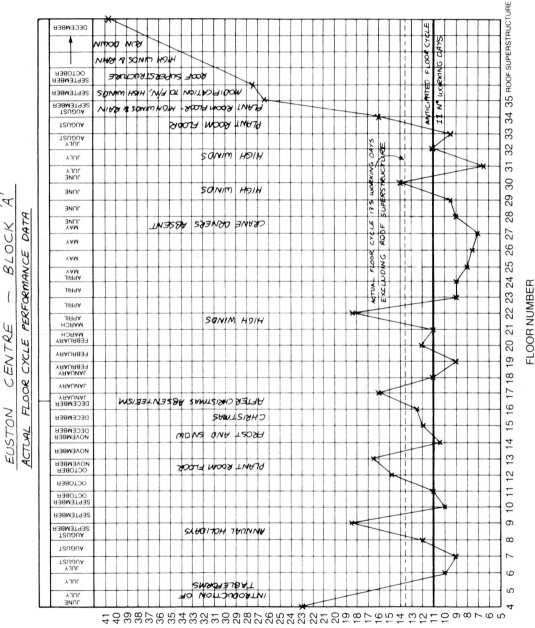

Figure 11.7 Construction cycle record chart against that planned.

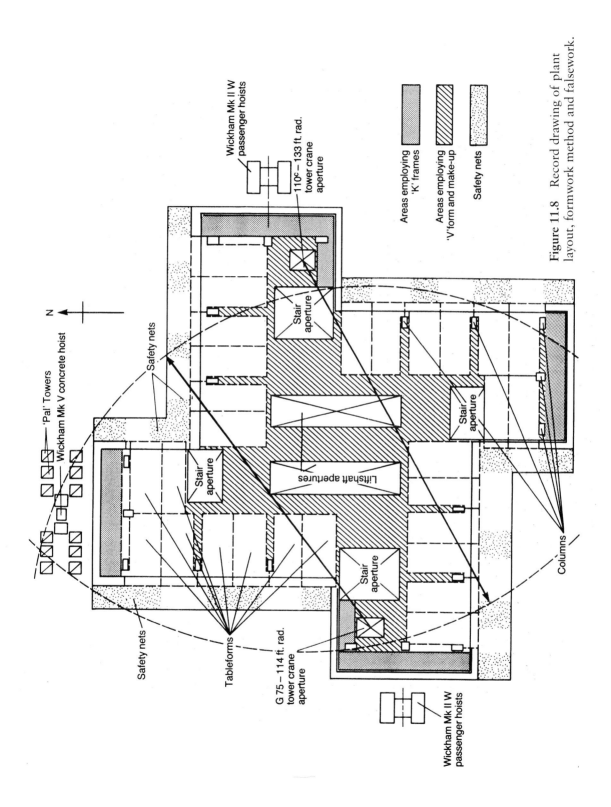

Figure 11.8 Record drawing of plant layout, formwork method and falsework.

Wickham Mk II W passenger hoists

110^c – 133 ft. rad. tower crane aperture

Areas employing 'K' frames

Areas employing 'V' form and make-up

Safety nets

N

'Pal' Towers

Wickham Mk V concrete hoist

Safety nets

Stair aperture

Stair aperture

Liftshaft apertures

Stair aperture

Stair aperture

Columns

Safety nets

Tableforms

G 75 – 114 ft. rad. tower crane aperture

Wickham Mk II W passenger hoists

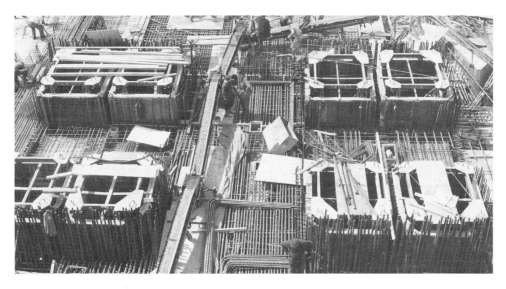

Figure 11.9   One piece lift shaft internal forms for crane handling.

in recent years[6]. Broadly it means design which creates the climate for rapid construction without any sacrifice to appearance or quality. See Chapter 5 for a more detailed consideration. The following factors relate to the speed and quality of construction:

1. In high rise structures, the cores control vertical rise speed. Core simplification is crucial to speed. If it can be done for slipform techniques (and it has to be), why not for traditional methods? One way this speed up of an in-situ core construction can be achieved is illustrated in Figure 11.9. The lift shaft forms are mounted on the steel template seen at the top of each shaft. Each side form is pivoted from the template and by means of push pull mechanisms can be collapsed inwards. Once struck, the whole unit can be extracted in one piece by the crane. When cleaned and re-oiled, the whole assembly is lowered onto levelled timbers bolted to the inside of the shaft. The wall forms are then jacked outwards to the top of the previously concreted shaft. All that remains is to plumb the assembly.

    To use this method, the structural designer

has to agree to dispense with projecting concrete thresholds and create a continuous unobstructed shaft. Thresholds are designed as bolt-on metal versions to be fixed after the shaft form has moved upwards.

2. The large area pouring of floors speeds construction, reduces cost and improves quality. The technical justification for this approach is given in a Reinforced Concrete Council guide[7]. Figure 11.10 and Table 11.6 show actual examples.
3. The introduction of pre-cast elements in horizontal locations can gain speed and improve accuracy and quality (Figure 11.11).
4. Simplification of floor details can save time and money.
5. Provision of aids to the maintenance of floor to floor and overall height accuracy – of great importance in relation to follow-up cladding.
6. Use of straight bar reinforcement.
7. Prefabrication of reinforcement in columns and beams – detailing which makes this as easy as possible.
8. Repetition as much as possible in formwork and falsework allows less labour-intensive methods.

Figure 11.10 View of large office block floor formwork, showing screed layout for large area pouring.

Figure 11.11 Pre-cast elements as edge beams to speed up work and guarantee accuracy and quality. Note pre-cast elements on floor formwork prior to concreting floors in background.

223

Table 11.6 Cost benefits of avoidance of stop ends (from actual pours)

Large volume pours	
Large concrete raft	2210 m^3
Poured in 50 m^3 bays	Stop ends = 10.27% of the cost of the concrete per m^3 in place.
Poured in 3 No. bays	Stop ends = 1.56% of the cost of the concrete per m^3 in place.
Large area floor pours	
Poured in 40 m^3 bays (as specified)	Stop end costs = 9.38% of the cost of the concrete per m^3 in place.
Poured in 120 m^3 bays	Stop end costs = 2.19% of the cost of the concrete per m^3 in place.

Slipform construction

It would be wrong, in a chapter on in-situ concrete construction, not to include a section on slipform techniques.

Slipforming is a specialized technique carried out by contractors who are properly qualified to provide the skills and equipment needed.

Slipforming systems have been progressively developed over the years from hand-operated jacking methods, to raise the forms at regular intervals, to sophisticated hydraulic methods, where the jacks move as one. The method can be employed in the construction of a wide range of structures from simple silos, chimneys and bridge piers, to large complex building cores and towers. Tapering profiles and reducing wall thicknesses can readily be accommodated.

Basic system

One of the most advanced systems is illustrated in Figure 11.12. It utilizes a jack climbing on a 48 mm diameter steel tube which has considerably greater stability than the alternative solid rod. This enables a jack of higher capacity to be employed with the consequent increase in yoke centres and platform area. Three levels of platform are commonly used with this system. The upper platform acts as a storage and distribution area and incorporates guide templates for the vertical reinforcement bars. It adds rigidity to the overall assembly which significantly influences the accuracy that can be achieved. Concrete, reinforcement, door frames, blockouts and plant items can be accommodated at the upper level. This reduces congestion at the middle level, the main working deck, which is level with the top of the shutter, where operations are carried out under sheltered conditions. The lower platform, suspended on rigid suspended scaffold frames, provides access to the concrete immediately below the shutter to enable finishes to be applied and blockouts exposed.

Advantages claimed for the use of sliding formwork

The use of sliding formwork, with its continuous casting process ensures the rapid, economical and accurate construction of tall structures. Construction joints occur only at the positions where the sliding process is stopped, perhaps to accommodate changes in section or for crane adjustments. The use of slipform construction, like all specialist methods, is dependent on two key factors:

1. Is it technically satisfactory?
2. Does it provide the least cost solution in **overall** terms?

What may be considered fashionable in construction methods has no place in the final choice – or should not have!

In evaluating slipform as a cost effective solution, the basis of comparison with alternative methods must be clearly understood. Once recognized, the comparative costing should follow the principles laid down in Chapter 7.

Categories of use

In order properly to compare slipform with alternative methods, the whole of the concrete process needs to be considered. If this is to be done accurately, it must be recognized that slipform can be used in three distinct categories, each of which plays a different role in the overall method of solution.

1. Where slipform is used for the whole structure – constructing silos, chimneys, communication towers and very tall slender columns in bridge construction, to name the

main examples. It is in this field that slipforming first began. Its use can be categorized as a **substitution** method.
2. Where the slipform element is a relatively small part of the overall structure. In this role, use relates mainly to multi-storey office buildings or similar structures with high core elements. In this case its use can be categorized as a **dominant** method.
3. The principle of the jacking system employed can, on occasions, be adapted to a lifting role where heavy lifts would be difficult by any other means. Here, an **unconventional** use.

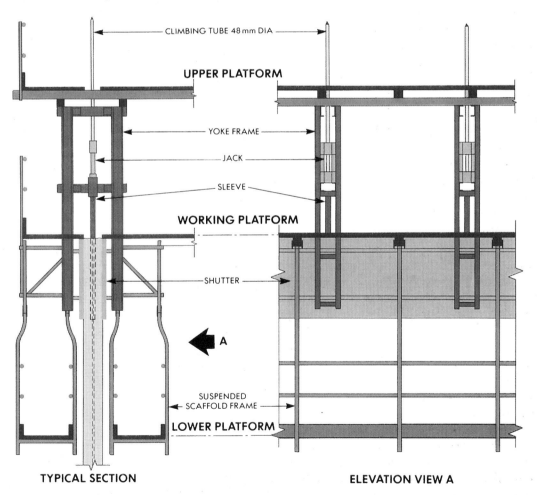

Figure 11.12 Slipform method. General arrangement of equipment. (Douglas Specialist Contractors Ltd).

Substitution method

The terminology used here should be self explanatory. Slipform is used as a direct substitution for other alternative solutions (Figure 11.13). When comparing methods, the direct comparison must cover all the items involved in the structure, as shown.

Direct cost comparison elements

Slipform
Provision of specialized equipment and formwork

Initial setting up time – usually about three weeks
Continuous reinforcement fixing
Plant and labour for multi-shift working
Supply of concrete
Access and safety needs

Cost element 'S'

Alternatives
Manufacture of formwork
Set-up cost per use

Figure 11.13 Slipforming the towers to the Humber bridge – a clear substitution example. (Photo Donald I Innes, Consulting Engineers Freeman Fox Ltd).

Reinforcement fixing
Plant and labour for single shift working
Supply of concrete
Access and safety needs

Cost element 'A'

To these costs must be added the alternative preliminaries and overhead items:
Time cost elements

| Slipform is faster over a minimum height. O/A Cost 'T$_s$' | Conventional methods slower O/A Cost 'T$_a$' |

The decision to use one method or another is dependent on, for slipform:

$$S + T_s < A + T_a$$

unless time is the critical factor.

Dominant method

In this situation, the approach is somewhat different. The use of slipforming for merely the core structure(s) of a building will dominate the structural planning, while not being the major part of the structure. It follows that the direct cost elements must now cover the whole of the superstructure when comparing with alternatives. In the slipform case, the core cost will need to be considered separately from the surrounding floor areas (Figure 11.14).

Assuming that the surrounding floor construction is the same for both alternatives, the direct comparison becomes:

Core cost + Floors cost (Slipform)
or
Core + Floor cost (Traditional)

Figure 11.14 Multiple slipformed core towers on a major building structure. The third one from the left is still under construction. Note the lack of other structural work during slipforming. See the text for economic factors. (Douglas Specialist Contractors Ltd).

227

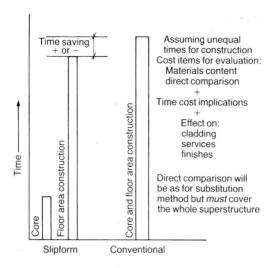

Figure 11.15 Dominant method – cost comparison basis with conventional in-situ construction

The result of the direct comparison may depend on how quickly the floor construction follows up the slipformed core. Time, therefore, may well be the key issue in this method for slipforming. Figure 11.15 makes the point clear.

The protagonists for slipforming always argue that the method is quicker in this mode of use. Convincing times for the core built in so many days are given but rarely does one see how long the surrounding floor took to complete and how long it was before the floor got started after the core had been completed. Many publicity pictures show a naked core, with no sign of the surrounding floors. It behoves the planner of construction methods to analyse this situation very carefully before opting for the slipform approach.

A further factor that needs looking into in dominant method use is the core design. Structural designers are well aware that, for slipforming, the core design needs to be as simple as possible. As a result, many of the awkward bits and pieces of a traditional core design are removed. One may reasonably ask why is this not done for traditional methods? How much faster would the traditional structure go if it was?

A final item that needs examination in the dominant method is: What effect does the method have on building cladding, service installation and finishes? Clearly an intangible item as defined in Chapter 7.

Planning factors

Slipforming is a specialist method. The firm concerned will supply, erect and dismantle the forms and jacking equipment, and give expert supervision and advice on mix design. The main contractor will normally provide the plant for lifting concrete and reinforcement, and all operative labour.

Operational performance (rate of progress) will be provided by the specialist for planning purposes, together with the methods of installing cores for openings and any temporary bracing that may be needed for overall stability in the dominant method.

The planner will have to assess concrete quantities and delivery rates required to meet the vertical rise rate planned, together with making of core box outs and any other items that have to be built in as the slide progresses. It must also be remembered that the labour force needs rapid access to the working level. This is best achieved by a personnel hoist erected either inside – if enough room is available – or outside and anchored to the completed core. Tower cranes are normally used for handling concrete and reinforcement.

References

1. *Health and Safety at Work etc Act 1974*. HMSO, London.
2. *Management of Health and Safety at Work Regulations 1992*. HMSO, London.
3. *Construction (Design and Management) Regulations 1994*. HMSO, London.
4. *Construction (Health, Safety and Welfare) Regulations 1996*. HMSO, London.
5. *The Lifting Operations and Lifting Equipment Regulations 1998*. HMSO, London.
6. Illingworth, J.R. (1984) Buildability – tomorrow's need. *Building Technology and Management*, pp. 16–19.
7. Goodchild, C.H. (1993) *Large area pours for suspended slabs – a design guide*. Reinforced Concrete Council, Crowthorne.

Chapter twelve

Pre-cast concrete structures

Pre-cast concrete structures are widely used, from single storey factory buildings to multi-storey structures for commercial, industrial and training use. They have also played a significant role in domestic accommodation.

Method of use

There are effectively only two types of pre-cast concrete structure:

1. Individual components erected in place;
2. Volumetric units.

As there are considerable differences between the two approaches, they are dealt with separately.

Individual components erected in place

The standard form of this type is the production of structural elements in a pre-cast factory, under controlled conditions. Component elements will normally be: columns, beams, walls, floor panels and staircase units. Separate units are then delivered to site for erection and joining to other units (Figure 12.1).

A further approach is load-bearing walls, supporting floor units – the basis of many 1960s high rise accommodation blocks. In view of the well publicized problems which have arisen with this method, it is not in favour as a structural

system. None the less, out of such origins has come the use of high quality decorative finish cladding elements for both pre-cast and in-situ structures. In some cases, such cladding elements may be extensions of the structural members – for example, in relation to columns, (Figure 12.2).

Pre-cast cladding is considered in more detail at the end of the chapter, and there is further mention in Chapter 14.

With these pros and cons in mind, the key issues in the planning sense become clear: those related to design details and those of an operational nature in relation to efficient handling and erection.

Design-related factors

1. Recognition that joints are critical. The method planner should examine joint details very carefully to establish:
   (a) Can the joint detail, after execution, be inspected to ensure that it complies in all respects with the designer's details?
   (b) Can the detail be properly dealt with in terms of access to the joint by those using levelling nuts, dry pack mortar or grouting up?
   (c) One leading pre-cast manufacturer[1] uses a patented form of structural connection between columns and beams

Figure 12.1   General view of pre-cast structure, showing component members – columns, beams, slabs, walls and stairs. (Trent Structural Concrete Ltd).

(Figure 12.3). It combines the strength and versatility of bolted steel to steel structural connections, with the design flexibility, inherent fire resistance and economy of pre-cast concrete.

2. Has stability in erection been considered by the designer and a sequence of erection put forward, even if the contractor may have other ideas? (Two authoritative documents [2][3] give advice in this area.)

3. Where possible, joints in columns should be above floor panels level. In this way, the floor provides a safe staging for those erecting the next level of columns and beams (Figure 12.4).

4. Overall, the proposed sequence of erection by the designer must be realistic.

5. Where metallic connections arise, is more than one metal involved? Are such metals compatible in the electrolytic sense?

Factors affecting operational methods

It will be clear that the erection of pre-cast structures depends on heavy handling equipment in the shape of suitable crange, in most cases. Therefore:

1. To be able to use the most economic crane, the loads to be lifted need to be made as equal as possible for the required radius of operation. In other words, the crane workload can be balanced to achieve the least number of lifts.

2. While formwork and falsework do not arise

Figure 12.2 Structural column and cladding combined. Note the need for edge protection.

unless joints are in in-situ concrete, pre-cast concrete erection still needs a degree of temporary works. In general such temporary works are related to the provision of temporary supports for stability purposes, lining and levelling, together with templates to ensure proper spacing, etc. Figures 12.5 and 12.6 illustrate some aspects of such temporary works.

3. Cost savings are being made in Figure 12.7 and 12.8 by utilizing other pre-cast members as kentledge to the raking support members. This method demands forethought in the planning stage. Additional cast-in sockets must be provided for, to anchor the rakers. The same will apply in relation to template members. Sockets for their fixing need to be allowed for in the casting of columns or other elements.

4. Safe access is obligatory when connecting units together. Two main solutions need examination: prefabricated scaffold towers, capable of moving as required by the erection crane; or mobile hydraulic access

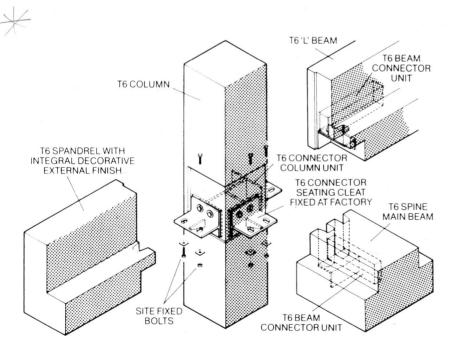

Figure 12.3 Structural joint between columns and beams by a leading manufacturer of pre-cast structures. (Trent Structural Concrete Ltd).

231

Advantages of pre-cast concrete construction

1. Complete manufacture of units carried out off site in normal circumstances, giving high quality control;
2. Minimizes erection time;
3. High quality of individual elements as factory made;
4. Does not require formwork and falsework on site;
5. Labour on site greatly reduced;
6. Greater capability for good Quality Assurance (for individual units);
7. With careful design, relating to the erection sequence, scaffolding can be minimized to a high degree;
8. With very restricted sites, the whole structure can be pre-fabricated off site and delivered as and when needed, avoiding the need for site storage.

Disadvantages of pre-cast concrete construction

1. Design needs to be complete before casting commences;
2. Last minute alterations to the structure are impossible;
3. Arguably dearer than in-situ concrete or structural steel;
4. Quite unsuitable if structural alterations might be needed in the future;
5. Units require care and protection while in storage, during transport and while handling on site, all of which involves cost;
6. Units can distort during the curing period;
7. If units are produced in the sequence of erection to save time, damage to one element can create havoc with the erection programme;
8. Prestressed floor panels or beams can create problems with uneven camber in different units.

Figure 12.4 Design allowing erection of columns from the adjacent pre-cast floor. Thus avoiding any temporary scaffolding to provide a safe place of work. (Trent Structural Concrete Ltd).

Figure 12.5 Levelling device for pre-cast columns – to hold plumb in position prior to grouting column. (See Chapter 8 for details of column bases for pre-cast columns).

Figure 12.6 Site A-frame for temporary storage of column and cladding element combined. Note also the need for protection on the architectural sections of the unit.

Figure 12.7 Minimizing cost by using the foundation to another base to provide kentledge to raking shores supporting columns. Planning must allow for the anchor bolts to be cast in when the base is poured.

Figure 12.8 Another save money idea – using pre-cast ground beam units as kentledge for the support of columns and their plumbing. Note also spacer bars between columns.

platforms, Figures 12.9 and 12.10. Mobile access platforms are well suited to pre-cast concrete erection as the working platform is only needed in one place for a relatively short time. The crane doing the erection is left to concentrate on that aspect alone – for which it is best suited. See also Chapter 4, scaffolding.

Planning the erection sequence and time scale

Scheduling of components

It is a first priority that a schedule of components should be established at the earliest possi-ble moment – if possible at the tender stage – and their sequence of erection. A good bill of quantities would list all variations, with key dimensions and the estimated weight. All this information should then be scheduled in a form similar to that in Table 12.1. In preparing such a schedule, it is wise to check the weights of units given. Pre-cast elements are normally more dense than in-situ concrete and may also con-tain more intense reinforcement. To be sure that adequate crane capacity is allowed, the concrete density used should be at least 2563 kg/m^3.

The last two columns of Table 12.1 provide data for establishing the crane rating that will be needed. Even from a small sample of weights, as shown, it will be clear that the floor units are the critical ones. They will control crane capacity needed.

Figure 12.9 Erection crane moving scaffold access tower to new location.

Figure 12.10 Using hydraulic work platform instead of scaffold tower, as mobile frees crane for pre-cast erection only.

Table 12.1 Form of Schedule for components (Reference numbers taken from Figure 12.11)

Ref. No.	Item	Cross-section area	Length	Location	Weight	Radius from centreline of crane	Crane rating in m/tonnes
A/1	column	$0.09\,m^2$	3.0 m	Gnd fl.	692 kg	15 m	10.35
C/6–7	int. beam.	$0.135\,m^2$	5.5 m	1st fl.	1.900 t	10 m	19.00
D/E (1 F)	Floor unit 1 of 4	$0.30\,m^2$	4.0 m	2nd fl.	3.075 t	12 m	36.90

Note: in compiling this schedule, one would list the items in the order of their erection. Above is an example of notation only. Weights based on concrete of $2563\,kg/m^3$.

A further value of this schedule is the ability to compare the units shown in the final working drawings with those given in the tender documents on which the choice of cranage was made. Variations may well occur in the final detailing and if the units become heavier, the planned cranage may not now have the desired load–reach ability and a heavier crane becomes necessary. Such a situation creates a clear item for a potential claim.

One final matter needs attention, in relation to crane rating. Once the schedule of weights has been calculated, crane capacity must take into account the additional weight of slings, spreader beams or other lifting gear necessary for the safe handling of any particular unit.

Sequence of erection

Once the schedules of units have been completed, the sequence for erection can be examined. In so doing, a number of questions need answers.

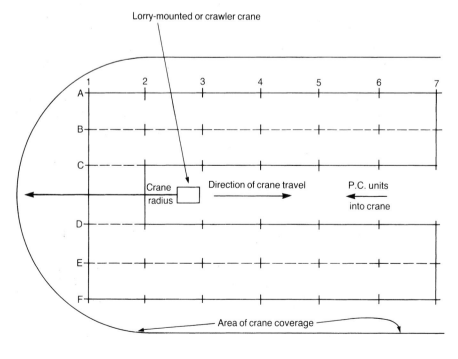

Figure 12.11 Plan layout showing erection based on leaving out the centre bay. In this way, the crane rating can be reduced as radius is minimized for erection.

1. Has the designer put forward a proposed erection sequence? If so is it based on minimizing the size of cranes needed. Figure 12.11 illustrates the point. The design allows for the middle bays of the structure to be left out initially for a crane to operate – thus greatly reducing the radius needed. As the structure proceeds, the crane fills in the middle bays behind itself.

2. If the above is the case, or if not, made so, the ground floor slab will need to be left out to avoid damage with cranes and delivery vehicles travelling over it. Only the sub-base material will be laid to provide hard access for crane and vehicles.

Criteria for crane selection

From the information derived previously, the required cranage specification can be decided, together with the temporary works needed for safety during erection. The procedure to be followed is tabulated in the box below.

Time-scale

The erection time-scale is a function of crane performance plus temporary works providing quick release from the hook.

Crane outputs will relate to the number of units to be lifted and an assessment of time per lift. For straightforward structures, companies experienced in the erection of pre-cast buildings will usually have their own performance data based on previous records. Makers of pre-cast units can either erect as subcontractors – giving a price and time or it can be done the hard way. Working out the crane's movement time scale from fix slings, lifting gear or spreader beams + slewing and lifting time + time holding while temporary supports fixed + slew back to pick up the next element. Due allowance needs to be made as well for unforeseen delays due to weather or general site inefficiency. Much will depend on the time of the year and the degree of control exerted by the site management.

Once a time scale is established for one crane,

Information needs for crane selection and associated temporary works

1. Establish the location of the heaviest units. Assess – how do they relate to site access availability? With big and heavy p.c. units can they be got into the site at all? Will they determine crane siting?

2. With multi-storey structures, any cranage erecting the structure will almost certainly be needed for follow-up activities, including the building cladding. In this case, tower cranes are the most likely answer. In settling the capacity, a comparison will have to be made between the pre-cast structure load weights and those (if known at the tender stage) of the cladding elements and any other one off heavy lifts – for example, lift motors, air conditioning chiller units, etc.).

3. Building area and height.

4. Weights to be lifted and at what radius. The answers come from Table 12.1.

5. The factors used in an adequate cost comparison between different crane type possibilities (Chapter 7).

6. The final item is to establish the temporary works needs to go with the erection process – especially those needed temporarily to support units while connections are made. Release from the crane hook is necessary as quickly as possible if the crane is to be operated efficiently (Figure 12.12 and reference [3]).

it will be necessary to see if this is going to fit in with the overall master programme envisaged or specified. If it does not fit, the planning will have to consider the implications of introducing a second crane.

Ordering units and delivery requirements

With the time scale settled, the basis for ordering and delivery requirements is in place when the contract has been won.

Decisions needed before final orders are placed are:

● Finalized delivery schedule and phasing;
● Discussion with supplier about, and agreement on, additional fixings required for assisting lifting and connection to temporary erection supports;
● Stockpile to be kept by manufacturer to ensure a damaged element will not stop erection (alternatively, what stock can be kept on site);
● Requirements in relation to temporary protection of units – mainly during storage, transportation and while erecting. If storage on site is planned, assessment of the cost of providing bearers, racks etc to avoid damage to edges, corners etc.

Once the orders have been placed, and before any erection is started, there should be a final check to see that all necessary safety needs have been foreseen, especially where the erection gang have to work in exposed positions. Have arrangements been made for safety belts and fixings for them to be anchored to the structure, or alternatively, adequate temporary barriers provided?

Volumetric units

Volumetric units present a quite different picture to the individual component approach. As the name suggests, the basic concept is to create pre-cast boxes which comprise a room or rooms of a larger building. Such boxes can then be fitted out at ground level, internally, and lifted into place within the overall concept to form part of

Figure 12.12 Pre-cast erection in progress showing push/pull props and wedging under the units to support the unit initially to release crane quickly. Used later for plumbing and lining, in association with the templates visible on the floor producing the curved elevation setting out.

a load bearing structure. In the case of hotel development, for instance, the method can be taken further, to the extent of making the volumetric unit equal to one bedroom and bathroom unit. It can then be fully fitted out, including the furniture, and lifted bodily into its final location.

The production of the 'box' can be realized in two ways: either casting the floor on a specially

---

Advantages of volumetric units

1. Large completely fitted out components possible;
2. All fitting out done in the manufacturer's yard;
3. Assembly in the best possible conditions

Disadvantages of volumetric units

1. System requires a large casting and curing area;
2. Need for heavy lifting equipment – both in yard and on site;
3. Heavy haulage necessary between yard and erection site;
4. Provision of good quality roads between the two, and on site is essential;
5. Jointing of large elements on site can be a problem;
6. As box units require stability, they are usually provided with a concrete floor and roof. This leads to redundant elements after erection completed. (Floor above on roof below, for example.)
7. If, for any reason, fitting out materials arrive late, erection on site is either delayed or has to continue without fitting out to clear the casting areas. Either way

Figure 12.13 Assembly of panels, cast flat, to make up volumetric unit. Note the need for a Straddle Carrier.

Figure 12.14 Completed shell of volumetric unit ready for fitting out.

prepared bed, followed by special formwork erected to allow the walls and roof to be poured in one, or casting the floor, walls and roof as separate slabs. Then lifting the walls to a vertical position, connecting them to the floor panel and finally connecting the roof panel on top. The connections are usually made by welding projecting reinforcement together (Figure 12.13), then pouring in-situ concrete to complete the joints. Once the box has matured, it is moved to a fitting out area by a stradle carrier crane (Figure 12.14). Fitting out covers partitions, plumbing and the fitting of baths, basins and internal decoration. Figure 12.15 shows the partitions under way. Once complete, the box has to be transported to its final resting place. The units are heavy and a lowloader is needed, together with a substantial crane at the erection site (Figure 12.16).

Figure 12.15 Skim plaster in hand, together with start of partitions and electrical carcass.

Figure 12.16 Transport and erection of volumetric unit in structure. In fact, the element has not been fitted out due to delay in supply of necessary materials.

Advantages and disadvantages

The disadvantages outnumber the advantages by a big margin. Studies have shown that the plant requirements are very expensive – straddle carriers at the plant and very heavy cranes in the field – due to the weights involved and the radii at which they need to operate. Originally thought of as a solution in underdeveloped countries, the road cost from plant to site alone can be very high.

The illustrations used to illustrate the volumetric approach were taken at a new development site in Venezuela. There was a shortage of finishing materials and the unit being positioned in Figure 12.16 can be seen to have no internal elements at all. Volumetric units have been tried elsewhere in the world – in the United States of America for rapid hotel construction and in Australia for bungalow construction in undeveloped areas. Both these examples have been successful, in their own way. The fact that, today, one hears little about continuing use anywhere in the world seems to prove the point that volumetric units have too many disadvantages, especially in relation to cost. It is not intended, therefore, to examine this method in any more detail.

The Tilt-up technique

In between the use of individual pre-cast components and volumetric units, there is a method involving pre-cast components known as the tilt-up technique. The method involves site casting of the walls of a building on its floor slab or on a separate casting bed adjacent to the building and then lifting the suitably cured panels by crane into their structural location. The panels form both the structural element and the cladding. Additional insulation can be catered for by using sandwich construction with insulation in the middle.

Tilt-up is widely used in New Zealand, Australia and the United States. The basic principle is illustrated in Figure 12.17. It will be clear that

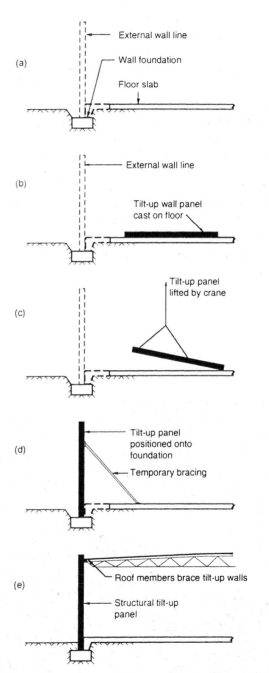

Figure 12.17 Principles of Tilt-up construction technique.

the floor slab must be cast as flat as possible for the wall panels to be accurately cast on top. If space on the floor cannot cater for all panels required, stack casting one panel on top of the other can be used or a casting bed prepared alongside the building and the panels cast there.

In either case, only edge forms are needed, with the external texture being achieved by form liners in sheet form laid on the floor (Figure 12.18). These also act as bond breakers between the cast and the floor. The same principle where stack casting is used. If a plain

Figure 12.18 Tilt-up unit being raised from casting bed on building floor. Note edge formers on left hand side of picture.

Figure 12.19 Completed Tilt-up building. While not apparent in the photograph, the entire surface of the concrete is painted. Paints used range from acrylic, high build elastomeric coatings to chemically bonded finishes. Brick finishes can be achieved with the panels cast on top of brick slips.

surface is desired, a bond breaker liquid is used instead.

Buildings up to three storeys are quite straightforward, but buildings in excess of this have been built. The method can also be used to provide cladding to a conventional in-situ or pre-cast frame.

The whole subject of Tilt-up construction is dealt with in considerable detail in a publication *Tilt-up Concrete Buildings: Design and Construction Guide*[6]. It is stated that the publication brings together worldwide experience in a form suitable for the UK.

It is interesting that the most common finish to Tilt-up panels is painting the concrete. Guidance in this respect is given. Figure 12.19 shows a finished structure with paint finish.

In a document said to be in a form suitable for use in the UK, it is perhaps regrettable that no references are given in relation to relevant safety legislation in the UK – only one brief mention in the text of the CDM regulations with little explanation of their contents. This is a pity, since it is a beautifully produced document.

Pre-cast concrete cladding

The cladding to structures is largely dealt with in Chapter 14, but, in the case where pre-cast panels are used, the subject is more readily dealt with in this Chapter. The requirements of pre-cast structures and pre-cast cladding, in the planning method sense, have much in common.

Contract conditions and decisions

Pre-cast cladding of a structure will normally be a critical path item. In consequence, the time taken to erect will be a key item in the assessment of contract time – if the contractor has to state – or a controlling factor in relation to time available for other works, if the client states a time for completion.

Important questions to be asked and determined are:

(a) Who will supply the units?
(b) Who will erect?
(c) How do the answers to (a) and (b) above fit in with the other elements of the construction process?

Pre-cast concrete supplier

If the contract conditions specify a nominated supplier, the main contractor will need to consider the reputation of such supplier in relation to:

(a) reliability in meeting required dates;
(b) accuracy and quality of the product;
(c) co-operativeness in relation to inspection, during manufacture, by the main contractor before accepting the nomination.

If the contract conditions place the pre-cast elements in the main contractor's bill of quantities, a decision has to be made as to who will make the units – the main contractor himself or a subcontractor selected by competitive quotations from firms of which the main contractor has previous knowledge. Today, it almost goes without saying that a specialist subcontractor would be adopted, from the point of view of expertise, convenience and cost – together with the plant capable of producing many sophisticated finishes in ideal conditions.

Erection of pre-cast units

It is tempting to sub-let the erection of units and thereby absolve the main contractor from any problems which may arise. What may seem, at first sight, to be a straightforward decision is, in fact, by no means so.

The interaction between basic structure and the cladding plays a major part in planning for contract efficiency. In particular:

● harmonization of crane workloads;

- influence of structural construction method on pre-cast unit fixing and vice versa;
- tolerances for the structure in relation to tolerances in pre-cast elements;
- conflicting requirements in relation to access and storage.

Experience shows that planning control and overall contract efficiency will best be achieved if the contractor erecting the structure also erects the pre-cast cladding. No arguments can then arise if problems occur due to tolerances on one side or the other not being met. This approach is true whether the structure is erected by the main contractor or a subcontractor. It also ensures that the organization of the crane(s) workloads and sequence are dealt with by one organization.

Planning and method assessment

To achieve a realistic planning and method assessment for the pre-cast cladding of a building the information supplied in the contract documents needs to be adequate. Key information items are:

- Details of the structural form;
- Adequate information illustrating the PC units to be handled; their location in the structure and the fixing method;
- Realistic assessment of the weights to be lifted;
- The degree of imbalance between various PC unit weights;
- Requirements in relation to tolerances;
- Any requirements in relation to the temporary protection of units once erected;
- Relationship of building to site and available access.

The listed items should be regarded as the minimum if the contractor is to price the work realistically. In too many instances, where these considerations are not met, delay and extra cost occur, resulting in unnecessary claims.

Pre-contract planning

When the contract has been won, the planning needs to be taken further. In placing an order for the PC cladding with a specialist supplier (as is usually the case these days), the following matters will need to be discussed:

- Agreement of delivery schedule and phasing required;
- Additional fixing or lifting inserts required for offloading, handling and lifting on site and fixing temporary support and adjustment items before the final fixing takes place;
- Information on method of casting and attitude of unit on delivery vehicle.

In addition, with the final cladding drawings available, the required cranage capacity needs to be re-checked, together with a re-check of the weights of the units themselves.

It is at this point that the needs for temporary works – in the shape of special lifting gear that may be necessary, together with all items of temporary support to achieve rapid release of a cladding unit from the crane hook – are decided. The final area of temporary works relates to the provision of storage racks on site to allow a buffer stock of units to be held, to avoid any delays in unloading or to provide replacement units if inadvertent damage occurs to one or more panels.

The need to release the crane from the panels quickly is essential for crane efficiency. The temporary fixings, apart from safety needs, need to be capable of adjusting panels in line and level to achieve 'eye sweet' appearance. This adjustment is often necessary due to panels warping during the curing period.

The whole subject of pre-cast concrete cladding was the subject of a conference organized by the Concrete Society and the Institution of Structural Engineers and others[5] at which the author presented a comprehensive paper on 'The Site' and its planning of the pre-cast cladding operation, together with considerable detail in relation to temporary works needed. On the temporary works side, those illustrated in the paper were published later in a book on temporary works[8].

Records of performance

The recording of performance achieved is an important part of a construction planners responsibility. Such records should be as detailed as possible and in a structured form. That is, with as much detail about the conditions prevailing as possible; plant used, labour force and illustrations of the panels, their fixing method, weights and installation performance achieved. The building structure type should also be recorded and the number of floors. A plant layout is desirable, too.

The two examples of performance records at the end of this chapter indicate the approach and the sort of detail that can be achieved.

Final alignment

The final visual result of a structure clad in pre-cast concrete will depend on the 'eye sweetness' of the whole. While cladding panels can be lined up from base lines, or in the case of curved surfaces by the provision of floor jigs, the result on the face of the building may still not be satisfactory due to vagaries in the surface of the panel. It is good practice to have a whole elevation to one floor level erected, to whatever setting out aid is used, but before final fixing eye lined externally, adjusting by eye if necessary, and only then permanently fixed. By so doing, a far better appearance will result as warping, dimensional errors in both structure and in cladding panels can be made to look less significant.

Protection

Where cladding is at ground level or in buildings of some height, where structural erection is still proceeding above, expensive finishes to pre-cast panels are vulnerable to dirt and mechanical

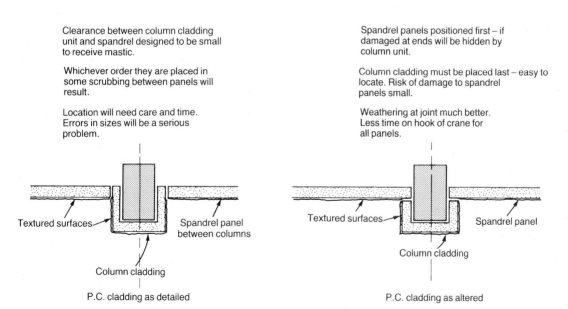

Clearance between column cladding unit and spandrel designed to be small to receive mastic.

Whichever order they are placed in some scrubbing between panels will result.

Location will need care and time. Errors in sizes will be a serious problem.

Spandrel panels positioned first – if damaged at ends will be hidden by column unit.

Column cladding must be placed last – easy to locate. Risk of damage to spandrel panels small.

Weathering at joint much better. Less time on hook of crane for all panels.

Textured surfaces — Spandrel panel between columns — Column cladding

P.C. cladding as detailed

Textured surfaces — Spandrel panel — Column cladding

P.C. cladding as altered

Figure 12.20 Sequence of erection changed to facilitate erection and avoid damage. A further bonus is improved resistance to water penetration.

damage. Cleaning on completion can be an expensive operation, while repairs to mechanical damage can never be carried out effectively in the long term. At worst, repaired areas may eventually fall off with danger to the public below. Recent years have amply demonstrated this view.

Adequate protection appropriate for the given situation will always be money well spent; both in relation to the quality of the finished product and avoidance of costs to the contractor.

Sequence of pre-cast cladding erection

At the earliest possible stage, when reasonably final details of the cladding arrangements are available, the planner needs to examine the proposed order of erection. Is the order the best both for ease of erection and avoiding damage panel to panel? Figure 12.20 illustrates this point from an actual contract. The Engineer called in the contractor for his advice, with the result that the sequence was changed as shown on the drawing. Not only was damage minimized as much as possible, a second benefit in the shape of a more weathertight joint also resulted.

References

1. Trent Concrete Structures Ltd.
2. Health and Safety Executive Guidance Note GS/28. *Safe erection of structures Parts 1–4* (1986) HMSO, London.
3. *Management of Health and Safety at Work Regulations* (1992). HMSO, London.
4. *Construction (Design and Management) Regulations* (1994). HMSO, London.
5. *Construction (Health, Safety and Welfare) Regulations* (1996) Stationery Office. London.
6. Reinforced Concrete Council. *Tilt-up concrete buildings: design and construction guide* (1998) RCC, Crowthorne.
7. Illingworth J. R. (1976) 'The site', paper in proceedings of one day symposium. *Pre-cast concrete cladding*, organized jointly by the Concrete Society, British Pre-Cast Concrete Federation, the Institution of Structural Engineers, and The Royal Institute of British Architects. Concrete Society. London.
8. Illingworth, J. R. (1987) *Temporary works – their role in construction* Thomas Telford, London.

Pre-cast Concrete Panels	**PERFORMANCE RECORD FORM**	Site:
		Date:

Description of building

3-storey reinforced concrete structure approximately 56 m × 28 m in plan. External concrete walls with a very limited window area (approx. 15% of elevation). Building incorporates a system in which heat recovery, lighting, and air conditioning are all integrated.

The pre-cast concrete cladding (which represents 60% of elevation) is in vertical bands (see Sketch No 1). String courses and roof edgings are also in pre-cast concrete. Mineralite is applied to the remaining exposed concrete wall areas.

Description of pre-cast panels and fixing methods

Panels manufactured by Portcrete Ltd using a Capstone with fine rubbed finish. Largest units weigh approx 2.3 tonnes. Panels are hung onto RC nibs cast onto the external walls. Panels anchored to wall by 4 No. 15 mm diameter Phosphor Bronze bolts. See Sketch No 2.

The roof edging and string course are also of Capstone, fine rubbed finish. Roof edging units fixed after wall panels to ensure no damage from crane ropes etc. See Sketch No 3.

Plant and equipment used

22RB I C D Power controlled load lowering — 21 m jib + 6.4 m fly jib.
Nylon webbing slings — 2 No. pairs — max load lift 2.3 tonnes.
Cabco Lifting bolts to suit Cabco Anchors cast into panels — 12 No. pairs.

Labour used

2 No. pre-cast unit fixers — 1 No. fixer + 1 No. labourer were subcontracted from Portcrete Ltd. This proved a good arrangement as the Fixer, being a Portcrete employee, had a greater degree of control and authority when organizing the delivery of panels.

Total labour in fixing gang

1 No. fixer
1 No. labourer } from Portcrete
2 No. labourers   main contractor
1 No. crane driver
1 No. banksman
—
6 men total fixing gang.
—

Sequence of actions and method of erection

Pre-cast panels arrive on site loaded flat and face up, 2 or 3 high. Deliveries were arrranged for 0800 hrs so as not to disrupt the daily erection programme.

247

Sequence of events was:

1. Off load pre-cast panel units with 22RB, using nylon webbing slings.

2. Tap out Cabco lifting sockets cast into panel.

3. Fix Cabco lifting bolts into panel.

4. Hoist panel to position with 22RB.

5. Pass panel down between independent scaffold and building. (A gap of 600 mm from the face of the bearing nib to the inner standard was allowed.)

6. When the panel is approximately in position, 4 No. 15 mm dia. Phosphor Bronze studs were screwed into Cabco sockets located in the back of the panel. 4 No. Neoprene packing washers were then placed over the studs. The thickness of the washer assembly was pre-determined by the engineer to achieve the correct panel position in line and level.

7. Packing pieces were then placed on the nib to levels determined by the engineer.

8. The panel is then manoeuvered so that the P/B studs align with the 38 mm dia. holes in the wall.

9. The unit is then pushed home. Two labourers inside then apply mastic around the bolt, place in position plate washer, nut and tighten.

10. The final action is to pour grout down the back of the unit to fill void between panel and the top of the supporting nib.

Performances achieved

Over a period of 5 weeks, 161 panels and 163 edge units were fixed.

Average performances achieved were:

      40 No. wall panels fixed per 5-day week

or   8 No. per day.

     163 No. roof edge units fixed in 6 days

or 27 No. per day.

Continuation sheet 3 of 5

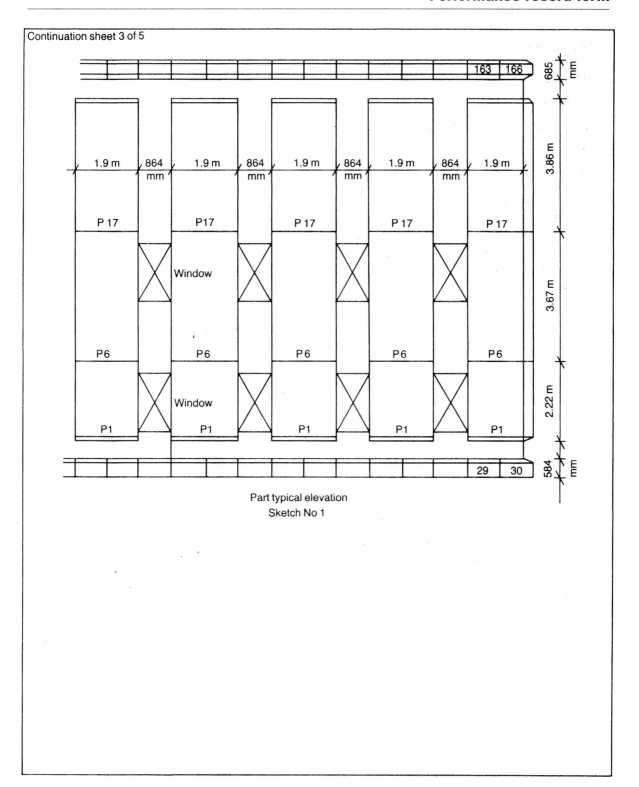

Part typical elevation
Sketch No 1

Pre-cast concrete structures

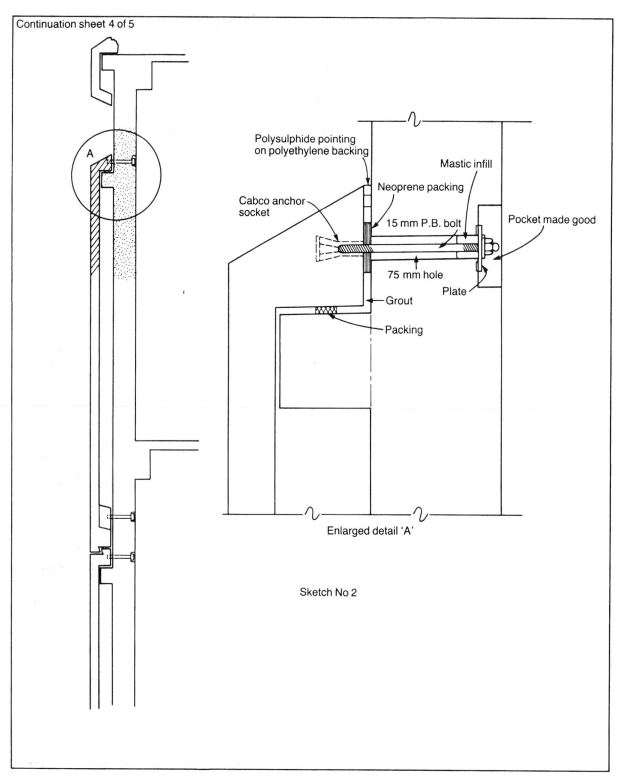

A

Polysulphide pointing
on polyethylene backing

Mastic infill

Neoprene packing

Cabco anchor
socket

15 mm P.B. bolt

Pocket made good

75 mm hole

Plate

Grout

Packing

Enlarged detail 'A'

Sketch No 2

250

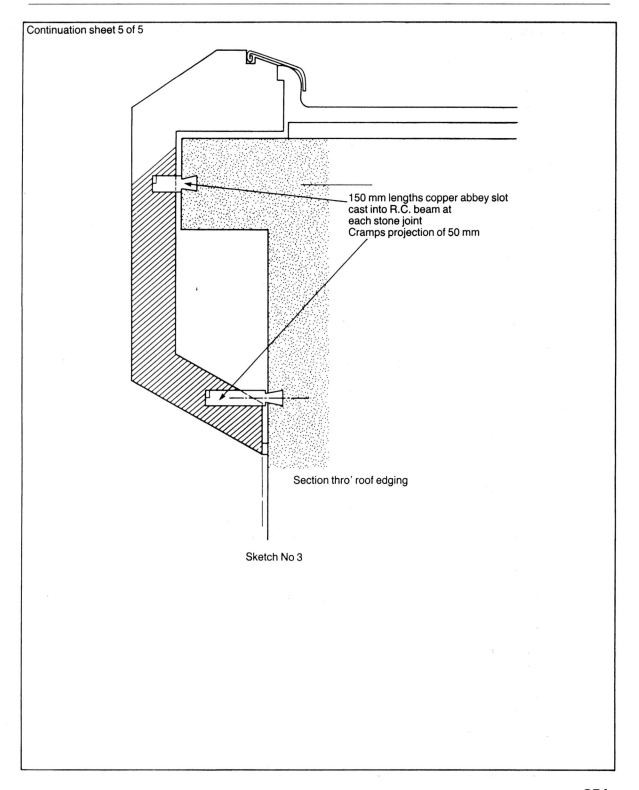

150 mm lengths copper abbey slot
cast into R.C. beam at
each stone joint
Cramps projection of 50 mm

Section thro' roof edging

Sketch No 3

Pre-cast Concrete Panels	**PERFORMANCE RECORD FORM**	Site:
		Date:

Description of building

10-storey steel framed office block. Concrete floors on profiled steel decking. Clad with pre-cast concrete panels, arranged to be both spandrel panels under windows and the head of the window at the same time.

Description of pre-cast panels

The cladding only involves the erection of three types of panel:

Type S1/S2 – 35% of total panels.
  – 3.60 m long × 1.90 m high × 127 mm thick.
  – Weight approx 2.2 tonnes.

Type S 13 – Used on north & south elevations – full storey height. They are same size as panels S1/S2 but are fixed vertically instead of horizontally.
  – Weight as above.
  – Have to be turned from horizontal to vertical before lifted into place.

Type S/ – Similar to S1/S2 but with a short angle return to pick up with end elevation pre-cast panels.
  – Slightly heavier, approx 2.4 tonnes.

Method of fixing to steelwork is shown in Sketch No. 1.

Plant and equipment used

Crane – Liebherr 75c (free standing with strengthened bottom sections).
Transport – Articulated lorry
  – carries 8 No. pre-cast panels standing on edge, chained to A-frame.
Lifting eyes – Screw in to cast in sockets for this purpose 4 No. min.

Labour used

1 No. Crane driver
1 No. banksman
1 No. Trades foreman (carpenter)
1 No. Labourer
1 No. lorry driver (assistance in unload to store only)

Sequence of actions and method of erection

Unload pre-cast panel

(a) Release chain mechanism securing panel to transporter.

(b) Screw in 2 No. lifting eyes to top edge of panel (cast in sockets provided for this purpose).

(c) Attach crane lifting tackle, hoist clear of transporter and place panel onto site storage A frame (Sketch No. 2).

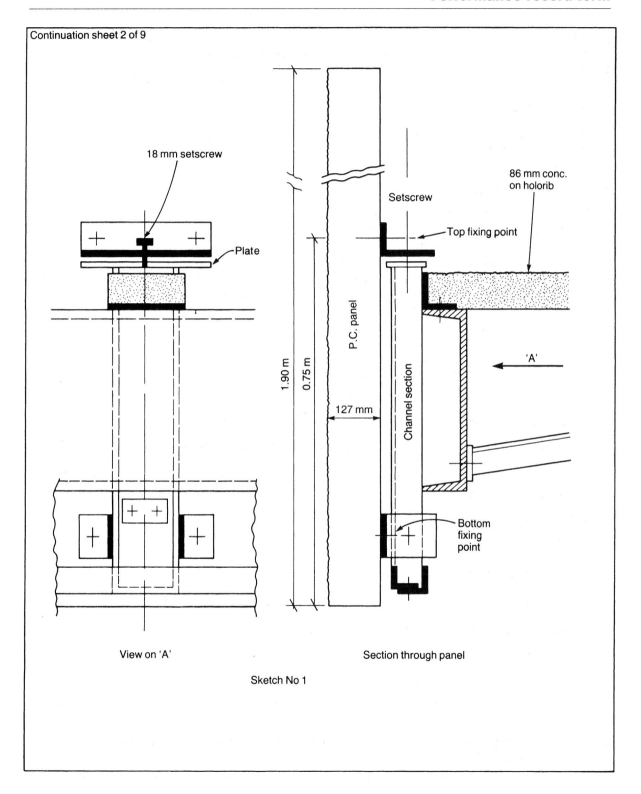

Continuation sheet 2 of 9

18 mm setscrew

Plate

86 mm conc. on holorib

Setscrew

Top fixing point

P.C. panel

Channel section

1.90 m

0.75 m

127 mm

'A'

Bottom fixing point

View on 'A'

Section through panel

Sketch No 1

253

Pre-cast concrete structures

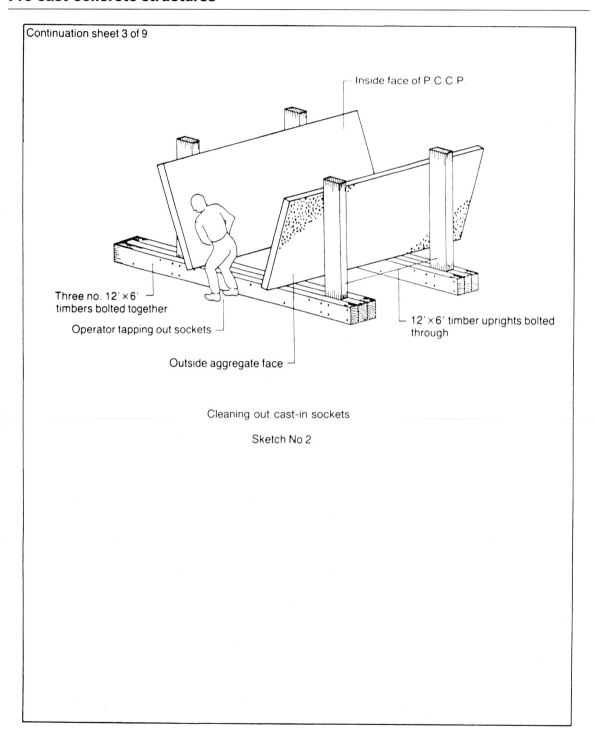

Inside face of P.C.C.P.

Three no. 12′ × 6′
timbers bolted together

Operator tapping out sockets

Outside aggregate face

12′ × 6′ timber uprights bolted
through

Cleaning out cast-in sockets

Sketch No 2

254

Observed time

Average observed time to unload and stack panel = 3.3 mins for panel S1/S2.

Carry out work to panel before initial erection

(a) Tap out the cast-in sockets to remove traces of grout and swarf. (There are some 20 No. 18 mm dia. sockets in a typical S1/S2 panel (Sketch No. 2).)

This work is best carried out whilst the panel is standing in the A-frame.

(b) Bolt on 2 No. fixing angles to panel (Sketch No. 1).

Observed time

This work in general is carried out by the crane banksman as a spare time job. Thus not a critical time.

Initial erection of pre-cast panels S1/S2

(a) Hoist panel from site A-frame and place in position, including inserting 'Servicore' jointing material, temporarily bolt at floor level connection (Sketch No. 1) and tie panel at top. For temporary tie in device see Sketch Nos 3 & 3(a).

(b) Initially eight or so panels should be erected at one time. This provides continuity and rhythm for the crane and erection gang.

Observed time

Average observed time to hoist and initially erect S1/S2 panels = 8.75 mins per panel.

Initial erection of pre-cast panels S/13

These panels are positioned on the North and South elevations and are full storey height. They are similar in size to the S1/S2 units, but instead of being fixed in the horizontal they are fixed in the vertical position.

They are transported, handled and stored in the same manner as the S1/S2. However, the initial erection demands another operation in order to bring the horizontal panel into the vertical position. It must be appreciated that to attempt to hoist a panel that is lying in the horizontal position to the vertical by using the end (which will eventually be the top) suspension points would cause the panel to swing on its axis, with disasterous consequences.

To overcome this situation, a swivel frame is used. In principal, it consists of a ground frame to which a triangular box is pivoted. How this works is shown in Sketch No. 4.

Observed times

(a) Average observed time to transfer S/13 panel from horizontal to vertical position = 3.9 mins per panel.

(b) Average observed time to hoist and initially erect S/13 panel to south elevation – 2nd floor = 10.5 mins per panel

Pre-cast concrete structures

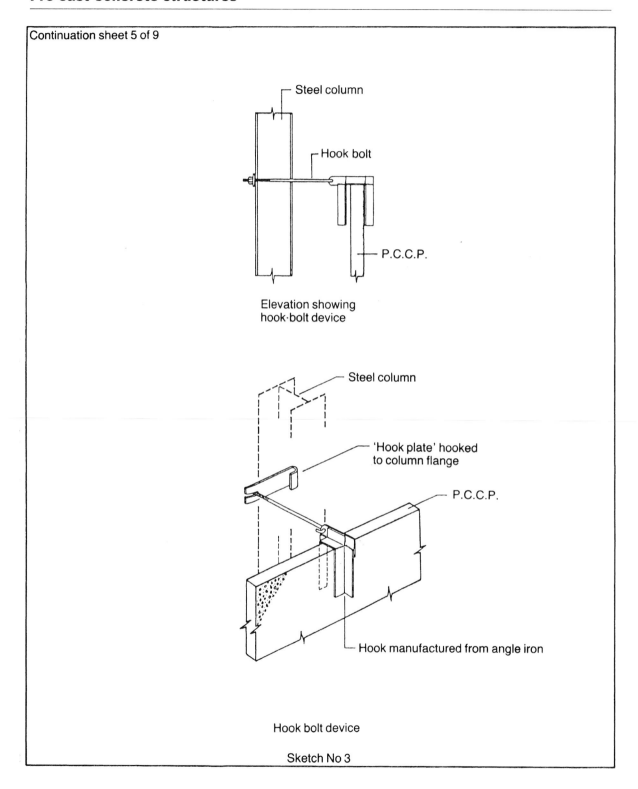

Steel column

Hook bolt

P.C.C.P.

Elevation showing
hook·bolt device

Steel column

'Hook plate' hooked
to column flange

P.C.C.P.

Hook manufactured from angle iron

Hook bolt device

Sketch No 3

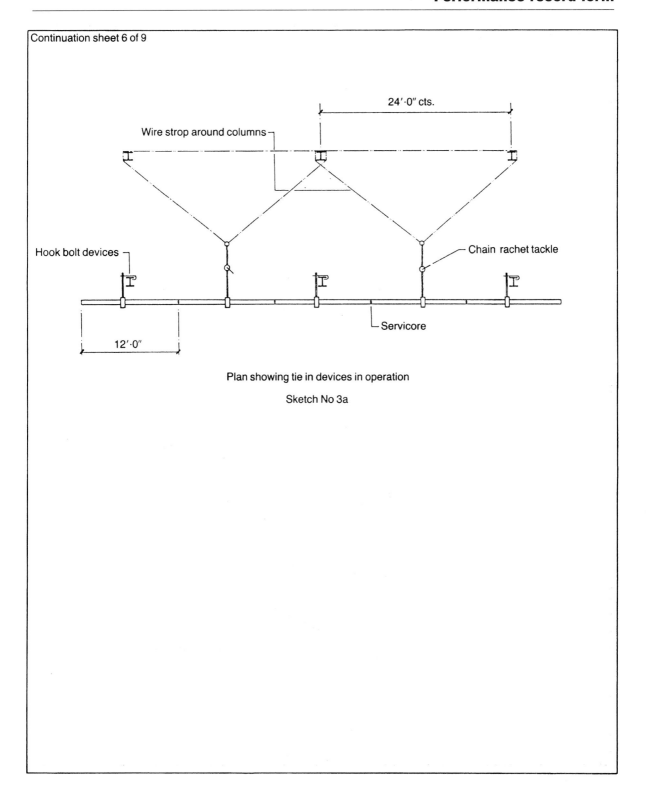

Plan showing tie in devices in operation

Sketch No 3a

Pre-cast concrete structures

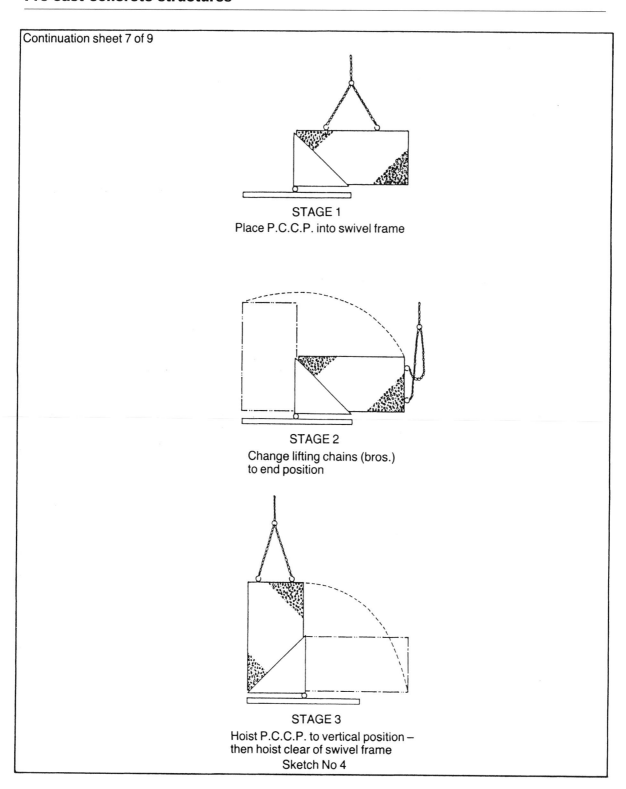

STAGE 1

Place P.C.C.P. into swivel frame

STAGE 2

Change lifting chains (bros.)
to end position

STAGE 3

Hoist P.C.C.P. to vertical position –
then hoist clear of swivel frame

Sketch No 4

Thus combined observed time to initially erect a S/13 panel is:

3.9

10.5
_____
14.4 mins
_____

Initial erection of pre-cast corner panel S/

Panel similar to S1/S2 but has short angle return to pick up with end elevation pre-cast panels.

Observed times

(a) Unload to store. As panels S1/S2.

(b) To hoist and initially erect S/ corner panel = 15.0 mins per panel.

Final erection

By final erection is meant the final lining and levelling process. Average times proved difficult to assess. Some panels had warped and the final erection called for eye sweet adjustment, which can take time. Other panels were dimensionally inaccurate, while the steel frame was found to be out of plumb and alignment, in some cases. Sketch No. 5.

The lesson here is to apply a quality control system from the start — factory inspections and dimensional checks before delivery. Steel structure to be adequately checked before erection starts and errors established.

Pre-cast concrete structures

FINAL ERECTION OF P.C.C.P. – SEQUENCE

1. Measure and mark centre line on top edge of P.C.C.P.
2. Measure and mark half column section to either side of c.l.
3. Place straight edge (spirit level) across column flange.
4. Adjust P.C.C.P. to align marks (achieved by tightening down adjacent P.C.C.P. then 'pulling' the P.C.C.P. in question to position by bottlescrew device).

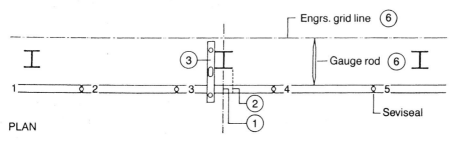

PLAN

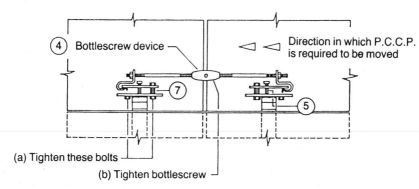

(a) Tighten these bolts

(b) Tighten bottlescrew

5. Level P.C.C.P. to datum mark (previously marked on columns by Engr.) achieved by adjusting set screw.
6. Adjust P.C.C.P. to line (chalk grid line previously put on slab by Engr.) achieved by 'easing out' or 'pulling in' P.C.C.P. to gauge rod.
7. 'Washer up' gap and tighten securing bolts.
8. Plumb P.C.C.P. on inside face (plumb rule) and tighten bottom securing bolts.
9. Once the P.C.C.P.s adjacent the columns have been finally erected the intermediate P.C.C.P.s (numbered 2–4) can be aligned to them.

Chapter thirteen

Steel structures

The use of structural steel, in the commonly recognized sections, dates back to the latter half of the nineteenth century. Allen, in a Technical Information Sheet for the Chartered Institute of Building[1], records that by 1887 Dorman, Long Ltd offered a range of 99 beam sizes as well as a vast range of channel and angle shapes.

In the years between the two world wars, steel structures dominated the major building arena. In a book published in the early thirties, *The Wonder Book of Engineering Wonders*[2], there is a chapter entitled 'The Engineer as Builder'. In the opening paragraph, the following statement is made, 'But when you see and study the erection of an office "block", a huge "store," or the headquarters of some great company, you discover that the stone or concrete front is only a shell, and that the real strength of the structure is in the network of mighty girders which are entirely hidden from sight when the building is finished'. The erection of County Hall and the buildings above Baker Street Station in London are pictured as examples. Elsewhere, most of the skyscrapers in New York were built in structural steel, the most notable, at that time, being the Woolworth Building in 1911–14, the Chrysler Building in 1930 and the Empire State Building, also built in 1930.

Structural steel was also widely used in industrial plants for its ability to achieve large open spans in manufacturing areas. It was also the norm for the boiler house structures in power stations where very heavy loads required support, together with the framing of turbine halls. Figure 13.1 illustrates the boiler house steel being erected at Fulham power station in London, the first phase of which was completed in 1936.

This particular picture is interesting for a number of reasons as it illustrates how things in construction can change over a period of some 65 years. All the steelwork is riveted and is being erected by distinctly unusual cranes. All the operatives wear cloth caps, men are seen standing on girders without any form of safety harness or other form of protection, while only one vertical ladder can be seen as a means of vertical access. In other words, operatives walked along the girders to the place of work.

Today, the Health & Safety at Work etc. Act of 1974[3] and the Construction (Health, Safety and Welfare) Regulations 1996 have to be complied with[4]. More detail on these Regulations is given in Chapter 1. In addition it is now compulsory for operatives to wear safety helmets in designated areas of work, certainly in a situation as in Figure 13.1. In the specific area of the erection of structures further guidance is provided by BS 5531:1988[5], Health and Safety Executive Guidance Note GS 28[6] and the Management of Health and Safety at Work Regulations 1992[4].

The unusual appearance of the two main erection cranes is historically interesting. Both are conversions from the creeper cranes used in the construction of the Sydney Harbour arch

261

Figure 13.1   Erection of boiler house steelwork to Fulham power station, circa 1935.

bridge, completed in 1932. Notice also that two beams are being lifted at the same time to speed erection by reducing the number of lifts.

Since the second world war, concrete construction, in the multi-storey role, developed rapidly in direct competition with structural steel. Today, competition between the use of steel or in-situ concrete is fierce. Much fast track development now uses structural steel and the planner and methods engineer, together with site management, need to be aware of the advantages and disadvantages of this form of construction.

The *Architects Journal*, in 1991, published a five part series of articles[8] on developments in steel and concrete construction. In Part 1 – Steel Frame Design, it was stated that, 'To-day, over 50% of all commercial buildings are of com-posite construction, including over 80% of building of five storeys or higher.'

Structural steel frames

The availability of steel sections in a variety of dimensional ranges provides the designer with a wide variety of options. The addition of hollow steel sections to the range widens even more the scope of steel in designing such items as space frame, lightweight roofs and other areas where the hollow section is a strong but light means of construction. Figure 13.2 illustrates the prod-ucts range of steel sections and where each sec-tion finds its main uses in steel structures.

While the steel structure takes the loads involved down to the foundations, the building's vertical cores which provide for stairs, toilets,

Advantages of steel structures

1. Can be totally pre-fabricated off site;
2. Complex structures can be trial erected at the point of manufacture;
3. Not subject to significant damage in transit and handling;
4. Capable of pre-erection into sub-assemblies on the ground to save erection time in the air;
5. Erection generally straightforward. With the use of profiled steel decking, welded to the beams, a safe place of work is created for the fixing of reinforcement and the concreting of the floors. All that is needed is perimeter protection;
6. Structural steel is far and away the best solution if the facility for future alteration of use is a built-in requirement;
7. Erection time for the structure arguably faster than the alternatives (the concrete lobby would dispute this);
8. Long term retention of strength properties;

Disadvantages of steel structures

1. Inaccuracies in manufacture not uncommon;
2. If fire proofing is called for, casing in concrete is time consuming and expensive. Against this, sprayed systems and box encasement, to name only two solutions, are much better and cheaper than the problems with casing in concrete, which involve significant expense in temporary works, in providing access and safe places of work, often at a considerable height;
3. Great care in detailing and erection are necessary when associated with pre-cast concrete cladding;
4. Safe working conditions during the steel erection may be difficult to achieve until a profiled

lift shafts and service rising ducts have a key structural role to play as well. It is such cores that provide a vertical structural element that gives lateral stability to the frame as a whole. Such cores can be constructed in steel or concrete. Where concrete is used, it is usually on the basis of slipform construction carried out ahead of the steel erection. One example of this approach is illustrated in Figure 13.3.

Floors usually are in concrete and can be in-situ or pre-cast. Today, the use of in-situ concrete, in what is known as composite construction, is becoming increasingly the most common method in high rise structures.

Composite construction

The original concept of composite construction was the use of profiled steel decking spanning between beams and tacked down onto them to achieve greater stiffness when pouring the concrete and, perhaps more importantly, to stop sheets being blown of the building in high winds. The profile types are shown in Figure

Section	Representation	Uses	Section	Representation	Uses
Joist		Stanchions beams, supports bracings, portal frames.	Structural Hollow Section (S.H.S.) Square		Roof trusses, stanchions, walkways, balustrading bracings, stiffeners, lattice work, light framed buildings.
Universal Beam (U.B.)		Supports, beams bracings, purlins, portal frames.	Structural Hollow Section (S.H.S.) Rectangular		
Universal Column (U.C.)	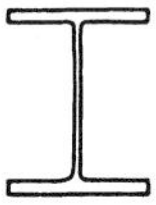	Columns supports, stanchions, heavy fabrications.	Structural Hollow Section (S.H.S.) Circular		
Channel		Bracings, purlins, walkways, roof trusses.	Plates		Plate Girders Box columns, and Box girders.
Angles Equal and Unequal	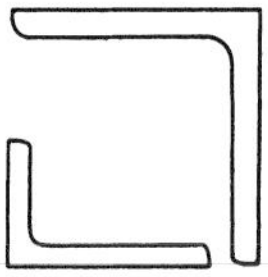	Roof trusses, purlins, bracings, stiffeners, light framed buildings lattice work.			
Tee's Short Stalk, and cut from U.C. or U.B.		Roof trusses, Roof trusses bracings, stiffeners, lattice work, hip rafters.	Universal Plates		

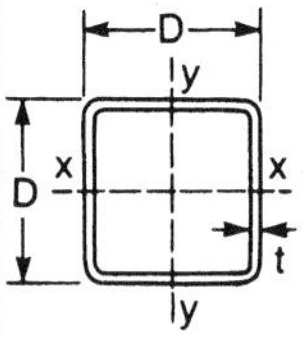

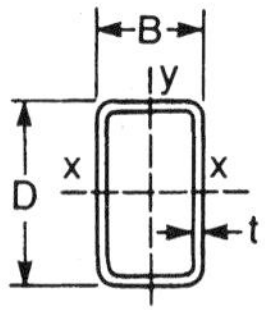

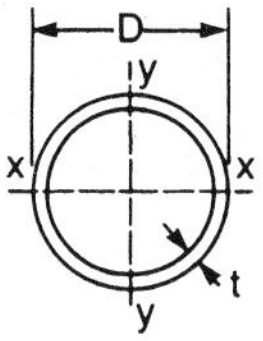

Figure 13.2 Illustration of product range of steel sections, together with main uses for each section. Note that each section has numerous dimensional options. (British Steel – Sections, Plates and Commercial Steels).

13.4. To ensure an adequate shear resistance between the concrete and steel, it was necessary to weld on shear connectors to the tops of beams. The general arrangement needed is illustrated in Figure 13.5, including edge details.

In further research, British Steel have now developed a new concept which not only does away with the shear connectors, but gives a much slimmer floor design. The method is known as Slimdek and is a registered trademark of British Steel (Sections, Plates and Commer-

cial Steels)[9]. This new approach became effective from May 1997. The Slimdek floor has the following components:

1. The asymmetric Slimflor beam. A rolled section with a nominal cross-section of a 190 mm wide top flange and a 300 mm wide bottom flange. It has a top surface embossment to develop composite action with the concrete encasement (Figure 13.6).
2. Slimflor beam. A registered trade mark of

Figure 13.3 Steel superstructure associated with slipformed concrete core.

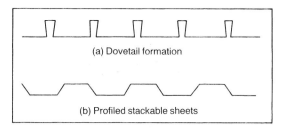

Figure 13.4 Common types of profiled steel decking. (British Steel – Sections, Plates and Commercial Steels).

British Steel (Sections, Plates and Commercial Steels) defining a universal column section with a plate welded to the underside of the bottom flange.

3. Rectangular hollow Slimflor beam. A hot finished rectangular hollow section with a plate welded to the underside of the section. Used primarily as an edge section.
4. ComFlor 210. A 210 mm composite profiled steel deck supplied by Precision Metal Forming[9]. It is planned to be phased out over the next two years and replaced by SD225 (below).
5. SD 225. A 225 mm deep composite profiled steel deck developed jointly by British Steel/Steel Construction Institute/Precision Metal Forming. This will be the standard deck profile for Slimdek.

Figures 13.7 and 13.8 illustrate the method and British Steel have produced a booklet detailing the system[9].

The advantages of the new design are:

1. A slimmer floor.
2. The profiled decking is capable of spanning 6 m between beams with the concrete loadings for which the floor is designed.
3. As detailed, the SD 225 deck is designed to allow extensive under floor servicing. The ASB webs allow round or flat ovals to be cut for continuous runs.

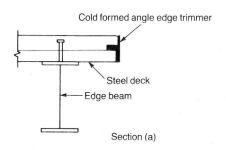
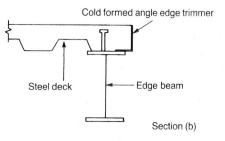

Figure 13.5 Edge trimmers to profiled steel decks in cold formed angles. (British Steel – Sections, Plates and Commercial Steels).

265

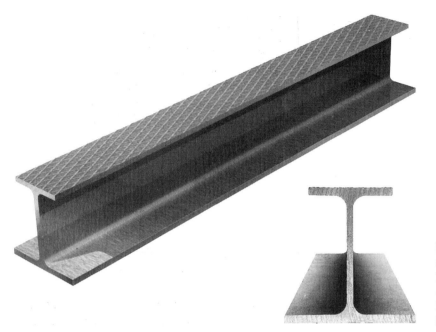

Figure 13.6 The asymmetric Slimflor beam. Note patterned top flange. (British Steel – Sections, Plates and Commercial Steels.)

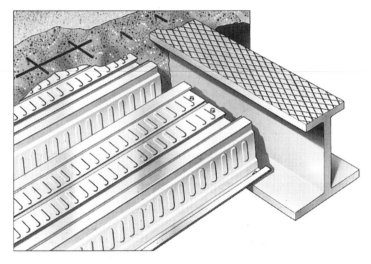

Figure 13.7 Details of Slimflor components. (British Steel – Sections, Plates and Commercial Steels.)

4. Vertical services can be catered for by boxing out on top of the profiled deck.

Technical matters apart, further practical advantages arise:

1. The decking can be laid and fixed well ahead and, once fixed to the steelwork, provides a safety shield to those working below the erection gangs.

2. The rectangular hollow Slimflor Beam forms the correct height for the edge form.

3. As the formwork to the floor is permanent, large area pouring is possible [10]. This in turn leads to economy in time and cost, with pumping of the concrete as the preferred option (Figure 13.9).

4. Temporary works are greatly minimized.

5. The profiled deck provides a safe place to

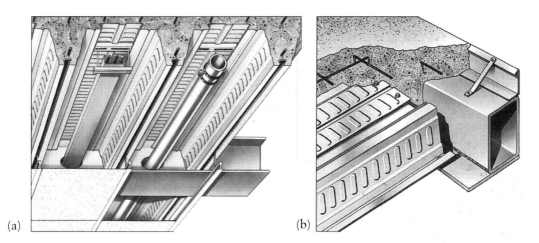

(a)

(b)

Figure 13.8 (a) Slimflor components including edge beam (rectangular hollow Slimflor beam). (b) Underside showing service installation ability. (British Steel – Sections, Plates and Commercial Steels.)

Figure 13.9 Large area pour on profiled steel floor decking by concrete pump. (British Steel – Sections, Plates and Commercial Steels).

work, with a means of access usually provided by the openings for stairs or from or from access scaffolding if used.

6. Safety provisions are completed by complying with the laid down needs for edge protection – hand rails and toe boards (Figure 13.10).

Sub-assemblies

The two examples given above of the value of composite construction also introduce the concept of sub-assemblies in the erection process. To erect the circular escape stair support members in mid-air would require temporary supports and access and safe working space for the welders which is time consuming and costly (Figure 13.11). The same would apply to the corner assembly shown in Figure 13.12. By carrying out the assembly of the supporting steel in the fabrication shop, time is saved together with the costs of access and temporary works, and a more accurate result achieved.

The sub-assembly concept is not only relevant to awkward areas. With adequate cranage, it can be used to speed erection of the main structural members. The principle is illustrated in Figure 13.13. Main connections between

267

Figure 13.10   Temporary edge protection – hand rails and toe boards as required by the Construction (Health, Safety and Welfare) Regulations.

Figure 13.11   Circular cantilevered escape stair Steelwork. Erection made easy by sub-assembly in works.

Figure 13.12   Further example of sub-assembly value. Complex corner detail erected in one piece.

beams and columns are fabricated in the works, together with parts of the beams themselves. As shown, the only joint made in the air is that between two halves of the linking beam. This junction is more simple than the connection of the beam to the column. In addition, only one joint has to be made instead of two.

In appropriate circumstances, where the lifting capacity allows, whole H-frames of pre-fabricated steel can be lifted into place as one unit. Greater accuracy of the structure can obviously be achieved as much of the sub-assembly has been carried out on the shop floor, where the checking of dimensions is much easier.

In recent times, the sub-assembly concept has been extended to include complex roofs needed for the popular atrium areas in buildings. The construction Press regularly shows complete roof assemblies being lifted into place by high reach and high capacity mobile cranes brought in specially for the one lift involved.

This concept reaches its ultimate conclusion when the whole of the steelwork is erected on

Figure 13.13 Sub-assembly in progress in works. (British Steel – Sections, Plates and Commercial Steels).

the ground and then jacked up into place. Figure 13.14 shows just such a situation. Heavy lifting jacks, mounted on only four corner columns, are being used to lift the entire roof assembly including pre-assembled services for a hanger at London Airport. The scale of the operation becomes clear when the size of the cars on the ground is studied. In an airport, minimizing the height of erection equipment can be crucial in relation to aircraft safety. Indeed, such heights may be specified in the tender documents.

Figure 13.14 Ground level roof assembly of entire hangar roof followed by jack lifting to final location. Jacks mounted on four corner columns. (Douglas Specialist Contractors Limited).

Steel with pre-cast concrete floors

Pre-cast concrete floor units have been widely used, in the past, in association with steel structures. While not so common to-day, with the advent of composite construction, pre-cast concrete floors still play a part as a solution to floor construction within structural steel frames. Where used, the pre-cast floor is non-composite with the steel structure in most cases. In these situations, the floor depth is usually deeper than for composite solutions.

The location of pre-cast floor units is in one of two options: 1. supported off partly pre-cased beams or 2. resting on the top flange of steel beams. Either way, a degree of in-situ concrete will be needed to stabilize the whole, together with grouting between the individual units and a surface topping to provide a finish suitable for the purpose of use (Figure 13.15). From the details shown in this figure certain points become clear.

1. If the pre-cast units are to be supported on partial casing, two things have to be examined: the length of the units to ensure that they can be put into position between the steel beams in order to sit on the casing and, second, the time that the casing can be carried out to minimize cost.

2. Concreting between the ends of the pre-cast units will provide stability against movement, but how is this achieved when the units sit on top of the steel beams? As illustrated, some form of concrete in-situ edge beam becomes necessary at the external edge parallel to the pre-cast units. If edge steel is

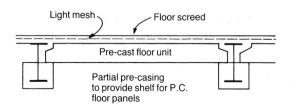

Figure 13.15   Pre-cast floors with partly cased steel.

required to be concrete cased, pre-casing will provide the stop against which the pre-cast units can be grouted. In the case where the beams come onto the edge steel beams at right angles something different is required. One solution is shown in Figure 13.8(b). Here the same edge trimmer as used in composite construction is used to provide the edge form for a half width in-situ fill-in to stop longitudinal movement of the pre-cast elements. Such a solution can be applied whether or not the steel edge beam is cased.

Casing a steel beam in concrete has always been an expensive item, if carried out on the pre-erected structural steel. Quantities are small in handling terms, involve formwork and a degree of reinforcement, while the means of access and provision of a safe working place can become very expensive. This tends to lead to working horizontally, so that a safe floor is provided for the next floor above and easy access to it, in much the same way as in composite construction.

Having to work horizontally is not always desirable either, especially when the structure and its contents would be better served by working vertically. Such a situation arises in the erection of a grandstand for race meetings or athletics. They are usually of some length and, on a horizontal erection approach, any cladding and internal fitting out is delayed – as demonstrated in Chapter 14, structures of most kinds need to be carcassed and finished upwards, with the exception of access towers, which need to be finished downwards. The economic solution to achieving vertical erection of the frame and floors can be achieved if any steel requiring casing in concrete is pre-cased before erection, on the ground.

Pre-casing structural steelwork

While structural steel requiring casing in concrete is declining in quantity, largely due to more modern methods of providing fire-proof casing to the steel, such structures as stands at sporting arenas need to be more concerned with the prevention of corrosion, as they are often at the mercy of the elements for many months of the year. In this context, casing in concrete provides a sound solution. The cost can be greatly minimized if the casing is carried out at ground level before erection.

The use of this method requires careful planning. The key points can be summarized as:

1. Agreement must be reached with the steel-work contractor to accept the heavier loads that he will have to lift.
2. Who will supply the erection crane? If the steel erector, he will need a heavier lifting capacity. Will the crane allowed for in his tender be able to cope? If not, the main contractor must be prepared to cover the difference in hire and operating rates, as the price for saving money and time on the pre-casing operation.
3. If the main contractor has priced for the crane on the basis that it is made over to the steel erector for erection and reverts to the main contractor for cladding and all follow-up work, it is merely necessary for the main contractor to have made adequate allowance in his tender for the appropriate capacity needed.
4. To be successful, pre-casing requires that steel deliveries have to be made earlier than usual, to allow time for the pre-casing operation. The overall programme must be based on a delivery time for the steel, the estimated pre-casing time and when erection can begin.
5. In assessing casing start times, it must be realized that the casing needs to be carried out in the sequence envisaged for erection. Space is needed for this operation, not only the casing, but also to allow proper sequential stacking of the cased members. The method has no merit on a restricted site. Pre-casing would only involve excessive haulage costs if carried out away from the site.

A publication on Temporary Works[11] looks at this approach in more detail than is possible here. Suffice to say that careful examination is necessary to determine which is the correct attitude of the steel for the casing to be carried out. That is, column units with sub-assembly beams attached are better dealt with the column unit laying on its side (Figure 13.16).

Establishing method and planning of steel structures

Steel structures as well as pre-cast and in-situ concrete or any other type of structure are covered by the Construction (Design and Management) Regulations 1994[7]. These regulations cover all parties to the construction process. They require co-ordination by all parties to maximize health and safety on site. The steel erector, the main contractor and the structural designer, in particular, must co-ordinate their ideas on the planning process, and establish with the designer areas of instability or other not easily recognizable possible areas of accidents. Appendix A gives a more detailed analysis of this legislation.

While it can be argued that it is up to the specialist subcontractor to provide time scales for his work, the method and planning staff have no means of judging performance if they have no records of previous work against which to assess current activities. It is also necessary to realize that both composite construction and the use of pre-cast floor techniques normally involve more than one party. As both methods differ to quite a degree they are considered separately.

Planning composite construction

When considering composite construction at the tender stage of planning a number of matters have to be considered.

1. Is the steel erection contractor already known?

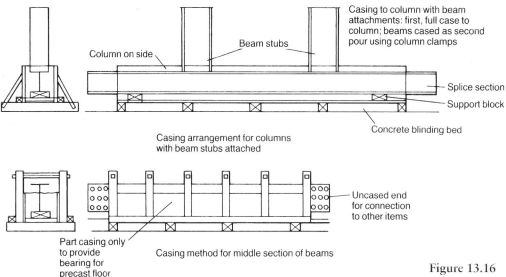

Casing to column with beam attachments: first, full case to column; beams cased as second pour using column clamps

Column on side

Beam stubs

Splice section

Support block

Concrete blinding bed

Casing arrangement for columns with beam stubs attached

Uncased end for connection to other items

Part casing only to provide bearing for precast floor units

Casing method for middle section of beams

Figure 13.16 Pre-casing of steel, method of approach.

2. If he has been nominated, information will be needed on his ideas of erection time. How will such times fit in with the overall contract period envisaged? If disagreement arises, how can it be settled amicably? Alternatively, erection times may be a contractual item.

3. The main contractor, or his subcontractor, will be responsible for the steelwork foundations. What is needed from the steel fabricator for building in? Holding-down bolts are the most obvious item. See Chapter 8 for more detail.

4. Clear lines of delineation are needed in the structure. Or who does what? Such delineation needs to be settled at the earliest possible moment. If possible at the tender stage. Items involved are (a) hoist and fix profiled steel decking, including all cutting to shape: provide and install shear connectors and edge trimmers. All these should be dealt with by the steel erector. (b) Fix reinforcement to floor slabs; pour concrete to floor.

Provide edge protection guard rails and toe boards and safe access to floor once profiled decking erected. All items here to be the direct responsibility of the main contractor,

and carried out either directly or by the main contractor's subcontractors.

5. Who will supply cranage? It is becoming increasingly common for the main contractor to provide the cranage and allocate periods to the steel erector before eventually using it for follow up operations. Careful agreement on both parties' needs in terms of radius and lifting capacity is clearly necessary. It will be clear that unless integration is agreed by all parties involved, arguments may develop which could result in claims for delay.

Planning for pre-cast floor construction

Many items are similar to composite construction, especially in relation to shared cranage and related matters. Where a degree of pre-casing is involved the items that need to be agreed between the steel erector and main contractor have already been listed. In addition:

1. It will be usual for the main contractor or his subcontractor to position the pre-cast floor slabs, but to agree with the steel erector for

him to hoist and deposit on the steel beams. This to be at his prerogative. When depositing it should be agreed to place several, one on top of another to save time, say 3 number in a pile.

2. Once at the required level, the actual final placement can be achieved as illustrated in Figure 13.17. Tirfor winches or similar are anchored to the steel beams above and pick up one floor slab at a time. This unit is then pulled sideways by other winches to the required degree for lowering to its final resting place.

3. The main contractor or his representative would be responsible for any make up concrete or grouting of the floor units, together with the final floor finish.

Planning – generally

Planning and method engineers who have had long experience in construction always build up their own data books of performance in all

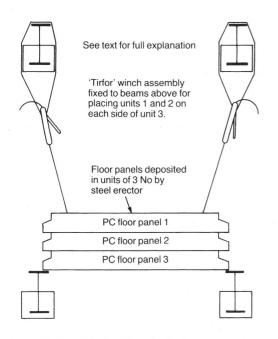

See text for full explanation

'Tirfor' winch assembly fixed to beams above for placing units 1 and 2 on each side of unit 3.

Floor panels deposited in units of 3 No by steel erector

PC floor panel 1

PC floor panel 2

PC floor panel 3

Figure 13.17 Method for final placement of precast floors with partly pre-cased steelwork.

areas in which they have had experience. In relation to steel erection, the yardstick used is often based on the use of x tonnes per hook per week. (That is per crane per week.) The tonnes of steel erected per crane will clearly be dependent on the type of steel erected. The number of lifts of lightweight roof trusses may be the same as heavy steel columns, but the tonnage handled will be quite different. Records of this kind are very useful if the steel firm has not been nominated at the time of tender.

Method engineers also need adequate knowledge of temporary works needs for achieving stability of steelwork during erection. Even if allowed for by the steel erector, site management must have adequate references to establish if the work is being carried out safely. Fortunately a number of publications provide excellent guidance in this respect[5][6][7].

The final area, in relation to planning steel erection, is in respect of attendances specified in the contract documents. These may include provision of services (electrical, telephones, accommodation and stores, welfare facilities, offloading materials and so on). The costs of such services will be borne by the main contractor and need to be estimated as accurately as possible. At the contract stage, a running cost of each individual item should be kept.

In industrial plants, much of the steelwork may be open to the elements and merely acting as a support to a process plant. Access is often by steel staircases, with floors formed by metal grid configurations. In one such plant, known to the Author, all the main contractor's work was priced on a schedule of rates for nominated types of work. Thus concrete to small areas of floors was measured as a given area and thickness, formwork and its support in area and height of support not exceeding a given figure, while reinforcement was measured in the usual way by tonnes. The contract documents emphasized that all items were described in accordance with the standard method of measurement for civil engineering works. So far so good! But what had not been stated was the location of the work in the otherwise bare steel structure. It turned out that the small areas of concrete were, in fact, 40 m up in the air. The taker-off had been working to the standard method of measurement for the formwork and falsework giving a figure for the supports not exceeding 3 m. This was because there was a steel walkway less than 3 m below from which the support could be taken. He had lost all sight of the actual height of the work in the steel structure! As the drawings had not been available when the schedule of rates was agreed, the work items were considerably underpriced. It was fortunate that the client understood what had happened and agreed to a re-price of the items in question.

References

1. Allen, P.H. (1985) *Steelwork today*. Chartered Institute of Building Technical Information Service paper No 51. CIOB, Ascot.
2. *The Wonder Book of Engineering Wonders*. Ward, Locke. 2nd Edition, London. Undated, but circa 1930.
3. *Health and Safety at Work etc. Act 1974*. HMSO, London.
4. *Construction (Health, Safety and Welfare) Regulations 1996*. HMSO, London.
5. British Standards Institution. *Code of practice for safety in erecting structural frames*. BS5531 : 1988. BSI, London.
6. Health & Safety Executive. *Safe Erection of Structures*. Guidance Note GS 28. Part 1 – initial planning and design (1984); Part 2 – site management and procedures (1985); Part 3 – working places and access (1985) and Part 4 – Legislation and training (1986). HMSO, London.
7. *Construction (Design and Management) Regulations 1994*. HMSO, London.
8. *Architects Journal*. Structural Developments. Five part series on developments in steel and concrete construction. Part 1 – Steel frame design. 24 April 1991; Part 2 – Steel frame buildings. May 1991; Part 3 – High speed concrete 8 May 1991; Part 4 – Visual concrete. 15 May 1991 and Part 5 – Steel or concrete? 22 May 1991.
9. *Slimdek – The System*. Effective from May 1997. British Steel, Redcar.
10. Goodchild, C.H. (1993) *Large area pours for suspended slabs – a design guide*. British Cement Association, Crowthorne.
11. Illingworth, J.R. (1987) *Temporary Works – their role in construction*. Thomas Telford, London. Chapter 11.

Further reading

1. Customer led – Construction led. Feature by a series of authors. *Steel Construction*, February 1991. pp. 11–20
2. Girardier, E.V. (1991) Construction led Chapter 2. Follow up article after the feature in (1) above. *Steel Construction*, August.
3. Taggart, R. (1991) *Steelwork erection – guidance for designers*, British Steel, Redcar.
4. *Structural sections effective from July 1995*. British Steel, Redcar.
5. Cardinal, H. and Bannister, A. C. (1995) *Aspects of steel selection for building and bridge construction*. British Steel, Swindon Technology Centre.

Chapter fourteen

Cladding, internal works and specialist services

Once the methods have been decided and the planning has been completed for all work up to and including the structure and roofing-in, the concept and approach for the remaining activities take a somewhat different path. Instead of relatively few clear cut items, the planning has to deal with a growing number of trades and specialist services. The main or management contractor's influence declines in relation to methods, his function now being more related to the following matters.

1. Where specified in the contract document, providing facilities for specialists and subcontractors. (Generally called attendances in the bill of quantities.) These may include: unload and store or elevate to the working level, materials, plant and equipment; provide storage facilities; provide use of scaffolding together with general site facilities.
2. Negotiating programme periods to integrate with other contractors and the overall master programme.
3. Obtaining detailed breakdown programmes from each subcontractor or specialist, showing time scales and labour content, to allow progress monitoring. Such monitoring

should also be carried out on a financial basis.
4. If required by the contract terms, playing a co-ordinating role in relation to the various contractors' drawings.

The importance of this overall monitoring becomes apparent when one realizes that the items in the area of cladding, internal works and specialist services represent a major part of the contract cost. The figures given below make the point very clearly. In fact, the figure quoted of 50% for all services can now reach 60% in high tech buildings.

A major problem, when dealing with the items covered in this chapter, is the ability to obtain adequate information on the content of the cladding, internal finishes and specialist services, at an early stage in the planning process. When tendering, it is unlikely that the main or managing contractor will have any real knowledge of content, unless some action is taken to coerce the architect to provide such information, even if only in the most general of terms. Experience has shown that the best way to proceed is to produce a check list related to cladding, internal finishes and specialist services for the type of

---

> ### Cladding
> Curtain wall type or pre-cast claddings can cost as much as the structure on which they are supported.
>
> ### M and E Services
> In a new office block, M and E services can represent 20% of the total building cost.
>
> ### All Services
> In some projects, the total cost of all services may be as much as 50% of the total building cost.
>
> ### Air conditioning
> In an office block of 100 000 square feet, fully air conditioned, the air conditioning plant will need something in the order of 8000 square feet of space.
>
> *(Construction News 9/6/88)*

building in question. Such a check list can then be sent to the architect with a request that he merely ticks off the items on the list most likely to be incorporated into the building. This approach offends no one and has been found in practice to work extremely well. Most architects see it as a useful *aide memoire* to their own future output of drawings and nomination of specialist firms. A typical check list of this type is given at the end of this chapter (Example 14.1).

From such bare information an experienced planner can get a fair assessment of the workload and the likely time involved. If the contract is won, more detail will become available. Unfortunately, such information is likely to come forward by degrees. For example, one may expect 10% on commencement, 70% by the first 3 months, 90% in nine months, with the remaining 10% spread to the end. While the example of a check list at the end of the chapter covers office blocks, similar check lists can readily be

developed for all other building types – industrial, hotels, schools, specialist buildings and so on. Experienced planners see such check lists as tools of the trade.

While the information starts as a tick on a check list at the tender stage, once the contract has been won and cladding and finishes and specialist services begin to be let, continual updating of programmes can be carried until, finally, the programme for any particular item becomes final. It follows that specific programmes, within the master programme, require continuous development by the site planning staff on behalf of the site management.

With the above general considerations in mind, more detailed examination of each of the main groups in the heading can take place.

External cladding

At the roofing-in stage of a structure, the claddings become the key to waterproofing the building, so that internal finishings and service installations can be pressed forward. As such, they should be programmed to be carried out as close to the structure as possible. In load-bearing structures, the load-bearing material may also be the cladding as well; or at least the waterproofing means. The so called cladding, here, would be merely decorative.

Types of cladding

The range of cladding materials, today, is very wide. They can be listed as brickwork (facings); pre-cast concrete (in many forms); masonry; blockwork; curtain walling; steel sheeting, self-finished or enamelled panels; types of plastic; glass fibre and glass reinforced cement; and, in low rise construction, rendering and tile hanging.

For low rise industrial structures, the use of self-finished steel sheeting in a variety of colours has become increasingly popular (Figure 14.1). With this material, the roof sheeting and the

Figure 14.1 Self-finished metal cladding draining to ground level gutters.

(a)

(b)

Figure 14.2 (a) Curved roof edge – roof continuous with wall sheeting. Run-off collected by secret gutters above any window or door head line. (b) Down pipe from secret gutter collection.

wall cladding can become one continuous element, doing away with gutters at the edge of roof line. The principle used is to provide continuous drainage down to ground level, or at some point above to clear door openings, etc. Collection is by secret gutters. Figure 14.2 illustrates the method. The resulting structure is enabled to show clean, flowing lines with a big improvement in appearance. Figure 14.3 shows a variation with brick cladding at lower levels.

With curtain walling, double or triple glazing with in-built insulation in the non-glazed areas combines the duties of keeping the weather out as well as keeping heat or cool air inside. Figure 14.4 shows mirror glass cladding, while Figure 14.5 illustrates large curtain wall elements being hoisted to building. In this case the crane used by the steel frame contractor was left behind to handle the cladding. Pre-cast concrete cladding was dealt with separately in Chapter 12.

Masonry claddings usually occur in high quality buildings attracting high rental values. At the most expensive end, granite, tooled or polished, and marbles are the most used. At the cheaper end, reconstructed stone finds a ready market, with sandstones and limestones somewhere in between.

Concrete blockwork, pointed or painted with cement paint, is common in low cost buildings such as sports centres and often used in access towers or walling in low cost industrial structures.

Glass fibre, resin bonded, together with glass reinforced cement have found roles as cladding

Figure 14.3 Variation on Figure 14.2, with facing brick cladding under secret gutter line.

Figure 14.4 Office building with mirror glass curtain walling.

Figure 14.5 Large curtain wall elements being lifted from delivery vehicle.

timber frame structures (Figure 14.6). An interesting variation is shown in Figure 14.7.

The efficiency of any cladding is much dependent on effective design and careful installation. The following listings deal with key issues and suitable construction methods.

From the purely method assessment angle, there are a number of matters which must be satisfactorily settled before erection starts. These are given below.

Case history

The importance of close designer/contractor relationships is well illustrated by an actual case history. A concrete structure in the Middle East was to be clad in heavy pre-cast concrete units. A good deal of discussion took place between the consulting engineer and the contractor concerning handling and turning problems, as well as the methods of fixing the cladding to the structure in a hot country. All these matters were settled amicably and to everyone's satisfaction. It

materials, though protection from the effects of ultra-violet light will be needed by means of suitable protective paints.

Housing and low rise flats or small office buildings in the low rise category mainly depend on the use of facing bricks or such added elements as tile hanging, rendered finishes on common bricks and plastic ship lap boarding on

Key issues in design and fixing of claddings

1. Accuracy of the structure – in relation to floor-to-floor dimensions, overall height of the building and, especially in tall buildings, avoidance of twist in the structural form. Curtain walling, especially, can be in great trouble from a twisted frame unless adequate tolerances have been provided.
2. Where the cladding is supported floor by floor on extensions of the floor slab, especially in relation to brickwork, constant checking is needed to ensure that each nib is vertically above the previous ones, and provides the bearing area needed.
3. With heavy claddings, brickwork and pre-cast concrete in particular, has the designer allowed for elastic compression of the structure, by detailing compressible material at the head of brick or pre-cast panels?
4. Tolerances and fixings – must be realistic in relation to the structural form used.
5. Materials in fixing components must be electrolytically compatible.
6. Detailing of fixings must allow inspection when completed.
7. Waterproofing in joints. There must be the ability to inspect after installation of inserted gaskets or compressible materials, to ensure that they are properly in position.
8. What is the relationship between the cladding and backup insulation?

Figure 14.6   Shiplap cladding above rendered brickwork on housing.

Figure 14.7 Facing brickwork cladding to low rise steel structure. Exposed steel work used as architectural feature over the main entrance.

Key items in method assessment for erection of cladding

1. One of the problems in planning the erection of cladding systems is that the necessary information has a habit of not being available soon enough. Of key importance is the knowledge of the weights of cladding units when pre-fabricated – curtain walling, pre-cast units and heavy masonry. If this information cannot be found at the time of settling cranage capacities for structural work, there is always the possibility that special cranage will have to be brought in at greater expense. Information is also needed on the sequence of assembly of curtain walling and how this will relate to structural progress.

2. Discussion is essential between designer and erector in relation to handling the units – do they have to be turned from flat to vertical or upended between arrival on site and positioned on the structure? Additional lifting points may be required or extra reinforcement for turning and handling moments. (See reference [1] for methods and solutions). Chapter 12 deals with pre-cast cladding.

3. Where it is necessary to leave areas of cladding until later to allow works access from hoists etc., always try to keep these in access areas of the structure. Cladding is often separate from the main cladding, apart from this area of the building being finished last. See later under Specialist services.

4. Where window frames are not integral with the glazing, they should be let as one contract. In the event of water penetration, only one party will be to blame.

5. What are the needs for temporary storage and who is responsible?

6. If the main contractor supplies the cranage, who supplies the lifting gear?

7. If connections are not easy, trial panels should be provided for testing the connections specified.

was not until the time for erection arrived that it was discovered that ten panels, on the side of the building away from the tower crane, were too heavy to be lifted at the radius involved. At that late stage, the only thing that that could be done was to hire in, specially, a suitably heavy mobile crane, as and when required, to lift one panel per floor. With hindsight, it was a taking-off mistake by the contractor, but a very expensive one.

Internal carcass and finishes

This work in a building is the most difficult to plan. Many specialist trades are involved and the final information in respect of their activities is received late, in planning terms. If, however, the check list system is followed, outline planning provides a start, capable of being progressively developed. In the case of internal carcass and finishes it is best to look at each separately.

Internal carcass

All internal carcass work has to be tailored to the type of structural frame and the cladding, services and finishes specified.

Likely items for inclusion
Back-up walls to cladding – including installation of smoke and fire barriers, together with insulation to the external wall (if not an integral part of the cladding system.) Partitions in brick or block: grounds for demountable partitions – which would be installed as finished items. In factories or warehouses or other industrial premises, firebreak walls: first fix joinery; fixings in ceilings for services or false ceilings; plasterer base coats and sometimes floor screeds. In planning terms, it is desirable to leave floor screeds until all other carcass work has been completed. The floor area can then be swept clean and the floor layer made responsible for clearing up any mess that he may make, at which point, the building should be clean for finishes of all kinds.

If this is not followed, the newly laid floor may need expensive treatment to remove plaster droppings and filler repairs to correct chipping and damage from previous trades.

Programming the order of work
Careful consideration of the best sequence for carcass trades should be made. Logical progression to avoid any carcass trade having to come back over other trades needs careful study. This is especially important as it needs to be made quite clear to all trades that they are to be responsible for clearing up their own rubbish or droppings. If one trade has to go back over another trade, arguments can arise over who made the mess. In this respect, any wet trades need to be programmed as early as possible in the sequence to give adequate time for drying out. As part of this philosophy, the main or managing contractor will need to assess the probable extent of rubbish removal involved and what methods he needs to adopt for its removal off site. It is often not appreciated the quantities involved. In a 28 storey office block, in London, a platform hoist was in use full time for barrowing out waste materials to ground level skips, once the full quota of trades were functioning inside the structure. Reference should also be made to Chapter 17 for ways of minimizing the cost of waste removal.

Handling and other facilities to be provided
It is preferable that attendances on other contractors should be defined in the main or management contractor's contract. It results in a clear situation at the tender stage as to what has to be allowed for. If the contract makes no provision in this respect, every specialist or subcontractor will have to negotiate his needs with the main or management contractor **after** the contract has been won. Given the preferable procedure, a table should be prepared from the contract documents showing the services to be provided by the main or management contractor and what provision is the specialist or trade

contractor's own responsibility. An example of a suitable table is given in Figure 14.8.

In the case of cranes or hoists that are provided for common user use, it will be important that agreement is reached with all contractors, preferably at weekly meetings, of the extent of use seen for the week ahead and approximate times needed.

Finishes

All second and, where operative, third fix items come into the finishes schedule. Where used, plaster tends to bridge the division between carcass and finishes. Backing coats are reasonably carcass, while floating coats are equally regarded as finishes. The situation is made easier by the emergence of one coat plasters, with quick drying characteristics.

It must also be remembered that finishes overlap with service installations on many occasions – decoration may have to wait for service work to be completed.

Likely items for inclusion

Second fix joinery, ironmongery and door furniture, plaster finishing coat or the whole process, electrician fit switch plates, light defusers, etc., plumber fit taps etc., fit doors and prefinished demountable partitions, floor screeds and/or tiling, tiling to walls and splashbacks, accoustic tiling to false ceilings and painter and decorator.

The above items may well need to be followed by client's furnishings – carpets, kitchen equipment, switchboard and hand sets, other communications systems, furniture and furnishings, blinds, together with alarm systems and security equipment.

The final act of finishings will be snagging all areas and, in particular, final test and balancing of air conditioning (once the building is occupied). These latter items may involve further decoration.

Programming the order of work

As with carcass items, the sequence of trades needs to be agreed by all parties as well as related to the sequence established for carcassing. It is also particularly important, prior to commencement of finishes, that adequate co-ordination has taken place in relation to all drawings provided by the various trades and specialists. Who is responsible for co-ordination should be established early in the contract, if not spelt out in the tender documents. Failure in this respect can have expensive repercussions in time and money, as the following case study shows.

Case study 1

The main circulation area of an airport terminal building, which was not air conditioned, was to be provided with an acoustic tile ceiling. In one section of the large area, snack bar facilities were to be provided with provision for the cooking of light meals. Fume extract ducts were provided above this limited area. As the duct work had not been completed, the acoustic tile subcontractor, to avoid delay, started work on installation at the opposite end of the circulation area. The tiling made rapid progress towards the snack bar section. It was not until the tiling was quite close to the cooking area that it was realized that the tiles were going to foul the extract trunking by 50 mm. That is, the trunking would project below the tiling by that amount. As so much tiling had been fixed, the only solution was to dismantle the trunking and remake it shallower and wider to pass under the acoustic tile line.

Clearly time and money was wasted. This was solely due to no clear responsibility being laid down in relation to the co-ordination of different contractors' drawings. It is at this point that arguments begin.

This particular example well exemplifies the point that all activities in the carcass and finishing stages also react to services as well.

SUB-CONTRACTORS ATTENDANCES

PROJECT :- - - - - - - - -

DATE :- - - - - - - - -

CONTRACTOR

	TYPE OF ATTENDANCE																										
A	SPACE FOR HUTTING																										
B	ACCOMMODATION																										
C	MESSING																										
D	LIGHTING FOR HUTTING																										
E	HEATING FOR HUTTING																										
F	PROVISION FOR CAR PARKING																										
G	STORAGE – OPEN																										
H	STORAGE – FENCED																										
I	STORAGE – COVERED																										
J	STORAGE – SECURE																										
K	PROVISION OF POWER – SMALL TOOLS/LIGHTING																										
L	LEADS ETC. FOR SMALL TOOLS/LIGHTING																										
M	GENERAL SCAFFOLD – FIXED ACCESS																										
N	– TOWERS																										
O	SPECIAL SCAFFOLD – FIXED ACCESS																										
P	– TOWERS																										
Q	LABOUR TO UNLOAD MATERIALS																										
R	LABOUR TO PLACE MATERIAL AT WORKSITE																										
S	CRANAGE – GENERAL																										
T	CRANAGE – SPECIAL																										
U	PROVISION OF EQUIPMENT (SPECIFY)																										
V	PROVISION OF POWER/FUEL FOR T & C																										
W	PROVISION OF HEATING FOR DRYING OUT																										
X																											

Figure 14.8 Form for recording required attendances on subcontractors by main contractor.

Handling and other facilities to be provided

As with carcass activities, requirements for finishing trades attendances need to be tabulated. The same sheet (Figure 14.8) would normally be used, and the same comments would apply.

Where specialist contractors want to use their own handling equipment, the main or management contractor must be consulted to ensure proper integration with the common user plant provided.

Specialist services

While it is desirable to separate specialist services from building carcass and finishes, it must always be recognized that co-ordination between all programme elements is essential.

Items for possible inclusion

Housing apart, specialist services in a building continually get more complex. The items involved in this area spread over a wide range of disciplines such as: water supply and H and C services; heating and ventilation; air conditioning process services in factories and hi-tech manufacture; process effluent disposal; communication systems – cabling, for telephones, telex, television; long distance computer communication, gas and electricity supply and lightning conductors; lifts, window cleaning cradles (on tall structures) and security equipment; fire alarms and sprinkler systems.

The services in a structure cross all the programme items boundaries (Figure 14.9). Co-ordination of service items with other programmes is essential. While primarily aimed at co-ordination between all parties to achieve site safety, the Construction (Design and Management) Regulations 1994[2] open the door to inter-communication between all parties to promote the greatest efficiency as well. Whoever is taking the co-ordination role must ensure that the specialist contractor programme fits the design programme – or vice versa – as well as the programmes of all other parties.

Programming the order of the work

Services need a different approach to the other areas already examined. As mentioned above, services cross all boundaries, so that their sequence of installation must necessarily relate to the building sequence as a whole. This starts at foundation stage with provision for service entries and drainage exits. In basements, there may be underfloor services of all kinds, including fixings for lift guides etc., in lift overrun pits in basements or under ground floors where there is no basement(s).

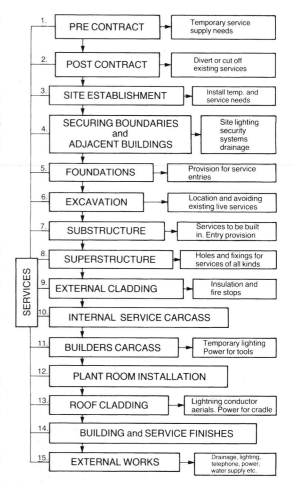

Figure 14.9 The impact of services on the construction process.

In these circumstances, the contractors involved in foundations and the structure are going to need small, but important, amounts of service information at a very early stage. Services programmes must reflect this, even though the bulk of their work may come much later in the overall programme.

With such divergency of services in a modern commercial or industrial building, their planning needs the services of an engineer with a wide experience of the type of building concerned. The sequence of activities needs to be correctly established and agreed by all services contractors at an early stage. This same engineer should then take over the responsibility for monitoring the service contractor's progress, both in time and cost.

While the sequence of events will vary according to the buildings content, there are key rules that will always apply if the most efficient result is to be achieved.

Key rules for services installation

1. Services installations should start with the rising mains and fan out from there into the office, production or other use areas.
2. In multi-storey construction, early installation in the core structures of services is needed, so that as each floor becomes available, services can radiate from the core area into the lettable space or production areas.
3. The importance of (2) above is twofold: (a) rising mains are in place in the event of fire during construction; and (b) as each floor services are completed they can be tested and accepted, when follow up decoration and final fixings can proceed. Once all finishes are complete, the area can be sealed off until handover.

These rules, in overall planning terms, create a planning sequence that is illustrated in Figure 14.10. Core structures, carcass proceeds upwards and their finishes take place downwards. Individual floors (lettable or production areas) by contrast, carcass and services and finishes proceed upwards.

Within the core structures which contain lifts, guides should be fixed upwards so that, at lift motor room level, the guides can accept counterweights and possibly prefabricated lift cars lowered in from the top of the shaft, before the lift motors are placed in position.

Where plant rooms are located on the roof of buildings, core carcass can still proceed upwards. Some testing will still be possible, electrical work and incoming communications systems in particular.

It is important to recognize that the core areas are also the main means of access at any particular level. As such, entry to the building from hoists or loading out platforms should aim to be sited in this area (Figure 14.11).

Case study 2

The following case study is a further example of the importance of core area service installation being able to proceed upwards. The layout of a small office building is shown in Figure 14.12. The office floor areas were designed as plate floor slabs, supported on columns. The one access/service core tower was designed in the traditional way. That is, columns and beams and in-situ stairs with half landings. The building had a small lift between stair flights, with enclosing concrete walls.

The site management were under pressure to speed up progress after a delayed start. As a result, the plate floor slabs proved irresistible with proprietary formwork and ready mixed concrete and proceeded apace to the pleasure of the client. To achieve this result, the core area was allowed to drift as it would have slowed down the rate of floor progress, having a much higher work content in a small area. This policy

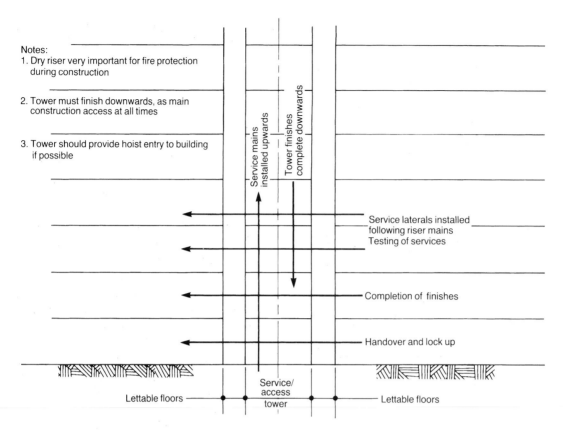

Figure 14.10   Sequence for service installation.

proved fatal in the end. Service core installation could not be started and, in the end, lettable area services could not be tested and the decoration could not be completed. All services and finishes came to a standstill, waiting for the core tower. Handover was late and a great deal of extra cost to the contractor was involved.

Commissioning of services

The Chartered Institution of Building Services defines commissioning as: 'The advancement of an installation, from the stage of static completion to full working order, to specified requirements.' Within the sense of this definition is meant putting a static system into motion – flowing of liquids or power availability to specified tolerances and, where appropriate, to temperatures and pressures laid down. Such

commissioning may relate to the building as a whole or in sections, if partial handover is called for.

The building services involved in commissioning can be extensive. The list of services, of course, will depend on a specific building's purpose. The higher the technology in the building, the more complex the services. Consider, for example, modern hospitals.

Commissioning is clearly something that needs to be properly planned and adequate time allowed for in the overall programme for its achievement. The main contractor needs to understand the services contractor's side and it is for this reason, that the overall planning and control should allow for a specialist services engineer who can both appreciate the specialist contractor's point of view and that of the

Figure 14.11 Entry from hoists to building. In this case, to escape stair landings.

numerous building contractors upon whom service installation will impinge. Above all, the main or managing contractor's staff must understand the need for allowing adequate time for commissioning to take place.

While most of the specialist services can be commissioned before handover, it is often not appreciated that air conditioning, while capable of commissioning before handover, cannot be finally balanced until the building is occupied. The reason is that final adjustments cannot be made until the heat generation effect of the occupants has been taken into account. Experience shows that such balancing may take as long as three months after the building has been occupied. The client needs to be made aware of this situation on the services programme, otherwise arguments can arise.

A more detailed examination of the back-ground to commissioning of services can be found in *The problems of commissioning building services*[3], *Co-ordination of mechanical and engineering services*[4] and *Services coordination on building sites*[5].

Provision of handling and other facilities

The same rules apply in relation to handling and other facilities as in the previous two sections of this chapter. The same form is equally applicable for tabulating attendances required by specialist services and their subcontractors.

It is in the area of specialist services that early knowledge of what has to be lifted into the structure becomes of crucial importance. Many items of equipment can be of considerable weight. If the main or managing contractor is required to provide cranage for common user purposes, the weights to be lifted must be known in the earliest method assessment stages. Failure at this stage may well mean bringing in a high capacity mobile crane to lift just one item into the structure – at considerable expense. A chiller unit for an air conditioned multi-storey office block may weigh in the region of 5 tonnes. While the weight may not seem all that great, the radius at which it has to be lifted is what determines the crane size.

An even more onerous situation may arise if heavy equipment, such as boilers, has to be handled into the basement of a structure but will not be available before the ground floor slab is placed and the structure above commenced. Such a situation needs to be brought to the client's attention as early as possible with a request that the contractor be advised of the action he should take – whether to delay construction, or leave out parts of the building until the item in question is delivered? In either case, completion will be delayed.

In these matters, much depends on the client's advisers having foreseen the problem and pre-ordered the equipment concerned.

Finally, in this section, if the specialist company wants to provide its own handling

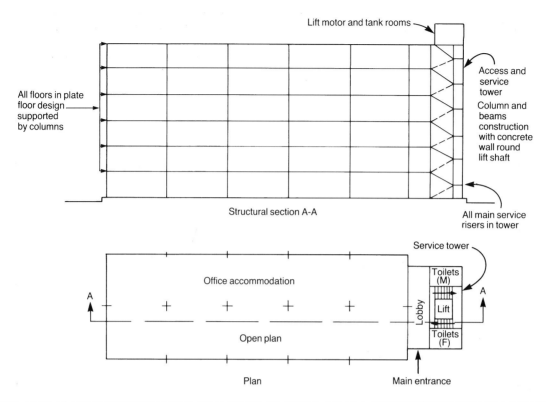

Figure 14.12   Small office building where the service core was left behind to give the appearance of more rapid progress by concentrating on the simple plate floor construction of the office areas.

equipment, close integration with the main or management contractor is necessary to avoid clashing in both plant location and programme timing.

References

1. Illingworth, J.R. (1987) *Temporary Works – their role in construction*. Thomas Telford, London.

2. *The Construction (Design and Management) Regulations 1994*. HMSO, London.

3. Roberts B. M. (1981) The problems of commissioning building services. *Building Technology and management*, May.

4. Barton P. K. *et al.* (1978) *Co-ordination of mechanical and engineering services*. Chartered Institute of Building occasional paper No 16. CIOB, Ascot.

5. Barton P. K. (1980) Services coordination on building sites. Chartered Institute of Building Site Management information service paper No 83, Ascot.

Example 14.1 Office blocks: Check-list for cladding, internal finishes and specialist services

EXTERNAL TO THE STRUCTURE

I.   Roof finishes
   (a) Asphalt
   (b) Roof felting
   (c) Tiles

II. External cladding
   (a) Brick/Block/Stone cladding
   (b) Pre-cast concrete/artificial stone facings
   (c) Window Frames
   (d) Cills
   (e) Emergency and secondary entrance doors
   (f) Steel staircases and handrailing
   (g) Glazing
   (h) Curtain walling

III. External finishes
   (a) Cement rendering
   (b) Marble/Mosaic/Granite facings
   (c) Painting
   (d) Main entrance doors

IV. External site works
   (a) Cement screeds
   (b) Concrete paving and kerbs
   (c) Asphalt
   (d) Garden flower tubs etc.
   (e) Trees, plants and shrubs
   (f) Brick paviors

INTERNAL WORKS

A. **Building trades** (Before service carcass)
   I. Internal brick/block walls

   II. Joiner 1st fix
      (a) Door Frames
      (b) Duct access frames
      (c) False ceiling carcass
      (d) Service duct carcass
      (e) W.C. partitions

B. **Services – carcass**
   I. Plumbing
      (a) Vertical pipework
         (1) R.W.P's, S.P's, V.P's rising and down mains

(2) S.P's
      (3) V.P's
      (4) Rising and down mains
  (b) Break tanks
  (c) Storage tanks
  (d) Switchgear
  (e) Horizontal pipework ('run outs' from rising mains)
      (1) Distribution mains
      (2) Hot, cold and waste pipe installations
      (3) Urinal stalls
      (4) Brackets and bearers for sanitary fittings
  (f) Cold water connections to mains supply

II. Heating
  (a) Vertical feed and return mains
  (b) Horizontal
      (1) Feed and return main
      (2) Radiator/convector installation and connections
  (c) Storage tanks
  (d) Boilers
  (e) Feed pumps
  (f) Heat exchangers
  (g) Flue exhaust system
  (h) Louvres and grilles
  (i) Switchgear and controls
  (j) Lagging

III. Fire protection
  (a) Vertical valves and pipework
  (b) Horizontal
      (1) Distribution pipework
      (2) Sprinkler installation
      (3) Fire resisting doors
      (4) Alarm and control conduit
  (c) Switchgear
  (d) Motorized control valves
  (e) Pumps and break tanks

IV. Gas installation
  (a) Vertical pipework
  (b) Horizontal
      (1) Distribution pipework
      (2) Fittings
  (c) Connection to mains supply

V. Ventilation/air conditioning
Vehicle exhaust/induced draught/forced draught systems
   (a) Vertical ducting
   (b) Horizontal
       (1) Ducting
       (2) Grill outlets
   (c) Fans
   (d) Motors
   (e) Switchgear and controls
   (f) Stand-by diesel motor
   (g) Filters
   (h) Heat exchangers/refrigeration units
   (i) Precipitators and humidifiers
   (j) Cooling towers
   (k) Pumps and compressors
   (l) Switchgear and controls
   (m) Lagging when needed

VI. Electrical installation
   (a) Conduits and wiring
       (1) Vertical ducts/Horizontal cable trays
       (2) Distribution to power and lighting circuits
   (b) Switchgear
       (1) H.T. switchgear/Controls
       (2) Mains to H.T. switchgear and landlord's meters
       (3) Transformers
       (4) L.T. switch gear/Controls
       (5) H.T. to transformers
       (6) L.T. to bus-bars
   (c) Wiring to
(1) Heating installation
(2) Ventilating installations
(3) Fire alarm installation

VII. Passenger/goods lift installations
   (a) Shaft guides
   (b) Door thresholds
   (c) Door frames
   (d) Unistrut inserts
   (e) Motors
   (f) Relay and control gear
   (g) Detail of lift pits

VIII. Lightning protection
    (a) Earth plates
    (b) Wiring and conductors

IX. Telephone installation
    (a) Service provider
        (1) Cabling
        (2) Control room equipment
        (3) Instruments and wiring
    (b) House
        (1) Exchange installation
        (2) Instruments and wiring

X. Window cleaning gear
    (a) Fittings required before roof finishes
    (b) Cradles, guide rails, motors etc.

C. Building trades (After service carcass)
    I. Wall and ceiling finishes
        (a) Cement rendering
        (b) Plastering
        (c) Dry lining
        (d) Special finishes
        (e) Wall tiling

    II. Wet floor finishes
        (a) Cement screeds
        (b) Granolithic
        (c) Terrasso
        (d) Quarry tiles

    III. Joinery 2nd fix
        (a) Skirting
        (b) Architrave
        (c) Doors
        (d) Door furniture
        (e) False ceiling carcass
        (f) Fixed screens
        (g) Timber panelling
        (h) Shelving
        (i) Notice boards
        (j) Special Fixtures
        (k) Laboratory or other special benches and cupboards, etc. May need several visits in sequence.

D. Final finishings
  I. Services final fix
    (a) Plumber – sanitary fittings
    (b) Heating and ventilating
       (1) Louvres
       (2) Radiator covers
    (c) Electrician
       (1) Lighting fittings
       (2) Incinerators
       (3) Switch covers and plates

  II. Painter and decorator
    (a) Painting
    (b) Wash down and clean out

  III. Dry floor finishes
    (a) Thermoplastic/Lino/PVC/Rubber tiles
    (b) Stair nosings
    (c) Wood block/Strip flooring
    (d) Carpeting

  IV. Demountable partitions – type and relationship to floor and ceiling finishes

  V. Final fittings clean out and hand over

Chapter fifteen

External works

The scope of External Works can be very large or quite small, or anywhere in between. It will all depend on the type of building or buildings and the surroundings.

The main areas of activity likely to be involved will arise from:

- Drainage requirements – surface water, sewers, trade effluents and any special process needs involving interceptors before discharge into sewers;
- Access roads, paved areas, storage parks and car parking facilities;
- Public Utility Services – water, gas, electricity, telephones and telex;
- Special industrial services – facilities for compressed air, oxygen and acetylene, inert gases (argon, neon usually), carbon dioxide for fire protection and liquid coolants for machine shops;
- Minor external buildings – sub-stations, compressor house, storage facilities for bottled gases and metering their supply, to name some possibilities;
- Security and lighting – security fencing and alarm systems linked to gate houses; external lighting of roads and paved areas and building security surveillance (closed circuit TV).

While the above activities span the whole contract period, the bulk of the work will fall into one or other of two main periods of the contract duration.

1. At contract commencement – drainage main runs, including connection to the designated outfall, whether surface water or foul sewers. In addition, the provision of access arrangements, hard storage areas, car parking for site personnel and hard areas for piling machines or other plant associated with foundation construction. And, whenever possible, service mains to building entry.
2. At or towards the end of the contract – completion of minor buildings, security fencing, surveillance equipment, lighting to roads and other areas where materials and equipment are stored. Final connections into the structures of service mains and final surfacing of roads and paths.

Where housing is involved, individual garden regrading, fencing and external stores come into the picture. Finally, site regrading and landscaping may be a major element (Figure 15.1), or quite minor on central re-development sites.

At contract commencement

Good planning and organization is required if the contract is to get off to a flying start, especially in relation to installing drainage and

Figure 15.1   Site landscaping on large scale for business park development.

access and hard areas for storage and site accommodation.

Drainage

Successful opening up of a construction site is very dependent on the rapid installation of drainage systems. If separate systems (surface water and foul) the surface water drainage needs first priority. It should be planned to work backwards from the outfall so that run-off is possible at all stages of its installation. Further reasons for this policy are that the site is helped to be kept as dry as possible for other follow-on works and, at the final clean-up stage, long waits for the ground to dry out for regrading etc. are avoided as much as possible.

The successful planning of the drainage installation requires adequate knowledge of what is specified and the site conditions which may affect decision making. Checklist 15.1 lists key issues. With the information arising from the checklist, the operational planning can begin.

Planning the method for drainage installation

Drainage method planning requires careful sequencing to achieve the most efficient use of the resources employed. To achieve this, the following matters require integration as each has an impact on the others.

Excavation machinery

Left to itself, a modern backhoe excavator is capable of running away from the other operations involved – trench support, pipe laying and testing and back-filling and reinstatement. Unless it is integrated into other activities a number of problems arise. For example, excessive trench support equipment becomes necessary at greater cost than need be. It taken off site because there is nothing for the backhoe to do, it will have to be brought back later on, to

Checklist **15.1** Information for planning drainage

1. Is the outfall to drainage inside or outside the site boundary. If outside, has permission been obtained to enter the land? If yes, are any special conditions laid down? (fencing requirements, reinstatement needs, removal of temporary access arrangements).
2. What temporary access arrangements are required for delivery of pipes, concrete and special backfilling material to other peoples' land?
3. Is a soil report available? Are there any difficult areas which may require special treatment? (running sand requiring well-pointing, Figure 15.2; well pointing dewatering, Figure 15.3).
4. What depth of excavation will arise? Areas over 1.2 m in depth will require support, unless in stable level bedded rock (Construction (Health, Safety and Welfare) Regulations 1996 [1].
5. Can surplus material after backfilling be spread on site? Or does it have to be removed?
6. What pipes are specified? Do they have patent rapid connection joints? Are they laid on concrete beds and with concrete haunch?
7. Back filling requirements. Are special materials and methods called for?
8. If large diameter pipes are involved, what weights need to be handled?

help in back-filling, or some alternative item of plant brought in.

Trench support methods

Most drainage trenches on site are capable of support by proprietary systems – that is standard solutions – unless of considerable depth (Chapter 4.). The method planner must recognize that such systems, while totally desirable, depend for their economic use on rapid turnover. That is, kept in one trench location for as short a time as possible. For this to happen,

Figure 15.2 Failure to assess problems with running sand. Proprietary support totally wrong in this situation. Should have planned and priced for wellpointing. See Chapter 2.

Figure 15.3 Well pointing in action for the construction of an outfall culvert in waterlogged dune sand. Note header pipe and suction pipes with the pumping equipment. Excavation totally dry.

it is necessary for pipes to be capable of testing in short lengths, not only between manholes as used to be the case when pipes were jointed with yarn and cement mortar. Except for large diameter concrete pipes, where older jointing methods may be required, it is normal to-day for quick action jointing systems to be the norm.

Integrated activities

With the above points in mind the approach to planning drainage begins to become clear. Thus:

1. The excavator must not only dig but assist in other activities as well (Figure 15.4). As such, its ability to be used as crane is covered by legislation. The excavator must have a Certificate of Exemption No: CON(LO)/1981/2(General)[4] which exempts it from the requirements of Regulations 28(1), (2) and (5), 29 and 30 of the Lifting Operations and Lifting Equipment Regulations 1998[5]. The Certificate must be signed by a competent person and set out the conditions and limitations of use. Certificates of this kind apply to excavators, loaders and combined excavator/loaders. A copy of the certificate is given in reference[4].

When such a certificate has been issued, the trench excavator can be used in a number of additional roles: unload pipes, handle into place trench support equipment, lower pipes into trenches as well as manhole rings. The converse also applies – assisting backfilling, removing trench support systems, trench sheeting and regrading.

Drainage: key items in making method decisions

1. The critical item will be speed of pipe laying and testing. Performance data can be provided from published data or preferably from company records of previous performance. The latter is always the preferred source as the figures can cover overall performance – not just achieved output when delay free. (For method of recording such information on site, see Chapter 6.)
2. Selection of excavator. Digging depth and lifting capacity (if heavy large concrete pipes are to be handled) more likely to determine size and power needs, unless the excavation is in hard materials.
3. The choice of support system will relate to the ground conditions. Where trenches are deep (usually in excess of 6 m) proprietary methods may not be adequate. If in doubt, consult temporary works specialist. A solution designed for the circumstances may be needed. This is likely to slow down the sequence of events.
4. Make sure that the excavator is used on as many activities as it can perform, in addition to the digging. See Construction Industry Research and Information Association publication *Trenching practice* [2].
5. If pipes to be installed are long, check that the support system is such that the pipes can be lowered into the trench between the strutting spacing!
6. Some authorities may specify special backfilling procedures. Consider the implications on time scale. See National Water Council/British Gas Corporation *Model consultative procedure for pipeline construction involving deep excavation*[3].

Once the above are assessed and answers obtained, the sequence of events can be established. When performance factors are applied, individual and overall time scales

2. Given that the pipes in use have approved coupling methods, testing can take place over short lengths, allowing support methods to move ahead quickly and back-filling to proceed much earlier. Thus the excavator time in use can be more concentrated, quicker turn round of hired support equipment can be realized and back-filling follow on close behind.

Temporary access, storage areas, car parking and site facilities

The second major item in the external works programme at the contract commencement concerns the provision of hard access to all parts of the site, hard storage areas for materials, site facilities and car parking etc. At the tender stage, every effort should be made to obtain details of the permanent access specification and the setting out of all roads, paved areas, permanent car parking areas and the like. With this information to hand, it is usually possible to minimize the need for temporary roads on site by putting in the base courses of permanent roads and paved areas as soon as possible and using these in the temporary role.

For these ideas to function properly, it is essential that their initial installation is linked to the early completion of the main site surface water drainage. By putting in gullies as the drainage proceeds, the site temporary roads can be kept as dry as possible (Figure 15.5).

With paved areas, the same principles apply.

Figure 15.5 Use of permanent road sub-base for temporary construction access. Note gully grating in temporary low level position to allow effective drainage.

Figure 15.4 Trench excavator handling drag box support as well as pipes and digging. (Details of this method of support are given in Chapter 4).

In the structural area, especially if the foundations are piled, the base course of the building floor should be installed before piling starts. This provides a piling carpet for the piling contractor to work on.

If situations arise where temporary roads are unavoidable, they will have to be removed during the site clean up period. With all the above factors in mind, and decisions made on plant and method, performance in each item shown in the table of activity sequence (Table 15.2) can be assessed and a time scale for each established, followed by an overall programme with that covering the drainage element.

(See p. 302 for Key items in making method decisions

Public utility services

Public utility services also need planning for installation prior to the superstructure commencement. To achieve this desirable situation often calls for great perseverance on the part of site management. The utility companies are well known for a reluctance to get their mains or, at least the ducts or pipes needed, in position at an early stage. Yet the advantages overall are considerable. If all underground activities can be completed at the foundation stage, the site can be cleaned up and allow a clear run for the superstructure. Scaffolding can be erected knowing where excavations have previously taken place. No clash need take place between superstructure needs and getting services into the building. Trench excavation at a late stage also baulks free access for follow up trades.

In planning, it is necessary to establish who will lay mains or ducts. It is most likely, today,

Table 15.1   Activity sequence for drainage

Item	Description
1.	Remove top soil, if not already removed in a general strip for roads, paved areas and over building area. Store on site for later reuse. Note: outside the site boundary, remove topsoil and store suitably for reinstatement
2.	Excavate trench, progressively
3.	Provide support to trench if more than 1.2 m deep
4.	Provide pumping facilities if needed (check with soil report)
5.	Spoil excavated material on one side of trench, not nearer than 1.5 m to the edge of the excavation
6.	Bottom up excavation and lay concrete base for pipes – or prepare any other specified base
7.	Lay and joint pipes
8.	Provide concrete haunch to pipes or other specified method
9.	Test length of pipes – minimum allowed
10.	Backfill progressively as testing completed
11.	Construct manholes and pipe benching. Install manhole cover
12.	Backfill round manholes. Remove all surplus material off site
13.	Surface water drains – install road gullies and provide with temporary cover. Gratings should be installed prior to final road base course and running surface

Table 15.2   Activity sequence for first stage road and paved areas construction

Item	Description
1.	Soil strip and take to soil storage tip.
2.	Excavate to reduce levels – take off site/use for landscaping/use as filling elsewhere on site.
3.	Lay and compact base course material
4.	Excavate for and lay concrete to kerb race. If required insert short lengths of scrap reinforcement to link kerb backing to kerb race concrete, when kerbs are finally positioned.
5.	Install road drainage gullies and connect to main drainage.
6.	Temporarily install gulley gratings at top of base course level.

that the public utility companies will sub-let most of the work to subcontractors who specialize in installing pipes, ducts or cables for the body concerned. Each will have to be identified and time scales agreed.

Negotiations regarding requirements should begin at the tender stage with the utility companies, especially in relation to delivery times for pipes, ducts, cables. Checks need to be made on any long delivery times that may arise, for instance in relation to sub-stations, special valves etc. At these early discussions, the planning engineer needs to press the advantages of early installation, both to the public utility company and for the overall efficiency of the contract as a whole. Once agreement has been reached on each utility's programme, copies should be issued to all involved for agreement. The time scale can then be integrated into the main construction programme.

In the case of central redevelopment work, it must be established early if new utility work is going to be involved outside the site. Permissions may be needed to open up the highway for new works, while the traffic branch of the Police will want a say in how much can be done at a time, to minimize traffic disruption.

The above matters form the group that needs to be completed as soon after contract commencement as possible.

Access: key items in making method decisions

1. To what extent will temporary roads and hard areas be needed? If the outfall for drainage is outside the site, temporary access will be needed and have to be removed and reinstated on completion. The access provision will need to cope with lorries delivering pipes and possibly ready mixed concrete vehicles.
2. If the ground is bad, will the use of geotextiles help to minimize the thickness of hardcore required? And also make the removal of temporary areas much easier? Get expert advice.
3. Where using the base course of permanent roads for temporary purposes, much argument can arise on the merits or otherwise of installing the permanent kerbs at the same time. Those in favour say it is necessary to do so to stop the base course material spreading outwards and causing unnecessary cost of material replacement when the final making up to levels is carried out. Those against rightly point out that the base course can easily be retained by using the kerb race concrete alone. If damaged by traffic running over it, it is much cheaper and quicker to bring up to standard than to remove, re-level and re-align the kerb stones. In most cases experience shows that the latter view is the correct one. With the degree of mechanization on the site today, it is almost impossible to control plant from crossing kerb lines elsewhere than nominated crossings.
4. The use of base courses as temporary roads and hard areas needs to be integrated with the drainage installation so that surface water run off can be achieved as soon as possible.
5. What method of excavation will be used? Soil strip will need a tip area on site for regrading purposes at a later stage. Is the rest of the excavation to go off site, be used as filling elsewhere or retained on site for landscaping purposes? Decisions on final use will affect the type of excavation plant used as well as the unit cost of the excavation. See Chapter 7.
6. Has a proper check been made conversion factors for filling materials where the material is bought by the tonne, but measured for payment by the cubic metre? See potential for waste in Chapter 16 in this situation.
7. Is compaction measurement method specified? What plant will be needed to achieve specification requirements?
8. If the kerb race only is to be positioned initially, it is desirable to insert short lengths of vertical reinforcement to stop kerb backing concrete moving when making up to

Towards the end of the contract

The remaining external works inevitably come into the second period – at or near the end of the contract.

Minor external buildings

Final details for such buildings usually come late in the day. Foundations details may have to wait until the equipment to be accommodated has been designed. For example: bases for transformers or supporting external plant and equipment. Installation of such plant may need to take place prior to completing a structure around it. Such specialist plant is often on long delivery and the planning must recognize that (a) orders have to be placed as early as possible, either by the utility company or whoever is going to install it and (b) failure in getting the equipment on time may result in the inability to commission the building as a whole.

Where domestic building is involved, the tenants' or owners' stores and garages will be the main external structures and will be on a one for one basis. An example, in relation to council housing is shown in Figure 15.6. Erection is dependent on completion of back garden regarding, with fencing as the final item.

Within the range of external buildings, sub-stations and transformers are the most important. They may be needed to provide temporary lighting, testing of electrical equipment as installed or for operating control gear for special industrial plant and equipment.

Figure 15.6 External small buildings, paths and fencing at rear of housing refurbishment (see also Chapter 21).

When planning the construction of minor external buildings, a number of factors have to be taken into account.

1. Small structures can create fragmentation of effort and increase cost.
2. The planning must try to create continuity of work by creating a trade team that can deal with a number of such buildings at the same time.
3. Minor works encourage waste. Accurate quantity take-off per building is essential so that the association of a number of small buildings can create full delivery loads of, for example, ready mixed concrete.
4. Excavation of foundations and disposal of surplus material should be linked to regrading needs.
5. What services are required in external structures? Includes drainage; water – toilets, fire precautions, drinking; electricity; gas; telephones; television monitoring and so on.
6. Are all details available in good time?

It should be clear from the above list that much more care is needed in the planning of external buildings than might be first thought, if costs are to be kept to a minimum at a time in the contract when a good deal of effort is needed on expensive clearing up. At the tender stage, it is likely that little information is available for such buildings. The onus for a satisfactory conclusion in financial terms will, therefore, rest with the site management team. Pressure has to be applied to obtain details from the client at the earliest possible date. Once to hand, an assessment of cost against what was allowed in the tender can be made, based on the available information, if any, at the time of tender.

Landscaping

On new ground sites, as distinct from redevelopment of existing sites, environmental considerations have assumed considerable importance. The landscaping needs will depend on the local situation, but a number of matters need to be kept in mind by the main contractor's planning and method staff.

1. The work of regrading and planting will be carried out by specialist landscape gardeners. Or, if not the regrading, certainly the planting.
2. The extent of such works will vary enormously depending on the local conditions.
3. In the planning overall, due regard has to be paid to the most suitable time of the year for planting trees and shrubs.
4. In addition to planting, much use is now being made of water as a feature. Water

recycling and aeration plant may well be involved.
5. It is very unlikely that full details will be available until late in the contract.

The way in which the demand for landscaping has grown is illustrated in Figures 15.1 and 15.8. This very extensive contour changing and large area planting is on what used to be old gravel workings that had been filled with domestic waste. Huge quantities of material have been moved and new material brought in to provide the conditions for high technology business units and the creation of a surrounding country environment. By contrast, Figure 15.7 illustrates small

Figure 15.7 External works, road realignment and tree planting at front of refurbished local authority housing.

Figure 15.8 Roads and lighting needs for large scale business park development.

external works and tree planting on a housing refurbishing contract.

It is often the case that schemes of this nature are let as separate contracts. As such they are not required to be complete when buildings are handed over.

Fencing and lighting

As any building progresses, security becomes increasingly important as services and fittings are installed. To assist in providing such security, it is desirable to create the final security fencing as soon as possible. In so doing, money can be saved on having to provide temporary fencing over long periods. Coupled with the permanent fencing should be perimeter lighting and surveillance equipment. In turn, for this to be effective, the security gate house needs to become available at the same time.

The security needs, early on, point another reason why the installation of power supplies should be pressed forward so that security equipment can function.

Final completion of roads and paved areas

Once the dirty stage of construction has been completed – work below ground in particular – every effort needs to be made to complete the first stage roads and paved areas, etc. The activity sequence is shown in Table 15.3. Figure 15.8 is a good example of this stage in road completion. Note also (Figure 15.9), that the work is being pushed forward even though the steel

Table 15.3 Activity sequence for final stage road completion

Item	Description
1.	Install kerbs and backing concrete
2.	Re-locate gulley gratings to final road wearing course level
3.	Make up sub-base (used as temporary road) to final levels and compact
4.	Lay tarmac base coat (or concrete to pavement if no tarmac (rigid construction))
5.	Lay wearing coat of tarmac (or ditto if concrete is to have topping)
6.	Clean out all gullies and check functioning properly

Figure 15.9 Road completion and regrading adjacent to new warehouse structure.

frame structure in the background has still to be roofed and clad.

Generally

Good planning and site control requires that the external works and site clearing up should be going on while the main earning activities proceed. They should not be left to the last minute when other work is coming to an end. To do so often means a last minute scramble in an uneconomic way.

Finally, remember the most important adage in construction:

'Clean up as you go – a clean site is a profitable site'

References

1. *Construction (Health, Safety and Welfare) Regulations 1996.* HMSO, London.
2. Irvine, D.J. and Smith, R.J.H. (1983) *Trenching practice.* Construction Industry Research and Information Association Report No. 97. Updated 1992, reprinted 1994. CIRIA, London.
3. National Water Council: British Gas. (1983) *Model consultative procedure for pipeline construction involving deep excavation.* NWC & BGC 1983.
4. Certificate of Exemption No. CON(LO)/1981/82 (General) Excavators, Loaders and combined Excavator/Loaders. Health and Safety Executive, London.
5. *The Lifting Operations and Lifting Equipment Regulations 1998.*

Chapter sixteen

Waste in the construction process

Some degree of waste of materials is inevitable in the construction process. All estimators allow wastage factors in pricing a bill of quantities. Experience shows, however, that unless site management control is tight, wastage can frequently exceed, often by a large margin, the figure allowed in the tender. With most of the building trade's work sub-let, today, the control of materials is made much more difficult, except where bulk materials buying is still done by the main contractor. Nevertheless, main contractor or subcontractor, financial benefit can be obtained by establishing control procedures to avoid excessive waste. At a time of financial difficulty when competition is fierce, any savings that can be achieved must be desirable.

This aspect is not the whole story though. Designers of buildings also need to appreciate that they, too, contribute to materials waste. In this case, the waste is avoidable through design and detailing knowing how such waste can arise. Quality Assurance is increasingly wanted or specified by the client and design or other waste can be a contributor to the failure of quality.

Waste of materials

The main reasons for the waste of materials in construction have been defined in Building Research Establishment Digest No 247[1]. Materials waste divides into four distinct categories: design waste; taking-off and ordering waste; supply waste and, finally, contract waste.

Design waste

1. Failure to relate all built in items to a common module. Excessive cutting to fit results.

---

1. It has been recorded on a contract that design waste of 14% was built into the brickwork before the job started.
2. Poor design on a housing estate in relation to brick modules changed a cutting waste allowance of 7½% into one of 25%!
3. In facing brickwork, the specification to colour match all bricks created an additional waste factor of 6–7% over all other allowances.

---

2. Sequence of assembly made over-complicated by the detailing specified.
3. Failure to recognize undesirable consequences of a design decision. Figure 16.1 shows a problem with pre-cast cladding detailing.

Take-off and specification waste

In this case, materials suppliers and the contractor or subcontractor involved both contribute. Such situations usually start with incorrect taking-off of materials. If excessive material is delivered, it is an open invitation to waste. If too little, waste can be created by delay to the contract.

> Drainage fittings often get little accurate take-off. Often ordered as and when required. If the site are aware of this, there is no incentive to worry about losses and breakages. Just order a few more!

The buyer has a part to play as well. Is he sending a copy order to site which clearly informs the site staff of the conditions under which the order has been placed?

> In one case, a site were cheerfully cutting surplus lengths off floor joists in the belief that they were only paying for the proper length plus a small tolerance. In fact they were paying for the lengths as delivered.

Delivery waste

Here waste comes about from a number of situations.

1. Supplier given an incorrect specification – failure to notify amendments.
2. Delivery of damaged items – poor packing causing damage or breakage en route.
3. Where delivery is self-unloaded, careless handling into storage.

> In 1979, the National Federation of Building Trades Employers, now the Construction Confederation, instituted a survey among 32 member companies, following complaints that many facing bricks delivered were unsuitable for facing work. Out of 253 000 bricks counted, 27 780 were unsuitable for facing work, or 10.9% of those delivered. The highest rejection rate was 47% and the lowest 0.2%.

Site waste

Site waste is squarely in the hands of the main contractor and his subcontractors. Such waste is usually in the following categories:

1. Poor storage and control; handling bad, leading to loss by spoiling. Also theft and vandalism.
2. Faulty conversion ratios for compaction of bulk items – filling etc.
3. Careless use – un-necessary cutting instead of looking for a suitable available length. Particularly true of timber.
4. Failure to protect items when installed.
5. Working to inaccurate dimensions causing over excavation, excess use of concrete, for example.

Poor storage and control

Materials at the mercy of the weather. No assessment of the need for covered storage in some items or secure stores for others.

> 'Whilst a lost or wasted brick might only cost about 10p (1982), at least £5 worth of work may have to be done to earn enough profit to pay for that brick'.[2]
> Perhaps a more telling comparison to members of the construction industry was made in a CIOB publication by Good [3] to the effect that 'a single brick can cost the same as a pint of beer'

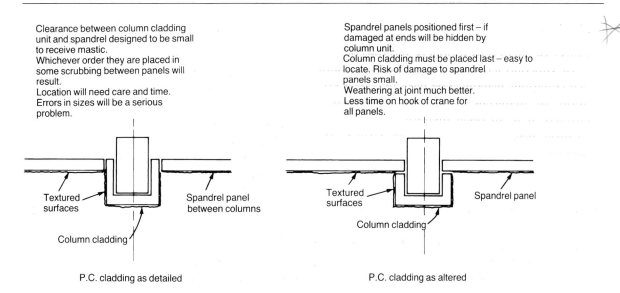

Clearance between column cladding
unit and spandrel designed to be small
to receive mastic.
Whichever order they are placed in
some scrubbing between panels will
result.
Location will need care and time.
Errors in sizes will be a serious
problem.

Spandrel panels positioned first – if
damaged at ends will be hidden by
column unit.
Column cladding must be placed last – easy to
locate. Risk of damage to spandrel
panels small.
Weathering at joint much better.
Less time on hook of crane for
all panels.

Textured
surfaces

Spandrel panel
between columns

Column cladding

Textured
surfaces

Spandrel panel

Column cladding

P.C. cladding as detailed

P.C. cladding as altered

Figure 16.1   Pre-cast concrete cladding as originally detailed and as finally agreed. Weathering detail improved and risk of damage during installation greatly minimized.

All building materials, today, are a magnet to the unscrupulous DIY fan. Proper precautions need to be taken to protect attractive materials in this field.

Site storage of materials

When planning the space required for the storage of materials, a balance has to be struck between storage cost and avoiding loss and damage. The two main considerations are their vulnerability to theft and how important it is to avoid contamination and damage. As a start, materials on site need to be divided into a series of categories as follows:

1. Heavy and awkward items. These include kerbs, paving slabs, drainage goods, gulley pots and manhole rings. Their weight makes them difficult to remove in quantity, which provides reasonable security. All that is needed is a hard level standing for offloading and stacking. As self-unloading is usual these days, such hard areas should provide room to accommodate delivery vehicles as well. Contamination by mud is not important.

2. Bulk items. Bricks, blocks, lintels, reinforcement and roof tiles, for example, all come in packaged form and are unloaded as such. Hard areas are again needed, but providing a clean, well drained surface. Many of these items are a target for DIY scroungers, but not easy to move in bulk. Brick and block packages should not be broken open until needed for use. Contamination from dirt has to be avoided. Reinforcement is best further supported on timbers allowing good drainage and easy drying.

3. Items not easily removed. Desirably covered against rain. Floor joists, roof trusses and carcasing timbers are examples. Storage should be on baulks of timber to keep the materials off the ground. Roof trusses need to be stored in the vertical position to avoid incorrect stressing. Tarpaulin covering is needed.

4. Attractive items. Second fix timber, doors and frames, chipboard and plasterboard, cast iron

goods and manhole covers, scaffolding and window frames need a secure compound. The plasterboard and chipboard should be kept in a hut within the compound and capable of fork lift handling. Doors and frames need dry storage within the compound.

5. High risk items. These include ironmongery, kitchen units, sanitary ware, electrical, heating and plumbing goods, and all require high security store within the compound. Steel shipping containers provide the ideal solution. They can be purchased second hand or hired and are now widely used.

The above examples give a good idea of the treatment for a wide range of materials. Different contracts may need different solutions to suit conditions. The costs of vandalism and theft should not be underestimated as replacement may take time and cause delay to the contract as a whole at considerable expense.

Faulty conversion ratios

Where bulk materials are involved, it is usually necessary to arrive at conversion ratios where material is bought by the tonne yet will be measured and paid for in m³.

> The original estimate for the conversion rate of stone filling from tonnes to m³, on an industrial contract, was underestimated. As a result, the contract lost 0.8 tonnes per m³ of stone every time that stone was used. Hence a loss to the contract – in this case £1442 per month at 1979 rates.

Careless use

Where the carpenter needs a piece of timber – skirting, architrave or the like, it seems human nature looks for a piece that is well in excess of the length needed and cuts to waste. It is too much trouble to search for a length much nearer to that required. How this attitude can escalate, in relation to tile battens, is shown below.

> A contracts manager for a company building houses was checking materials ordered against measured on one of his housing sites. To his horror, he found that 45% more tile battening had been ordered than the bill measurement. Investigation showed that no attempt was made to use verge offcuts, however long. More surprisingly, he discovered that the plasters found tile battens made excellent trammel rods, the battens broke across the knee easily for starting fires in the cold weather and, finally, made excellent slatted shelves in all the subcontractors' site offices.

Even when materials have been finally installed, waste can easily come about.

> In a military barracks, wash basins were installed, plumbing complete, tile surround in place and encasement to plumbing installed and painted. Someone drops a hammer and shatters a basin. In this case, waste is not just the cost of a new basin. To replace, tile surround has to be removed, plumbing disconnected and encasement to plumbing removed. Delay occurs while delivery of new basin awaited. Once the new one is installed, everything else has to be re-assembled. The cost is effectively doubled – all for the sake of no hardboard protection over the basin, when trades still have to work overhead, being provided. Bad planning!

Working to incorrect dimensions

Wastage in this area usually arises in work involving bulk materials. Examples are: excavation, filling and concrete. All occur in the ground works stage. Excavation in excess of the dimensions specified not only means more excavation to be removed and possibly back filled, but in foundations and floors also involves more concrete.

A concrete reconciliation carried out on completion of a factory floor showed a waste of 14%. Investigation showed that this was entirely due to a poorly prepared sub-base, the underside of the concrete being too low.

Over dig on bases for a factory, when checked, was found to vary between 0% and 71%. The average came to 38%. As a result, the waste of concrete in these foundations came to 36%.

Waste of manpower

Much of the published work on waste confines itself to materials. Examinations seem to stop when materials reach the point of final positioning. Yet this activity is as much materials handling as anything leading up to it.

Once the desired rate of progress for any activity has been established, all contributing activities have to be geared to achieving and maintaining the necessary flow of materials. This, however, is only one part of the planning process necessary at this stage. The appropriate method of handling, together with the correct labour force, needs to be established on the basis of maximum output at least cost. That is, the establishment of a balanced chain of events. For this to be achieved, two items not yet considered have to be examined.

Efficient place of work

If the final assembly is to be as efficient as possible, the work place must be both safe and suitable for the activity in question. Points are: adequate space to manoeuvre and temporarily stack the material being used, together with the working platform being arranged at a level best suited to minimize effort on the part of the operatives. Finally, compliance is necessary with the appropriate Construction Regulations which relate to the operations in question.

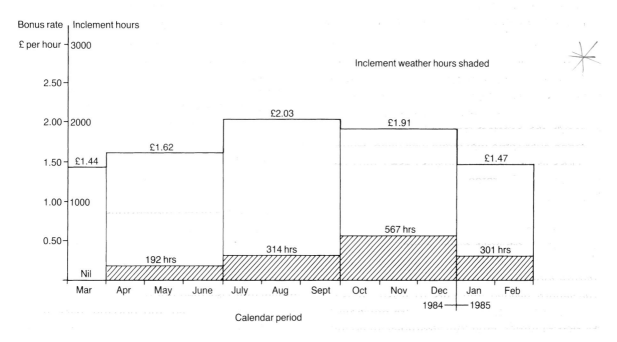

Figure 16.2   Histogram (1 of 4). Influence of rain, wind and cold on bonus earnings.

Influence of the weather

The weather plays a significant part in the efficiency of the labour force. The winter of 1963 was the worst in living memory in terms of cold – snow and frost were almost continuous for three months. The disruption to construction was great at a time of vigorous construction activity. The Government of the day brought over from Canada an expert on such conditions, to advise on what action should be taken to prevent such a disastrous stoppage in the future. As a result, much money was spent on the concept of covering the work, or providing covered areas into which operatives could be transferred if the weather was bad.

In the event, these ideas were an expensive failure. While winters like 1963 occur every year in Canada, they do not occur very often in the UK. Operatives disliked working in heated enclosures which had high humidity creating oppressive atmospheres. The enclosures were

costly and did not pay for themselves in normal winters. As a result, the whole concept died a natural death.

This is, in many ways, an object lesson in method planning. The views of an expert, who had no real experience of the British climate, were taken rather than examine the situation in depth. If a group of experienced construction site managers had been asked to consider their view of what was most likely to upset site activities, it is very doubtful if frost and snow would have been given as the prime cause in the south of the country.

With hindsight, too much attention was given to frost and snow. If a careful study is made of the main causes of lost operative production, it is only too clear that wind and rain are the major causes. And at all times of the year. To confirm this view, studies were made on the level of bonus earnings over a whole year. Four sites, in the North of England, agreed to co-operate.

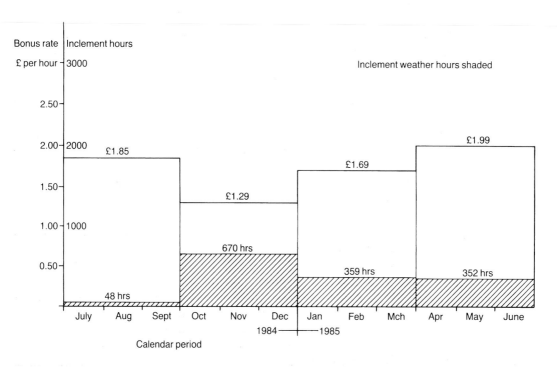

Figure 16.3 Histogram (2 of 4). Influence of rain, wind and cold on bonus earnings.

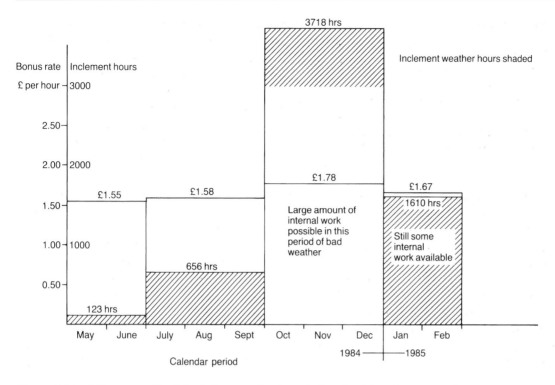

Figure 16.4 Histogram (3 of 4). Influence of rain, wind and cold on bonus earnings.

The sites in question still had the main contractor directly employing the traditional building trades, so records were readily available.

In the histograms in Figures 16.2–5, the activities covered in calculating bonus earnings were: drain layers, concreters, external works activities, together with bricklayers, carpenters, decorators and scaffolders. Three of the sites were housing, while the fourth was a sports centre. All the figures are based on plotting bonus earnings (at the rate per hour), on a quarterly basis against the months of the year. Also shown are the hours paid as inclement weather time, together with any notes of significant relevant matters that affected the bonus figure.

Figure 16.2 illustrates the more usual situation: highest bonus earnings in the summer months and least in the winter. Further inspection shows that the warmer the period of the year, the less effect inclement weather has on bonus earnings. October/November/December have more inclement weather hours, but bonus earnings are greater than in January/February/March, even though the inclement weather time paid was greater.

Figure 16.3 is the same sort of situation as Figure 16.2, but overlapping in the calendar period.

Figure 16.4, while taken over the same calendar period as Figure 16.3, has had a much wetter location. The graph clearly shows the value of available internal work, when the weather was at its worst.

Figure 16.5 illustrates the effect of all the inclement weather factors. Rain, high winds and the residual effects of wet weather. At the same time, work available under cover maintains a bonus level in the latter months of the year. Overall, a depressed level of earnings occurs over the whole year.

In more statistical terms, the figures used above can also be used to assess operative performance in the Work Study Rating sense. (Table 16.1).

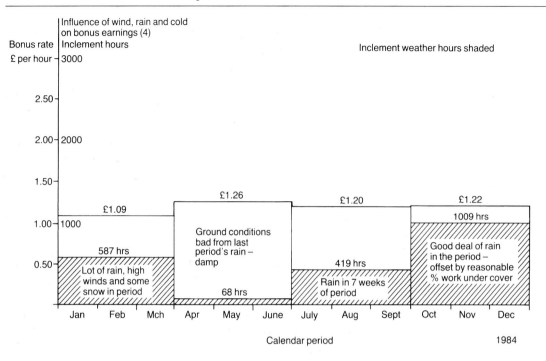

Figure 16.5 Histogram (4 of 4). Influence of rain, wind and cold on bonus earnings.

The figures show:

i) Oct–Mar performance averages 86. An increase of 33% on target rates would be needed to achieve the 114 performance in the Apr–Sept period.

ii) The norm set of £1.70 was achieved over the whole year.

The same figures can be re-examined on a more seasonal basis:

Winter months Nov–Feb (incl) – £1.20
= 71 performance

Summer months Mar–Oct (incl) – £1.88
= 111 performance

This comparison indicates a more pronounced effect of inclement weather on productivity. In this case, targets in the winter months would require an allowance of 56% for conditions.

If any further reason is needed for a more thorough examination of means to offset the effects of wind and rain, it is interesting that the Building EDC Report *Faster building for industry*[4], in Table 14.17, p 47, lists causes of delay

Table 16.1 Assessment of operative performance from Figures 16.2 to 16.5

Period	Hours worked	Bonus paid	Bonus rate £ per hour	Perf. rating norm. £1.70/hr
April–June	11348	22632	1.99	117
July–Sept	7755	14262	1.84	108
Oct–Dec	9112	11725	1.29	76
Jan–March	7097	11995	1.69	99
Totals	35312	60614	1.72	100

on construction sites. Of the 56 sites studied in depth, 27% of case studies listed bad weather as a main cause of delay, while 25% of case studies listed late materials delivery.

Combatting the weather

While some areas of construction activity will remain at the mercy of the weather, unless the client specifies full protection to be provided and pays for it, or totally uneconomic protection measures are taken, much more consideration needs to be given to areas where remedies are possible with minimum cost. Housing and building work lend themselves to the provision of cost effective protection against wind and rain. What is so surprising is that this area has not had more attention in the past.

Research into making the workplace more pleasant, in rain and windy conditions, shows that the construction industry would do well to look at their counterparts in agriculture and horticulture. Their approach to the protection of livestock and tender plants provides lessons that the construction industry would do well to study[5].

The use of plastic mesh drape netting is now increasingly common on buildings. It is called for to stop debris falling from buildings to the street below – hence its common name: debris netting. In fact such netting can reduce wind velocities as much as 50%. When draped on the scaffold, significant reduction in wind force can be achieved. This effect is well known in agriculture and horticulture, as is the ability of this same netting to keep frost from plants some distance below. What is really surprising is that no research has been done on the ability of the mesh to stop significant penetration of rain when hanging vertically. Equally, how would it work overhead? Or would a plastic sheet be better?

When one considers the drop in human productivity illustrated above (human waste), a research project in this field would pay handsome dividends.

The role of planning in preventing waste

Those who are involved in the method planning role have an important part to play in avoiding waste of all kinds. While the management of materials during construction rests with site managers, important contributions can be made at the tender stage. The matters on which the planning engineer needs to concentrate are listed below and each is then examined.

Design and detailing

The method planner is well placed at the tender stage to examine the drawings provided, and any others seen at the architect's office, as this will be a necessary preliminary to the tender planning. In so doing, the experienced planner would also be looking for any sign of bad detailing, problems that could arise in erection, and detailing which does not fully consider the implications of differing modular dimensions – hence creating excessive cutting of materials. Detailing may also make erection, in respect of connections and joints, very difficult to carry out efficiently. In an age of growing Quality Assurance requirement in construction, such matters become very important – especially if the detailing deals with protection against water penetration. Any such matters found should be discussed with the architect and engineer, so that problems can be settled before construction commences.

Specifications

Specifications can drastically affect planning the work and its cost. Many are carried along from year to year and do not get updated in line with the latest research and methods. In turn, they can drastically affect the method planning. For example, specifications should allow full height pouring of walls and columns, provided certain requirements are met. Failure to do so will increase time and cost.

Key areas for planning attention in avoiding construction waste

1. Examination of all available working drawings for erection difficulties, which would lengthen construction time and increase cost.
2. Does the design and detailing require excessive cutting?
3. Are specifications in relation to materials clear.
4. Are the methods of handling materials suitable to the materials involved? Has a cost comparison with other alternatives been carried out. (The obvious does not always turn out to be the best).
5. Have you planned a suitable scaffold for the operation in question?
6. Have you considered the time of year and means of providing weather protection against wind and rain to maintain output as high as possible.
7. The disposal of waste can be expensive. Have you considered the possibility of being able to bury waste on site? Or at least such items as old concrete, broken brick and block. Can timber be burnt on site?
8. Is valuable scrap likely to be generated? Planning should allow separation of such scrap with a view to selling to a suitable dealer. Where demolition is involved, considerable quantities of structural steel or broken out reinforcement may arise, of considerable value as scrap metal.

In a more trades-oriented sense, the toilet areas of a block of offices had a tiled finish. The planning engineer concerned discovered that the tiling subcontractor would have to come back four times after the first visit as the detailing meant that items had to be installed before more tiles could be fixed. Discussion with the Architect enabled the tiling to be carried out in one operation – with savings to the client and a faster programme for the contractor.

Handling of materials

It has already been seen that construction is primarily to do with the handling of materials. The bulk of materials delivered today come as packaged in one form or another. Many are offloaded by the delivery vehicle. To avoid damage, good level storage areas are needed and the offloading adequately supervised.

Experience has shown that packaged materials are far less liable to damage than those delivered loose. In planning the handling on site, it follows that the package should be kept intact as long as possible – certainly to the point of reaching a loading out tower near the point of final installation. In planning the handling method on site, therefore, it is necessary to consider the most suitable equipment for the job, and if necessary, cost it out against any alternatives. **The obvious is not always the best.** The following case study makes this point only too clearly.

Case study

The site comprised 170 dwellings, 84 No. 3-storey flats and 86 No. 4-storey maisonettes, plus roads and sewers and external works. The site was generally level. The layout is shown in Figure 16.6.

A previous, similar, site had used hoists to distribute materials into individual dwellings. With the number of blocks involved, the movement of hoists from block to block on this contract would require 21 erect and dismantle operations. Because of this, it was decided to compare the possibility of using fork lifts against the cost of the hoists – both to see which

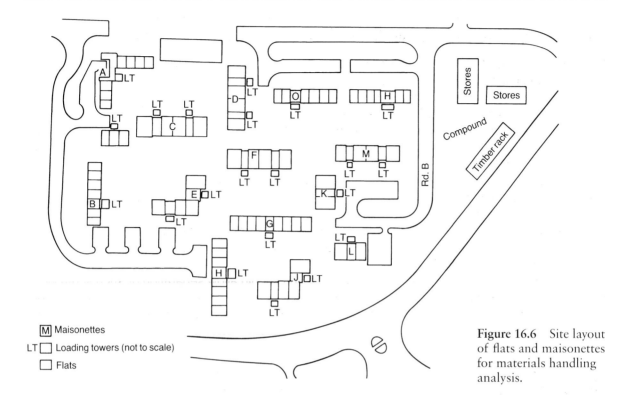

Figure 16.6 Site layout of flats and maisonettes for materials handling analysis.

was the most cost effective as well as improving handling efficiency. To reach the second floor of the flats and the ground floor of the upper maisonette, high reach fork lifts were needed.

The resultant cost comparison is given in Example 16.1.

To every one's amazement, the use of the high reach fork lift turned out to be £6476.95 cheaper. Or, in more statistical terms, a saving of 38.75%.

For the completeness of this study, it must be recorded that no plant feeding the hoists has been included in the cost comparison. The reason is that, on this site, the workload called for one fork lift with the hoists – dealing with unloading and storage. A standard machine would still be needed, even with the high reach model.

Having decided to use the high reach fork lift, it was then necessary to consider the structural format to achieve greatest efficiency in use. Figures 16.7 and 16.8 illustrate this aspect of the planning need.

In the case of the flats, it is necessary to consider how materials will be handled to individual flats once delivered by the fork lift to the required level – either direct to a floor or onto a loading out tower adjacent. It is clear that the best approach is that shown on the left of Figure 16.7. Distribution around a stair well is not at all happy.

The maisonettes present a quite different configuration. An access balcony exists at second floor level, and by locating a loading-out tower over the entrance, distribution to the upper maisonettes is readily achieved without the need for any scaffolding. Because the units are maisonettes, access to the fourth floor is achieved by means of the internal staircase. If the maisonettes were replaced by four-storey flats, the method would not work as the high reach would not cope with the fourth level.

Example 16.1 Cost comparison of platform hoist with high reach fork lift truck. All figures relate to 1976

Platform Hoist

Erect and dismantle complete including safety gates

(Per S.G.B. quotation)	=	£112.00 fixed sum
Hire of Hoist (Wimpey Ant Queen)		7.00
Fuel cost @ ¾ gal/hr for 40 hr week @ 23p	=	6.90
Allowance for maintenance @ 25% hire	=	1.75
Lubrication at 10% fuel cost	=	0.70

Total Site Requirements

21 No. Hoist erect and dismantle @ £112	=	£2,352.00	
21 × 13 = 273 weeks hoist hire @ £8.75 inc. maintenance	=	2,388.75	
Assuming hoist in use 50% of 40 hr week			
Driver cost = 273 × £80 × 50%	=	10,920.00	
Fuel cost = 273 × £7.60 × 50%	=	1,037.40	£16,826.50

High Reach Forklift Truck

Hire of machine	=	£73.50 p.w.
Add 25% for maintenance	=	18.40
		91.90
Fuel and lub cost	=	15.20 = 1½ gal/hr @ 23p gal + 10% lub
		107.10
Driver @ £80 per week	=	80.00
	£187.10 for 42 weeks = £7,858.20	

Cost of loading out towers

Maisonettes 7 No. @ £120	=	£ 840.00
Flats 14 No. @ £75	=	£1,050.00
Additional Saga scaffold hoists required for dashing		
2 No. used @ £2.50 per week for 42 weeks	=	£ 210.00
Fuel for above = ¼ gal petrol @ 65p per gal per hour per hoist		
@ say, 50% usage	=	£ 273.00 £10,231.20

Saving on F.L.T. over Hoist	=	£6,595,30

Scaffolding needs

Where scaffolding is necessary, one needs to be sure that it is not only safe (complying with Construction (Health, Safety and Welfare) Regulations 1996 [6]) but provides a suitable working environment for the operation to be carried out – adequate manoeuvring space for the materials involved, adequate access and safety in stacking, where this arises.

Weather protection

Bearing in mind what has been said previously, that most delays arise from wind and rain, can any economic measures be taken to ensure that work can continue in such conditions?

Disposal of waste

The amount of waste material that builds up on a site is usually far greater than one would imag-

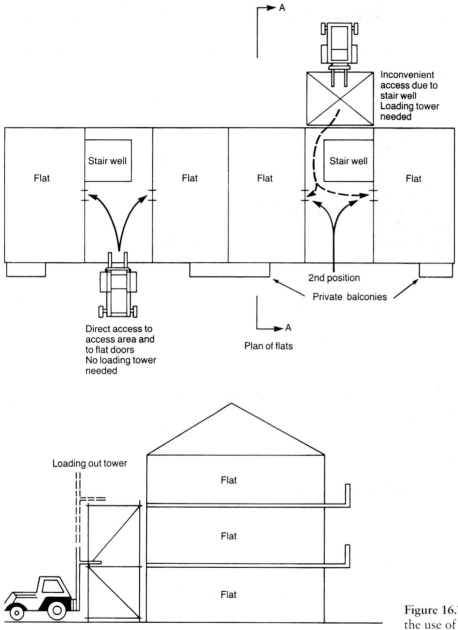

Figure 16.7 Handling layout for the use of fork lift, showing temporary works, for flats.

ine. It all costs money to dispose of. Table 16.2 illustrates this fact very forcibly. In clearing site at the end of a town centre redevelopment, 10 No. rubbish skips were needed each day. Examination of the skip contents shows that a high percentage of waste material was demolished brick and blockwork, broken concrete and waste mortar. All this waste is being taken off site at some cost by a waste contractor. Why? Could not the materials be buried on site to

319

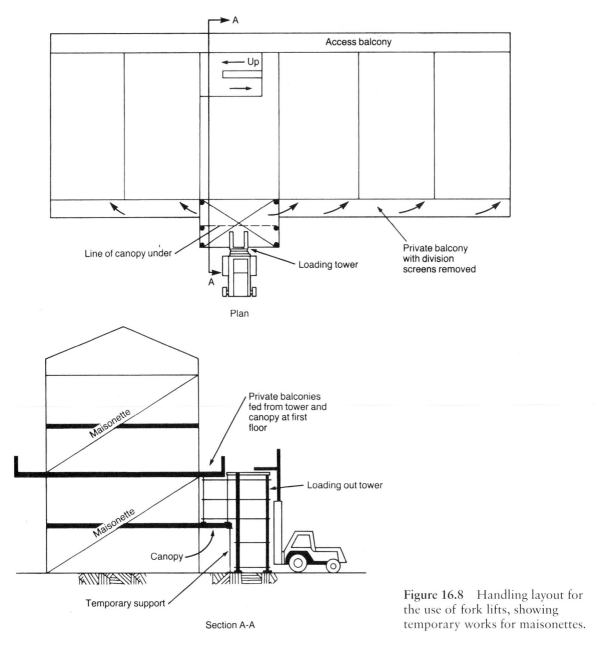

Figure 16.8 Handling layout for the use of fork lifts, showing temporary works for maisonettes.

make up levels? Good planning would have considered this aspect early on in the contract. The only real thought given was in relation to scrap reinforcement. Here a designated skip was placed in the steel bending yard for scrap. The contents were sent to a scrap merchant and the value credited to the contract.

Burning of materials

In respect of timber, plywood, wrappings and such like, the simplest way of disposal, without taking off the site, would be by burning. If one wants to consider this option, liaison with the local authority is needed. They may not co-

Table 16.2 Contents of ten waste skips on day one of study

Skip No	Contents
1.	Quarter full. Mainly demolished brickwork and broken bricks.
2.	Half full. Nearly all plywood rips 75–150 mm wide and max. one metre long. Small pieces of timber 100 × 50 and 75 × 50. Small quantity of brick mortar.
3.	Quarter full. Polythene sheeting, a few bricks, polyurethane boarding in broken sheets and broken out concrete.
4.	Almost full. Blockwork from demolition. Large cardboard boxes full of packing material. Ex insulation subcontractor.
5.	Half full. About equal amounts of timber and concrete. Ply rips as above, short lengths of 100 × 50. Concrete end-of-pour lumps, wheelbarrow size. Some plaster, waste paper and woodwool.
6.	Full. Broken out concrete, cardboard boxes, smashed wheelbarrow, damaged shutter panel, broken asphalt and polyurethane sheets, mortar and small pieces of timber.
7.	Half full. All timber, mainly badly damaged plywood sheets which appeared to have been used as a road. A broken ladder, split scaffold boards and short lengths of 100 × 50 riddled with nails.
8.	Almost full. Concrete and mortar lumps. Small amount of brick rubble and small pieces of timber and plywood – all less than 400 mm long or 400 mm square.
9.	Half full of rubbish similar to skip 8. But also contained broken fibreglass shutter and small amount of reinforcing rod less than 1.5 m long.
10.	In steel bending yard. Half full of reinforcement in short lengths less than 800 mm long. Mainly bent off-cuts from reclaimed steel due to schedule changes. Sent to scrap merchant. Value credited to contract.

operate in a liberal interpretation of the Clean Air Act 1968[7] and the Clean Air (Emission of Dark Smoke) (Exemption) Regulations 1969[8]. The latter exempts 'timber and any other waste matter (other than natural or synthetic rubber or flock or feathers) which results from the demolition of a building or clearance of a site in connection with any building operation or work of civil engineering construction, (within the meaning of section 176 of the Factories Act 1961)' [9]. The exemption is subject to three conditions being met. Of these, only the first is a problem. This states that the exemption only applies if 'there is no other reasonably safe and practical method of disposing of the matter'.

On the face of it, that would seem to be the end of it. In practice, many local authorities seem to allow burning of timber and other waste, provided that dark smoke does not arise, nor ash material rise into the air. In fact, it is not too difficult to devise a crude furnace to achieve such ends.

Communication and control

If materials handling and control are to mean anything, tight control needs to be exercised between the parties involved. This is especially true where the main contractor does the procurement of materials for labour-only subcontractors. Inadequate supervision and control means that the main contractor does not know where the materials are used or how efficiently – yet he is paying the bill. The subcontractor has no incentive to avoid waste, or protect materials, as he does not have to pay for them. The importance of close control will become much more important when or where clients insist on the main contractor operating a Quality Assurance scheme as part of the contract requirements.

Early meetings with subcontractors are needed to discuss responsibility for handling materials. To avoid cost, some main contractors put the onus for handling materials on the subcontractor involved. Site management needs to look closely at this situation as, in the end, a

321

totally uneconomic result can develop. More plant than is needed gets onto the site and a shared handling solution could save money overall. Unless resolved, it will be the main contractor who foots the bill in higher prices from the subcontractors.

Specialist service contractors

The comments above apply to subcontractors in the traditional trades field. The situation is different with specialist service contractors. Many now operate sophisticated control systems, usually computerized, to make sure that their own profit margins are not being eroded by waste of expensive materials and equipment. None the less, clear agreements are needed in relation to the main contractor's responsibility for offloading and storage of high value materials and possible onward handling.

Conclusion

To conclude this Chapter, the whole situation is summed up by Example 16.2. It is self explanatory.

The whole materials control and management aspect, rather than the planning function, was originally covered in three reports published jointly by the Construction Industry Research and Information Association with the Chartered Institute of Building[10][11][12]. These are still available. More recently, CIRIA has produced a new publication *Waste minimization in construction – site guide*[13].

Example 16.2 Replacement material needed due to waste

The following replacement materials were needed on a 50 house site built in 1980–1981. Assuming that the profit on the contract was 5%, the above losses, totalling £4878 at that time, would represent an additional 0.5% profit if they could have been avoided.

9.098 No. Facing Bricks.
  226 m² 100 mm Concrete blocks.
   6.6 tonnes Cement.
   54 No. Sheets of Paramount dry partitioning. Damaged during off-loading from delivery lorry, the JCB with forks being unsuitable for this type of material.
    6 No. Catnic lintels replaced. 2 No. damaged, 4 No. stolen.
   21 No. Window boards. Stolen.
   10 m Glazing bead. Stapled to door. Falls out when extensive handling.
   15 No. Internal doors because of damage on site.
    5 No. Kitchen units and worktops. Damage on site.
    2 No. Nosings. Damaged on site.
    2 No. Porch gallows brackets. Stolen.
1,200 m 50 × 25 P A R timber. Schedule errors, off-cut waste and theft.
    2 No. Window frames and 1 No. door frame. Stolen.
      Extensive ironmongery due to theft.
  400 m Skirting. Scheduling error.

References

1. Building Research Establishment. (1981) *Waste of building materials*, Digest No 247 HMSO, London.
2. Building Research Establishment. (1982) *Material control to avoid waste*. Digest no 259. HMSO, London.
3. Good, K.R. (1986) *Handling materials on site*. Chartered Institute of Building Technical Information Service paper No 68, Ascot.
4. Building EDC. (1983) *Faster building for industry. A Report*. Crown copyright. HMSO, London.
5. Ministry of Agriculture, Fisheries and Food. (1986) *Windbreaks for horticulture*. Booklet No 2280, MAFF. Crown copyright.
6. *Construction (Health, Safety and Welfare) Regulations 1996*. HMSO, London.
7. *Clean Air Act 1968*. HMSO, London.
8. *The Clean Air (Emission of Dark Smoke) (Exemption) Regulations 1969*. HMSO, London.
9. *Factories Act 1961*. HMSO, London.
10. Illingworth, J.R. and Thain, K. (1987) *The control of materials and waste*. Published jointly by the Construction Industry Research and Information Association and the Chartered Institute of Building. CIRIA Special Publication No 56 or CIOB Technical Information Service paper No 87.
11. Illingworth, J.R. and Thain, K. (1988) *Handling of materials on site*. Published jointly by the Construction Industry Research and Information Association and the Chartered Institute of Building. CIRIA Special Publication No 57 or CIOB Technical Information Service paper No 92.
12. Illingworth, J.R. and Thain, K. (1988) *Materials management – is it worth it?* Published jointly by the Construction Industry Research and Information Association and the Chartered Institute of Building. CIRIA Special Publication No 58 or CIOB Technical Information Service paper No 93.
13. Construction Industry Research and Information Association (1998) SP 133 *Waste minimization in construction – site guide*. CIRIA, London. At a later date, two additional documents will become available: SP 134 *Waste minimization in construction – design manual* and SP 135 *Waste minimization in construction – boardroom handbook*.

Chapter seventeen

Method statements

In the previous Chapters, consideration has been given to the method planning of the main elements comprising the construction process. These individual elements now require bringing together as a coherent whole, culminating in the production of a Method Statement, one of the most important documents in the planning process. Before this can be done, an analysis needs to be made to determine the logical sequence of the activities and least time for execution, the latter leading to establishment of a contract time, if not already specified in the contract documents. Such a Method Statement should not be confused with the Submission Method Statement, which is discussed in detail later in the chapter.

Determination of the logical sequence of events (tender stage)

The precise stage that this operation is carried out depends on the individual doing the construction method planning. Many experienced planners would opt for carrying out the logical sequence study before considering the individual method assessment for main elements in the bill of quantities. A much more thorough understanding of what has to be built is achieved prior to making method decisions and resource balancing takes place. If left to the time when individual methods have already been decided upon, it is highly probable that alterations will need to be made to resources previously assumed.

To achieve the above analysis, a network diagram provides the best method. Such a network needs to be hand drawn, as the information on what has to be built is extracted from available drawings, specifications and bills of quantities. In so doing, the construction planner develops a detailed appreciation of what is involved in considerable detail – and the extent to which information is going to be lacking at the tender stage. It is at this stage that the check lists given in Chapter 14 can be presented to the architect for him to complete as far as possible. Even so, information on specialist services, cladding and finishes is likely to be fairly sketchy. At the same time, rough ideas on contents provide programme items and sequence, even if time scale cannot be established other than an informed guess, based on experience.

Once completed, the network can be examined and corrected or adjusted to provide what seems to be the best sequence of events. Each item on the network can now be studied in relation to the appropriate method for adoption in relation to plant and labour time and content. In other words, the operational method statement is beginning to be built up. By producing method statement items, one by one, each can be checked against the network to see if the timings allowed fit in. If they do not, one of two options arise: (a) an adjustment of resources for one or more items to make things balance or (b) the resource balance is adjudged to be the best that can be achieved, altering the network to suit.

With the above approach, the network can become a time scaled version, either within the contract time specified, or the contractor's view of a realistic time for the contract. At the same time, the network diagram will now be capable of showing which items are critical if a given completion time is to be achieved.

Pre-qualification stage (prior to being invited to tender)

To limit the number of contractors asked to tender for a project, a client will often select a number of likely contractors and invite them to pre-qualify before issuing papers for tendering purposes. From the submissions made, a smaller number will be actually invited to tender or, in the case of appointing a managing contractor or contract manager, to state the terms on which such work would be undertaken.

Pre-qualification can take a variety of forms, depending on the particular client's requirements. The following is an example of what may be asked for.

1. Outline drawings to illustrate the outline scheme. Potential tenderers are to provide an appreciation of the likely construction problems and how these might be overcome by use of the tenderer's experience being utilized at the final design stage.
2. A proposal for the conditions of engagement for the working with the design team.
3. A statement setting out the experience of the firm in projects of a similar nature together with the ability to provide staff of proven experience in this type of work.
4. Proposed scope of conditions of contract and the basis of determining the tender sum.

In presenting a pre-qualification document, it is important that the contractor shows himself in the best possible light with high class presentation of both written matter and visual aids.

The pre-tender planning method statement

This is a document to be produced by the pre-tender planning team. It is designed to provide the estimator with the planning team's assessment of the plant, method and labour needed for each major item in the bill of quantities, or unit rate in a schedule of rates format. It will also give a time scale for the work. At the same time, it will provide a record of how the tender was made up for the benefit of site management if the contract is won. As such, it should include the following:

1. A clear and succinct description of what has to be built, the methods on which the pricing was based. In turn, such information provides the basis for the formulation of a tender programme for submission with the tender. It will also provide the basis of the contract master programme if the contract is won. See Chapter 18 for details.
2. A record of the pre-tender analysis from the method point of view.
3. Plant requirements and their duration on site.
4. Labour requirements and durations.
5. Information on temporary works needs, methods and details for pricing by the estimator.

Method Statements are confidential to the contractor, especially in relation to cost make-up, unless the contract is of the cost plus variety, when such information has to be disclosed. In the case of the main contractor or managing contractor, method statements should be required from all specialist or subcontractors, either at the pre-tender stage or as soon as possible after the parties are known when the contract has been won.

A particular feature of pre-tender method statements should be the inclusion of any safety measures required for the safe prosecution of the works. (The obvious inclusion here is scaffolding needs to comply with the Construction

(Health, Safety and Welfare) Regulations 1996[1]. Failure to allow enough money for safety items can lead to taking chances to save money.

All other temporary works needs should be detailed by the method planner as far as the available information at the tender stage will allow. Where it is apparent that complex temporary works will be needed, but the information at the tender stage is inadequate properly to assess the cost and time, reference to this situation needs to be made in the Submission Method Statement (q.v.).

A typical method statement of this type (pre-tender) follows. While this is for a major foundation, the approach is exactly the same for other sections of the construction process.

Example pre-tender method statement for a large concrete base in a plant producing chemicals

1. Introduction and description of the works

The contract is for the construction of the foundation for a fermenter vessel, in a chemical works. The vessel in question is 73 m in height and weighs 600 tonnes.

The foundation for this vessel is 18 m square and 2 m deep. On top is a plinth 10.6 m by 16 m and 450 mm deep. The whole of the base is supported on 81 No. bored concrete piles, 12/14 m deep and 750 mm in diameter (Figure 17.1).

The fermenter vessel is to be held down on the base by 48 No 40 mm diameter Macalloy bolts, 2.200 m long.

The top of the base also houses a further 112 No. smaller bolts to hold down a cladding structure, erection masts and other equipment foundations etc.

The client has specified that the entire base structure must be poured in one continuous operation to avoid pour planes within the concrete.

To match in with other plant erection, the base pour must not take place later than the end of October 1977. To that end, design drawings were specified to be available early in July 1977.

2. Method statement

(In preparing a method statement, a standard format is desirable. Each item needs to be identified by the bill of quantities reference number and the calculations given supported by any additional notes that are necessary to provide the estimator with a clear picture of what is in the planner's mind.) In this particular example, the key method item was the support of the large number of holding down bolts, which had to be very accurate. The method used was complicated and is inappropriate for this example of a learning method statement. It may be studied in detail in the author's book on temporary works[2].

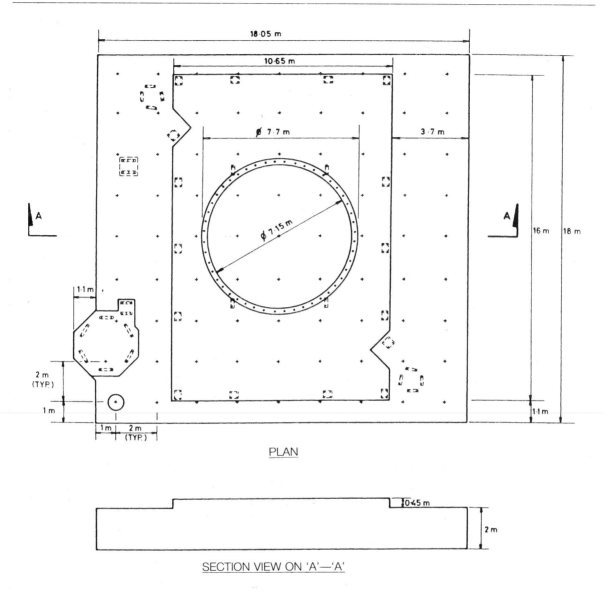

Figure 17.1 Foundation for large pressure vessel.

Bill item	Description, plant and labour, performance and time.
	Excavation to pile cap: Quantity involved 850 m³ *Plant to be used*: Hymac 580c and lorries. *Performance of machine*: Average output of 11.4 m³ per hour. This figure reduced by 20% to allow for excavating between close spaced piles. Therefore use 9 m³ per hour. Duration: $\dfrac{850}{9}$ m³ = 95 hours or *12 No. 8-hr days*
	Formwork to perimeter of foundation (Consideration of the problem led to the concept of constructing a weak concrete retaining wall against steeply sloping open cut excavation, as in Figure 17.2. 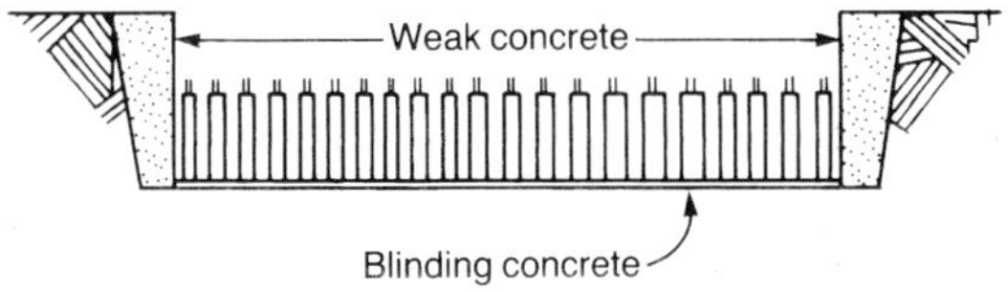 **Figure 17.2** The advantages of this approach were (a) minimizes excavation; (b) clean square tank for cutting down piles; (c) no movement risks when pouring concrete; (d) stable perimeter for concrete pumps; (e) second-hand formwork could be used, with finish not important. *Formwork*: Quantity involved 144 m² *Duration for erection*: 144 m² ¥ 2.5 man-hours = 360 man-hours Allow 4 No. men in formwork gang and 8-hr day thus: Duration: $\dfrac{360}{4 \times 8}$ = 11.25 days *say 12 days*
Not in bill	*Place weak concrete*: Estimated quantity 66.5 m³ *Plant*: Direct pour from ready-mixed concrete vehicles. Allow 15 m³ per hour = 4.43 hours *say 5 hours*

Bill item	Description, plant and labour, performance and time.
	Concrete gang: Allow 4 No. men – no reinforcement involved or need to vibrate concrete. *Duration*: Allow 8 hours to include prepare and clear up at end of pour.
	Reduce piles to specified 'cut-off' level Of the various alternatives for cutting down piles, it has been decided to use conventional compressed air breakers. Quantity to be broken off piles 168 m^3 *Plant to be used:* 4 No. heavy breakers with 2 No. two-tool compressors *Performance rating*: 4 man-hours/m^3 *Duration*: 168 m^3 × 4 man-hours/m^3 = 672 man-hours or $\dfrac{672}{4 \times 8}$ = *21 days with 4-man gang*
	Reinforcement to Foundation Hand fixing, in situ. Quantities, performance and man hours: <table><tr><td>*Dia*</td><td>*kg*</td><td>*rate*</td><td>*man-hours*</td></tr><tr><td>40 mm</td><td>40360</td><td>0.008</td><td>323</td></tr><tr><td>32 mm</td><td>15511</td><td>0.008</td><td>124</td></tr><tr><td>25 mm</td><td>35291</td><td>0.010</td><td>353</td></tr><tr><td>16 mm</td><td>2000</td><td>0.041</td><td>28</td></tr></table> Total man hours 828 *Duration*: Allow 5 man gang, 8-hr day = *40 man-hours/day* $\dfrac{828}{40}$ = 20.7 days *say 21 days* *Plant*: Allow a mobile crane of nominal 10 tonnes capacity for distribution of steel into the base area. *21 days hire*

Bill item	Description, plant and labour, performance and time.

	Concrete to base and plinth
	
	Quantities: Main base $648\,m^3$
	                  Plinth 77
	                  Sundry minor plints <u>24</u>
	                                      $749\,m^3$
	
	Overall period allowed for the pour 10 hours. To include set up time and clean up and wash down time at the conclusion.
	
	Actual pour time planned as *$100\,m^3$ per hour or 7.5 hours* to be carried out by concrete pumping. *Rate per pump $25\,m^3/hr$, therefore 4 No. pumps required.* (With such a large pour, the pump contractor agreed to provide a standby pump on site free of charge.)
	
	Ready-mixed concrete: Supplied from three separate plants to avoid any risk of breakdown. To meet demand, the number of ready-mixed vehicles has been carefully calculated as follows:
	
	  Plant No. 1 Turn round time 40 mins
	
	  Plant No. 2 Turn round time 50 mins
	
	  Plant No. 3 Turn round time 65 mins
	
	Study of the journey times and capacity of the trucks give:
	
	  17 trucks on the road, $6\,m^3$ capacity
	   8 trucks planned as standby
	
	17 trucks at $6\,m^3$ capacity will deliver $100\,m^3$ per hour with a built-in reserve from the nearer plants to allow for road delays.
	
	Sundry small plant
	
	Petrol vibrators: 4 No. plus 4 No. standby
	
	Welding set: 1 No. (for stiffening up any movement in rebar or bolt support system).

Bill item	Description, plant and labour, performance and time.

	(concrete pour continued)
	
	Labour requirements
	
	Concrete gang: 4 men per pump = *16 men*
	
	Supervision: 2 managers + pump foreman = *3 staff*
	
	Quality control: Concrete quality engineer for spot checks at supply plants and site throughout the duration of the pour.
	
	Sundry labour: Welder 1 No.
	                        Standby scaffolder 1 No.
	                        Standby carpenters 2 No.

	Notes relating to the Method Statement 1. With a large concrete pour of this type, it is essential that all those involved participate in the planning – design engineers, contractor, ready-mixed concrete supplier and the concrete pumping specialist. In Figure 17.3, showing the master programme for the base, notice the programmed meetings of all parties prior to the actual pour. 2. Large pours cannot readily be stopped due to failure in concrete supply or plant breakdown. For this reason, standby facilities in all aspects have to be studied and provided for. 3. Pours of this magnitude normally take place on a Sunday: (a) minimum traffic problems; (b) concrete plants can be used on single pour only; (c) better negotiating position with ready mixed company with regard to price per cubic metre, plant has earned its bread and butter potential during the week and covered its overheads.

To further bring the method statement to life, Figure 17.4 illustrates the organization needed for such a large pour. Figure 17.5 shows the pour in question in progress.

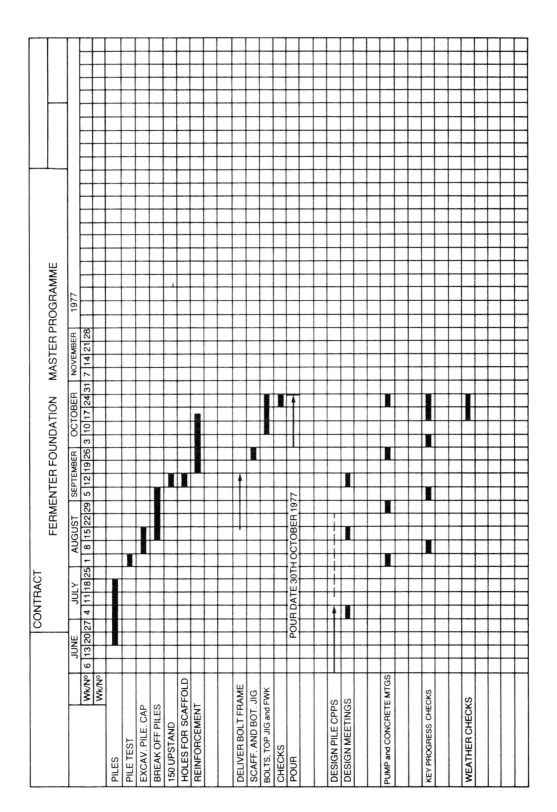

Figure 17.3 Programme for base construction.

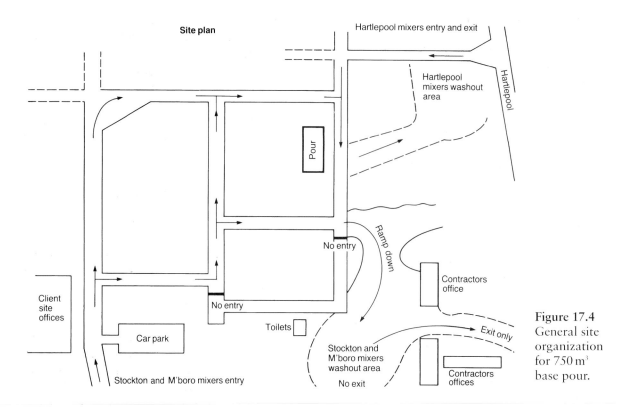

Site plan

Hartlepool mixers entry and exit

Hartlepool mixers washout area

Hartlepool

Pour

Ramp down

No entry

No entry

Client site offices

Car park

Toilets

Contractors office

Exit only

Stockton and M'boro mixers washout area

No exit

Contractors offices

Stockton and M'boro mixers entry

Figure 17.4 General site organization for 750 m³ base pour.

Figure 17.5 Actual pour in progress. (ICI Agricultural Division).

Submission method statements

The submission method statement is a totally different document from the Pre-tender Method Statement. Indeed, while the latter is confidential to the contractor, the former is specifically directed to the client. It can be likened to a public relations exercise, designed to sell the contractor's competence and experience to the potential client, how he sees the contract being carried out and drawing the client's attention to possible changes to save time and money.

This type of method statement would normally include the following:

1. A description of the contractor's intentions, in broad terms.
2. Drawing attention to key items for the contract's success.
3. A programme showing what will be necessary in relation to design input for the progress desired to be achieved, to meet the completion needed.
4. Any potential cost savings that could be made by alterations to the designer's intentions. This calls for tact.
5. Site layout arrangement and any problems that may arise in relation to access
6. Facilities and skills available from the contractor to assist the client and his advisers, should they be needed.

In more general terms, the aim should always be to pin-point, at the earliest time, issues that may affect time/cost if they are not met.

Many contract documents specifically state that no alterations to the documentation are permitted. (The reason being that the client wants to be able to compare all submissions on an equal footing.) Where a contractor sees the ability to make savings by doing some aspects of the work differently from that specified, the Submission Method Statement opens the door to putting these ideas forward, while pricing the works as specified. A sample statement of this type, for Dollar Bay in London's Docklands, is presented on the following pages. The text and illustrations are published by courtesy of Tarmac Construction, in a slightly abridged version of the original.

Notes on the Submission Statement

1. The contractor is selling his expertise to the client.
2. He is drawing attention to assumptions that he has had to make due to lack of information.
3. He is defining the plant and methods he intends to use.
4. He draws attention to the need to use large area pouring to the suspended floors for rapid construction with attendant savings to the client in cost.
5. The first stage tender programme sets out the contractor's overall concept of time. As no information is available about specialist subcontractors, the tenderer records that he is assuming that such subcontractors will comply with the programme.
6. It is indicated that the tender programme will be developed into the contract master programme in the light of any additional information available, if the submission is successful.

References

1. *Construction (Health, Safety and Welfare) Regulations 1996*. HMSO, London.
2. Illingworth, J.R. (1987) *Temporary Works – their role in construction*. Thomas Telford, London

DOLLAR BAY

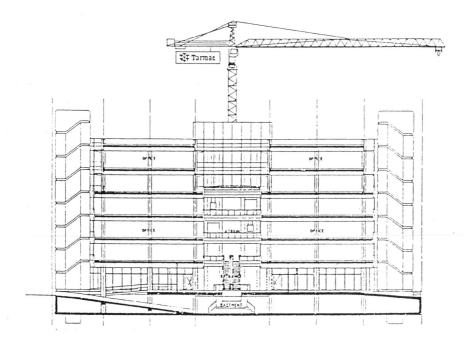

FIRST STAGE SUBMISSION

Tarmac Construction

INTRODUCTION

Tarmac Construction has the expertise and resources to carry out this prestigious development and would welcome the opportunity of joining the Professional Team during the pre-contract stage. From our investigations we have planned this project on a 'fast track' method in order to provide our client with the completed contract at the earliest possible date together with the high quality of finish that Tarmac Construction provides. We trust the enclosed information meets your requirements. Our staff have been put on standby should you require further information.

METHOD STATEMENT

Site Establishment

On taking possession of the site, a survey would be carried out of the pre-contract works. Prior to the commencement of the permanent works, any discrepancies found would be notified to the client's representative.
During this period a thorough investigation will be made, in conjunction with the Statutory Authorities, to identify and locate existing services.

Site Hoardings

The existing hoardings previously erected during the pre-contract works will be taken over by ourselves as indicated within preliminaries page 1/43 item D. They will be maintained until completion of the contract.

Site Accommodation

Discussions have taken place with the London Docklands Development Corporation regarding the possible siting of our accommodation complex on floating pontoons in the Dollar Bay dock. Indications at this stage are that our alternative would not be acceptable due to the congestion of moorings within the locality.
Our tender proposal for siting of temporary accommodation is therefore indicated on site establishment plan TP.001 attached (Figure 1).
A limited set up would also be located within the basement area for the early stages of the sub-structure works.
On completion of part of the suspended ground floor slab, the full accommodation set-up would be installed as follows:

Client's requirements	As specified in the preliminaries bill page 1/30A;
Tarmac accommodation	Provision for: Project Executive, Construction Project Manager, Site Agents, Engineers, Planning, Surveyors, General Foremen and Administration;
Welfare Facilities	Canteen, Drying Room, Toilets and First Aid.

Additional requirements, i.e. main subcontractor's supervision offices and secure facilities, would be located within the formed basement area.

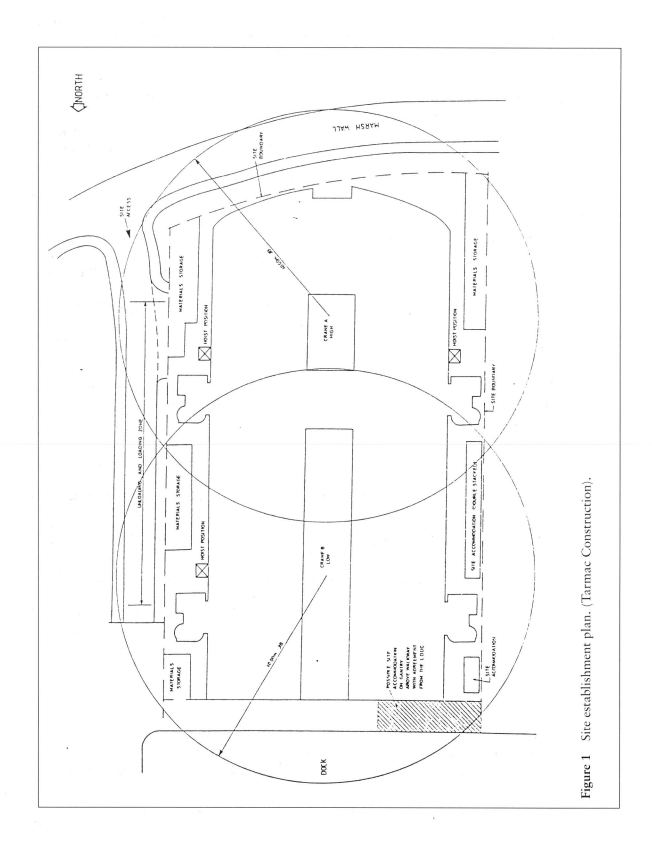

Figure 1 Site establishment plan. (Tarmac Construction).

Site access

During sub-structure works, the main site access will be via Marsh Wall, with entry to the site from the Southern boundary.

We anticipate that an access ramp to the basement area for vehicles and equipment will have already been formed in conjunction with the piling mat during the pre-contract works. The area to the East of the building, which will form the access road in the permanent works, will be used for loading and unloading etc, to avoid the congested Marsh Wall.

Materials Distribution

In order to meet the programme period and gain the maximum coverage of the site, we propose the use of two tower cranes for vertical and horizontal movement of materials. It would be our intention to install the tower cranes at an early stage in order to utilise their access advantages during sub-structure works.

Both cranes are to be located within the Atrium voids as indicated on the Site Establishment Plan TP 001 attached.

Loading platforms are to be tied into the main reinforced concrete structure for the landing of materials by our tower cranes.

Passenger/goods hoists are to be installed on completion of three floors of the main frame. This arrangement will provide additional back-up to the tower cranes during superstructure works and will be utilised for the vertical distribution of materials once the tower cranes have been removed.

Sub-structure Works

Basement construction would commence with the piling operation. Access for the piling and equipment will be via the previously formed ramp.

We have assumed that pile testing will be carried out *either* prior to the main contract works or concurrent with the piling programme. We have not allowed additional time for pile testing.

We anticipate that the successful piling contractor would use at least three rigs. Two would progress the driven piles with the third handling the bored, cast in-situ piles straddling the existing Thames Water Authority sewer.

We have assumed that pre-boring of piles will not be necessary and that the progress of the piling contractor will not be affected by obstructions.

The piling works would commence from the North elevation and progress to the South elevation. This sequence would dictate the progression of subsequent works through to the superstructure.

As soon as sufficient piles were complete the remaining sub-structure works would commence in the following sequence:

1. Excavate pile caps and foundations;
2. Form concrete pile caps and foundations;
3. Establish underground drainage;
4. Make up to formation levels;
5. Form basement over site slab (blinding, reinforcement and concrete);
6. Form concrete perimeter walls (formwork, reinforcement and concrete; together with inserts, openings etc.).

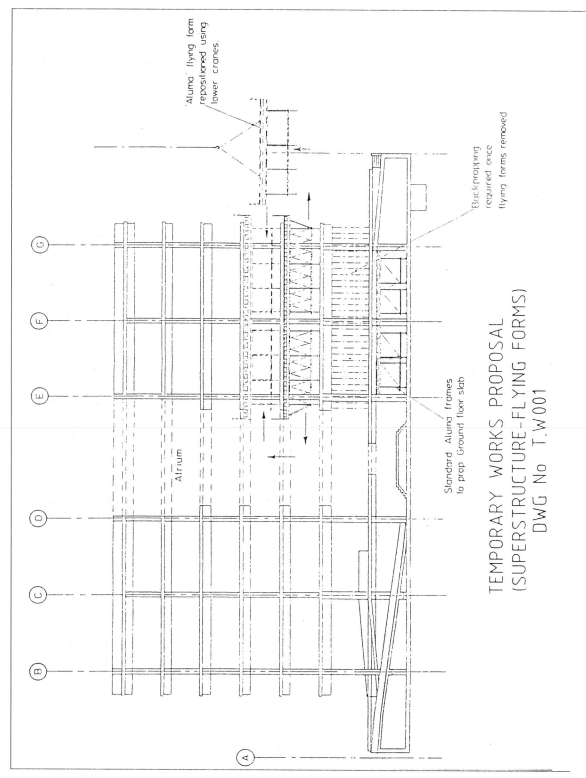

'Aluma' flying form repositioned using lower cranes.

Backpropping required once flying forms removed

Standard Aluma frames to prop Ground floor slab

Atrium

TEMPORARY WORKS PROPOSAL
(SUPERSTRUCTURE-FLYING FORMS)
DWG No T.W001

Figure 2 Temporary works proposal – superstructure formwork. (Tarmac Construction).

To allow completion of the piling works, the access ramp would require relocating. A new ramp would be installed in an area already piled and would subsequently be removed for completion of the basement slab.

For speed and efficiency, the retaining wall forms would be fabricated off site. They would be delivered to site to achieve immediate erection, thus saving time and working space. Concreting operations will be carried out with the tower cranes for retaining walls, columns and core walls. Mobile concrete pumps will be used for pile caps, slabs pours and car park ramps.

Superstructure

A careful study of the available formwork systems in relation to rapid progress has been made. The selected methods recognize that careful design and planning provides the key to achieving the desired speed whilst maintaining quality.

Our scheme for the ground floor slab uses the Aluma System. This system is quick to use and is ideal for restricted basements. Its lightweight construction makes it easy to strike and manhandle. For the subsequent upper slabs we have planned on the use of flying forms. This method employs a series of trusses and aluminium beams bolted together as a complete unit of crane handled formwork. Each unit incorporates a self-contained edge access, thus eliminating the need for an external scaffold rising with the frame (Figure 2).

We have assumed that the table forms can be struck in three days with continuing support being provided by back propping.

Circular formwork to columns would be manufactured off site in glass reinforced plastic or steel to provide the high quality finish required.

Structural steel perimeter stair towers are to be erected on completion of the fourth floor slabs, in conjunction with the metal stairs installations.

Distribution of concrete, reinforcement and formwork for the frame would be by tower crane. However, to maintain the rapid construction of the frame, we propose the use of mobile concrete pumps for concreting the suspended concrete slabs. We intend to cast these slabs in approximately $150\,m^3$ pours as indicated on the attached drawing No TW 003 (Figure 3). This approach will minimize construction joints and saves time and cost.

External Envelope

Temporary weatheright barriers will be installed (as specified on page 1/40 item A) progressively as the frame is constructed. This would give the earliest possible commencement of internal trades.

Due to the curtain walling type, and the manufacturer not being specified during the tender period, full research has not been possible for prospective design, manufacture and installation periods. Indicative time periods have been incorporated within our submitted programme based upon our knowledge of market trends currently being experienced.

The method of installation of curtain walling will be dependent on the type selected. Various methods of access for erection can be employed and include:

1. Traditional scaffolding;
2. Cradles supported from roof level;

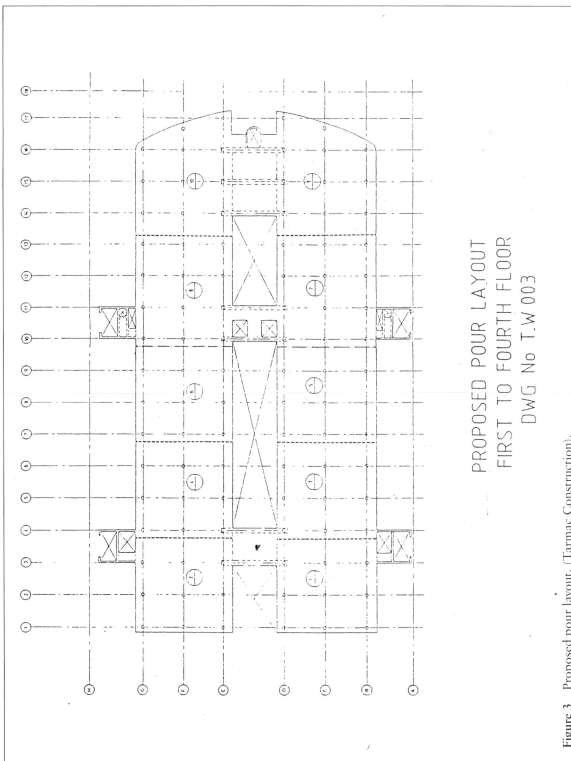

PROPOSED POUR LAYOUT
FIRST TO FOURTH FLOOR
DWG No T.W 003

Figure 3 Proposed pour layout. (Tarmac Construction).

3. Most climbing platforms rising from the ground floor;
4. Internal fixing of curtain walling requiring no external temporary platforms.

In our prelims we have allowed for the installation to be carried out as Item 4 above. We are currently using this method successfully in Docklands.

Mechanical and Electrical Services

Tarmac Construction has within its organization an Engineering Services Department. It would be our intention to allocate for the duration of the contract a Services Co-ordinator from this department. He would be responsible for ensuring close liaison with the Specialist Services Sub-contractors and the Consultant. He would co-ordinate and supervise the mechanical and electrical works and witness the commissioning.

Whilst the services requirements for each individual contract vary, we have found from our experience that successful co-ordination is about good communication and this is achieved by holding regular building services subcontract meetings.

Mechanical and electrical services would be installed in three sequenced operations. Main ductwork and electrical conduits at ceiling level and in risers would commence as soon as the progress of the frame permitted.

The proposed toilet pods should be installed prior to the Atrium roof steelwork. As this is prior to the watertight date we suggest that these pods are sealed with a watertight membrane prior to delivery to the site.

Installation of these pods would be by our tower cranes, which have a three tonnes lifting capacity.

Lighting boxes would commence on the progressive completion of the suspended ceilings.

This would be followed by the installation of the raised flooring.

Surface outlet/switch covers and cover plates would be installed during the final stages, after the decorations have been completed on each floor. The testing, commissioning and final balancing of all services would follow and would be of a progressive nature.

PROGRAMME BAR CHART

The first stage tender programme attached indicates our overall approach to this development. It includes knowledge accumulated during the tender period. We assume that Specialist Subcontractors will comply with this programme (Figure 4).

Our Senior Planning Engineer responsible for the preparation of the preliminary programme would, if the company are appointed, expand on the information obtained at the tender stage and develop a master contract programme, which will form the basis of monitoring progress through the contract period.

For the various elements of work undertaken by the Specialist Subcontractors, information concerning their performance data will be further investigated during the second stage. The requirements for subcontract order dates, and design, manufacturing and delivery periods will be reflected in our master programme. We have based our preliminaries on the tender programme.

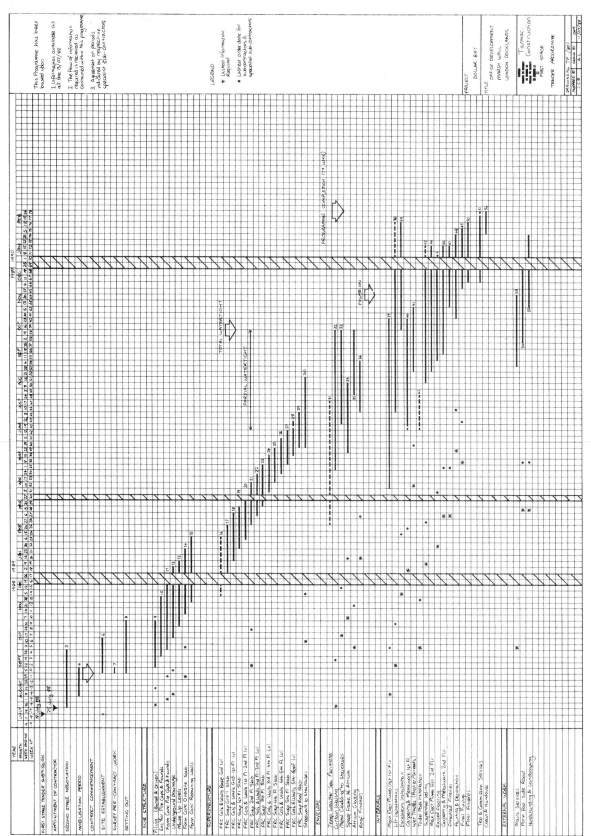

Figure 4 First stage tender programme. (Tarmac Construction).

SITE ORGANIZATION STRUCTURE

Key personnel for this project would be provided from within the Tarmac Construction organization. These managers and engineers have, in many cases, been trained by the company and have established their positions on varied projects over a long period.

TRAINING AND EMPLOYMENT OF LOCAL PEOPLE

The divisional policy is to promote the formal training of employees at all levels. Most of this training is carried out in-house in the division's own fully equipped training school.
Participation in external training schemes is supported and encouraged, particularly for young people.
At more senior levels, the training programme is geared towards management development.
We are currently establishing a training centre in the Docklands in conjunction with the Construction Industry Training Board.

HEALTH, SAFETY AND WELFARE OF WORK PEOPLE

The division recognises and accepts its obligations to ensure a safe system of work for all employees and others who could be affected by its operations, in accordance with the requirements of the Management of Health and Safety Regulations 1992 which will be backed up by a team of experienced full time Safety officers.
Further, the Company is fully conversant with the Construction (Design and Management) Regulations 1994 and will provide experienced staff to work in association with the Client, his Planning supervisor, the Professional team Specialist contractors, subcontractors and others who may be obliged to work on site, for the furtherance of Health and Safety on site for all parties.
Finally, the implications of the Construction (Health, Safety and Welfare) Regulations 1996 in superseding the Construction (General Provisions) Regulations 1961, The Construction (Working Places) Regulations 1966 and The Construction (Health and Welfare) Regulations 1966 are fully understood and incorporated in the Company's own publication *Guide to Accident Prevention and Statutory Compliance.*

Chapter eighteen

Contract planning control on site

The previous chapter reached the stage where construction methods, having been evaluated, were translated into specific plant and labour to create time/cost information. In turn, the resulting time elements were able to be analysed and integrated into a tender programme. Given that the tender is accepted, the planning now requires re-examination in relation to any additional information available and whether, in the light of such information, any changes will result in relation to the methods to be adopted. With such a re-examination the needs of the site in relation to planning control documents can also be established.

Before doing so it is desirable to establish a clear picture of what is meant by a planning control document, more frequently called a programme.

There is a surprising variety of interpretation as to what a programme is and what it should do. A man well versed in Critical Path programme techniques, who should have known better, once said to this author, 'You tell me what plant and methods you want to use and the sequence of their operation and I will do the planning!' What this author said to the gentleman in question would be inappropriate here. Nevertheless, it does make the point that supposedly experienced people have confused ideas about what a programme is and its relation to planning.

In this author's view, a suitable definition which properly reflects the purpose of a programme is: 'A programme can be defined as a time/sequence presentation of the results of planning the construction methods to be adopted, for management implementation in the field.'

In other words, the programme develops from the original method statement, either for the contract as a whole or a specific part. Nevertheless, while this is the basic function of a programme, its value needs to be seen in a much wider context. Amongst other things, it should provide management with the capability to:

1. Be the vehicle for analysing the most efficient sequence of operations;
2. Provide a means by which progress can be recorded against projected;
3. Be a tool of management for highlighting areas which need attention in the long or short term;
4. In certain special forms, provide a method of communication to subordinates in the field.

Varieties of presentation

Programme presentation is capable of almost endless variation – from specific basic types to varieties of such basic methods developed for special situations.

Basic types of programme

The best known basic methods are: Gant (bar) Charts; Line of Balance; Critical Path Networks; Precedence Diagrams and S-curves. All these types are described in detail, with examples of their function in the Chartered Institute of Building publication[1] already referred to in the Introduction. This publication also gives many references which cover the detailed use of the methods listed.

Of the above basic methods, the Line of Balance approach has today largely fallen out of use. Designed primarily for repetitive construction such as housing, it did not readily provide clear visual appreciation of the situation at a glance.

In addition to the basic methods and their derivations, further specialized forms are available for short term planning, resource balancing and time/cost recording.

The experienced planner in construction methods needs to be aware of the techniques available, together with the value of each, both in the planning analysis role and how they can be applied to give effective control and communication to management on site.

To make the right choice, it is necessary to examine what is wanted in a particular situation from a programme. The following are the key points:

1. Does the method aid initial analysis of the situation and especially in establishing the best sequence of events?
2. Are deficiencies in progress shown up as the work proceeds?
3. Can visiting management obtain an immediate view of the contract situation on arrival on site?

4. With the preponderance of subcontractors on site today, does the programme method give assistance in cash flow control?

From these questions, it will be obvious that one programme presentation is unlikely to be the answer to all requirements. It is also better to consider the situation in two stages – in respect of tender planning on the one hand and the operational planning after the contract has been won, on the other.

Pre-tender stage

When a construction method planner begins his assessment of the most cost effective approach to the building or structure at the tender stage, his mind must go back to the definition of 'planning' given in the Introduction to this volume: 'Understanding what has to be built, then establishing the right method, the right plant and the right labour force to carry out the works safely and to the quality required, in the least time and at least cost.'

In the time available, the planning staff have to achieve an appreciation of what has to be built, together with the sequence of events that will need to take place for earliest completion. To achieve this desirable result, the use of a hand drawn Network or Precedence diagram is far and away the best method to adopt. The action of drawing and linking of arrows, for example, keeps the planner's mind continually assessing if anything is missing from the sequence, whether all the information available has been included and where are any items which conflict with each other. In this latter case, further analysis, in more detail, may resolve the problem.

Once completed, the network or precedence diagram can be considered in relation to method and time scale. In so doing, the basis of the operational method statement will begin to emerge. This, in turn, develops into activity times and points the way to the correct sequence or overlap of events and provides the first assessment of overall time scale. As has been shown, in the previous chapter, the final product is the tender programme, best shown in bar chart form.

What is important here is that the network or precedence diagram is hand drawn and no attempt is made to develop it, at this stage, any further. The two methods are used as analysis tools only.

Post-tender (contract) programmes

At this stage it is necessary to consider two distinct situations – programmes for industrial and commercial projects, and those for housing and residential construction.

Industrial and commercial projects

In order that the minimum planning requirements can be satisfied for industrial and commercial projects, construction programmes need to be established in a number of stages. The reason for this is due to project information becoming available only as the contract progresses.

The main stages that are usually required to tie in with the flow of information are:

1. master programme;
2. sectional programmes;
3. weekly programmes;
4. preliminary value programmes;
5. subcontractors programmes;
6. materials procurement programme.

Contractors generally agree that the basis of presentation of these programmes is best carried out on a bar chart format. The generation method can be a matter of preference – direct development from the tender programme, producing a computerized network and, from this, using computer graphics to provide bar chart presentations of the network in what is known as linked bar charts or cascade diagrams. The varieties are endless and can be tailored to suit an individual contractor's preference. Whatever

means are employed, the resulting bar charts are the most easily understood at a glance, both by all levels of supervision on site and by visiting managers or client's representatives.

Notwithstanding the remarks above, there is considerable merit in treating the short term or weekly planning in a more communicative form. The method is described later.

Master programme

Under the Joint Contracts Tribunal form of building contract, JCT 80 rules, (Cl. 5.3.1.2), a Master Programme has to be submitted to the architect as soon as possible after the contract award. In GC/Wks/1, the type of contract for Government building and civil engineering works, clause 6 does not require a programme as such, but that works be carried out in an agreed sequence up to the date of completion. To achieve this, a programme would need to be prepared for a project of any size. Such a programme would effectively become the master programme for the contract.

Whether called for contractually or not, the preparation of a master programme prior to the commencement of the works is an essential need in relation to progression of the works and in the evaluation of delays and claims.

Derivation
It is essential that the master programme is derived from the programme that was drawn up and agreed as part of the tender submission. It needs to be drawn up by the planner who is to be responsible for the project planning. It must be agreed by site management and Regional/Head Office management before any issue is made to the client or his representative.

Format and content
The master programme is best prepared in bar chart form. As such it provides a generalized view of the duration of activities which is immediately understandable to all parties.

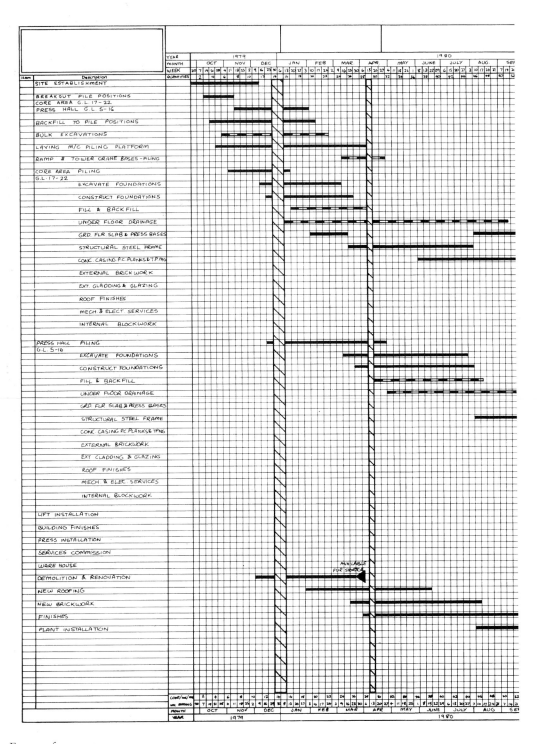

Figure 18.1   Format for master programme.

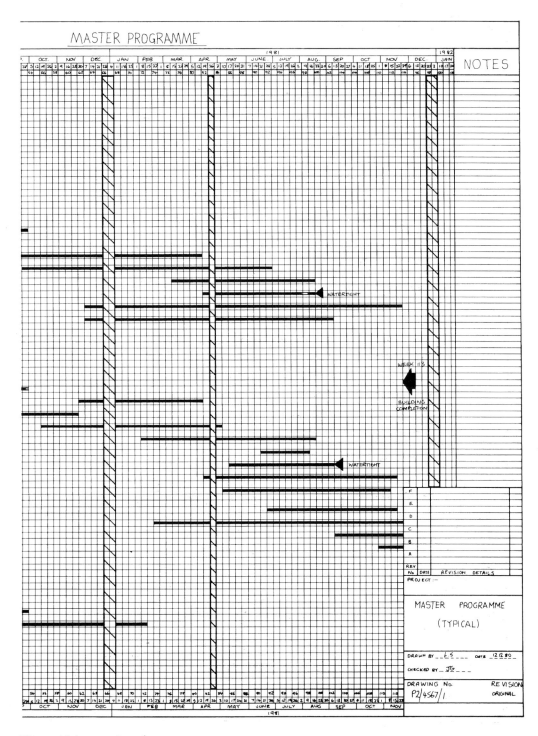

Figure 18.1   continued

Except on the largest and complex projects, the master programme should be confined to a single sheet (Figure 18.1).

Where the architect insists on the use of a network programme, it will usually be advantageous for a bar chart to be prepared from the network for the recording of progress. While a network programme is ideally suited to pretender analysis, as already mentioned, it is not suitable as a readily understood operational presentation.

The same can be said of Precedence Diagrams. The difficulty, in either case, is that, while those who prepare such programmes understand their content, they do not present a readily visible appreciation of a situation in the way a bar chart does. The situation tends to get worse when either method is computerized. As the contract proceeds and more activities become involved, the printout gets more and more voluminous. For a site progress meeting it is useless; apart from the time taken to fit the figures to the network or precedence diagram. As a major subcontractor once said, 'I have discovered the real value of computer printouts. They provide me with an endless supply of scrap paper!'

Sectional programmes

Requirement

Sectional programmes are required for the detailed site planning of the contract. They are prepared for specific parts of the work. That is, sectional programmes are the detailed work schedules for a particular area or stage of operations. As such, it is important that their preparation ties in with the master programme.

Sectional programmes will be an on-going activity as detailed information becomes available from the architect and subcontractors. Such activity will take two forms: (a) refinement of programmes as more information becomes available; (b) the preparation of subcontractor

programmes, once they have had time to assess their work loads, but using milestone dates provided by the main contractor, so as to fit in with the master programme.

Format and content

Yet again, the format is best kept to that of the master programme, using bar chart methods. The time scale used should normally be weekly, to line up with the master programme. Where critical areas arise, there may be advantage in using a daily time scale, especially if many short term activities have to interrelate. An example of a sectional programme is given in Figure 18.2.

Relationship to master programme

All sectional programmes must tie into the dates shown on the master programme and relate to sections of work shown on the master programme.

As sectional programmes are prepared, the cumulative resources involved must be checked to see that they remain within those envisaged for the master programme, and on which the pricing was based.

Supply of information

While the master programme will, inevitably, be prepared without all the information being available, sectional programmes need to be as accurate as possible as soon as possible. The JCT 80 form of contract requires that the contractor shall advise the architect of the dates by which information will be needed, such dates not to be unreasonably in advance of requirements or unreasonably close to the date stated (Table 18.1).

Once the master programme has been issued, the planning staff on site should start serious consideration of dates by which additional information will be needed. In evaluating such dates, account needs to be taken of the advance activities necessary before construction can take place. For example, if materials are long delivery items, information will need to be available

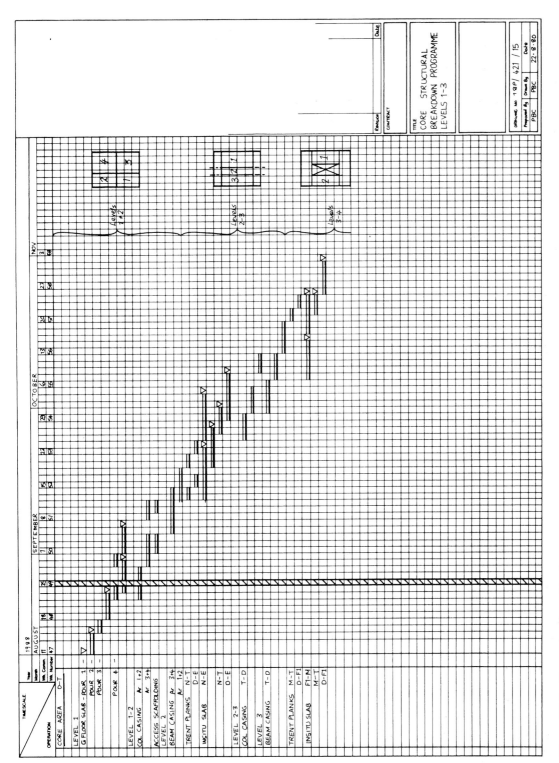

Figure 18.2 Example of sectional programme.

Table 18.1 Sample form for drawings and information required list

Typical contract
Drawings and information required list

	Date required	Date received on site (major part (90%))	Date of receipt of final details
Instruction to place order for service division and connections:			
Gas			
Electric			
Water	Immediately		
Telephone			
Existing Service Drawings	"		
Foundation drawings:			
(Architects) Setting out			
(Engineers) Piling			
Do. Retaining Walls and G.F. slab	30.1.81		
(Architects) Drainage			
Instruction to place order with Piling Contractor	27.2.81		
Frame Drawings:			
(Engineers) Plans			
Do. Sections			
Do. Details (breast panels etc)	17.4.81		
Do. Bending Schedules			
Do. Reinforcement Dwgs			
(Architects) Plans			
Do. Sections			
Do. Window Details	26.6.81		
Do. Elevations			
Do. Block and Brickwork			
Instruction to place order for Lifts	30.1.81		
Instruction to place order for Electrical installation	27.2.81		
Instruction to place order for Mechanical Ventilation	20.1.81		
Instruction to place order for Fire Resisting Doors	28.8.81		
Selection of and instruction to place order for facing bricks	28.8.81		
Details of roof finishings, Asphalt tiles, copper or aluminium	26.6.81		
Details of Glazing – internal and external	26.6.81		
Details of Joinery: (All items)	26.6.81		
Details of special wall finishes (i.e. mosaic)	26.6.81		

Detailed drawings of main entrance and foyer	26.6.81		
Details of and instruction to order false ceilings	26.6.81		
Co-ordinated services drawings – complete	15.5.81		
Details of and instruction to order W.C. pans	23.12.81		
Selection of and instruction to order sanitary fittings	25.1.81		
Instructions to place order for telephone installations	16.2.81		
Selection of wall tiles	23.2.81		
Details of instruction to order terrazzo	23.2.81		
Selection and details of Quarry tiles	23.2.81		
Details of and instruction to order kitchen equipment	27.3.81		
Colour schedules for decoration	23.12.81		
Details of and instruction to order floor finishes	23.12.81		
Details of and instruction to order demountable partitions	23.12.81		

External Works Details:
  Pavings
  Road Finishes
  Bollards
  Lighting
  Signs
  Road Marking
  Walls
  Seats
  Flower Tubs
  Stores			

for going out for quotes, obtaining approval for a price and the delivery time once the order has been placed.

To enable correct time scales for the supply of information to be made and no key items missed, the experienced planner will use a tabulated format so that a comprehensive issue of information required dates results, rather than piece-meal requests. An example of such a format is given in Table 18.1. While largely self-explanatory, it should be noted that three date columns are used: the first for date required, the second for recording when the major part of the information has been received and the third when final details have come to hand.

The reason for this last column is that often the architect thinks that all information required has been completed, but has overlooked the fact that some apparently minor detail has not been settled or issued. It is often the case that such lack of small details can stop completion of an item so that hand-over is not possible.

Such items as the drawings and information required lists lend themselves to computerization. Monthly printouts sent to the architect or resident engineer form a valuable record for potential claims, especially if not contested in writing.

In the same way, computerized drawing registers allow much greater control of amendments and when received. More especially to record the history of often endless amendments with minimum trouble. Dates of receipt of drawings are essential so as not to rely on date issued on the drawings.

Weekly programmes

Weekly programmes provide the ability to control activities on a short term basis and allow regular short term reviews of progress. In other words, prevent site management discovering that the contract is running weeks late! There is also a second and important value to the weekly programme. Communication on construction sites to the lowest levels of site management is not notably very good. The weekly planning presentation shown in the example in Figure 18.3 provides such communication on a day-by-day basis over the span of a week. At the end of the period, the site construction planner can review the situation and re-assess the needs for the following week to keep to schedule.

Weekly programmes should be produced by the site management team.

Format and content

As will be seen from Figure 18.3, a double format is used. The top part is in the normal bar chart form, while the lower half is a visual presentation of where the work is to be carried out on a day-by-day basis. By showing the major quantities of work to be achieved, management can see quickly if there is likely to be an unacceptably uneven spread of work throughout the week. If so an adjustment can be made to bring progress back to programme.

At the same time, the visual presentation shows foremen, gangers and charge-hands the location of work day by day for the period of

the programme. There is no need to seek instructions on where to go next on completion of a day's work.

Preliminary value programmes

The cost of preliminaries is an area of the project that site management needs to keep firmly under their control. A preliminary value programme provides a means of achieving this desirable situation. It is designed to establish and monitor the spending of the preliminary monies to achieve the best effect.

Format and content

A number of possible methods of presentation exist for preliminary value programmes. The method shown in tabular form (Table 18.2) is a means by which the preliminary value programme is integrated into the weekly cost system. Preliminaries costs are shown for all items included in the costing system. To compare actual cost against against value available, the cost accumulated can be plotted to date. In addition, changes to major site plant can be costed and the effect of such changes evaluated. Any agreed amendments to the preliminaries during the course of the contract can also be monitored.

The above approach is limited by the extent of the items included in the weekly system. A more visual system is shown in Figure 18.4. Being independent of any costing method, it can be used solely to monitor the main or management contractor's input to provide preliminary items, where the bulk of the measured work is subcontracted, or where the main contractor also has a labour input as well as subcontractors.

As with the first method described, its purpose is to provide site management with an accurate cost control of preliminaries.

The operation of this method is very straightforward. The programme is based on the estimator's split up of preliminaries by spreading

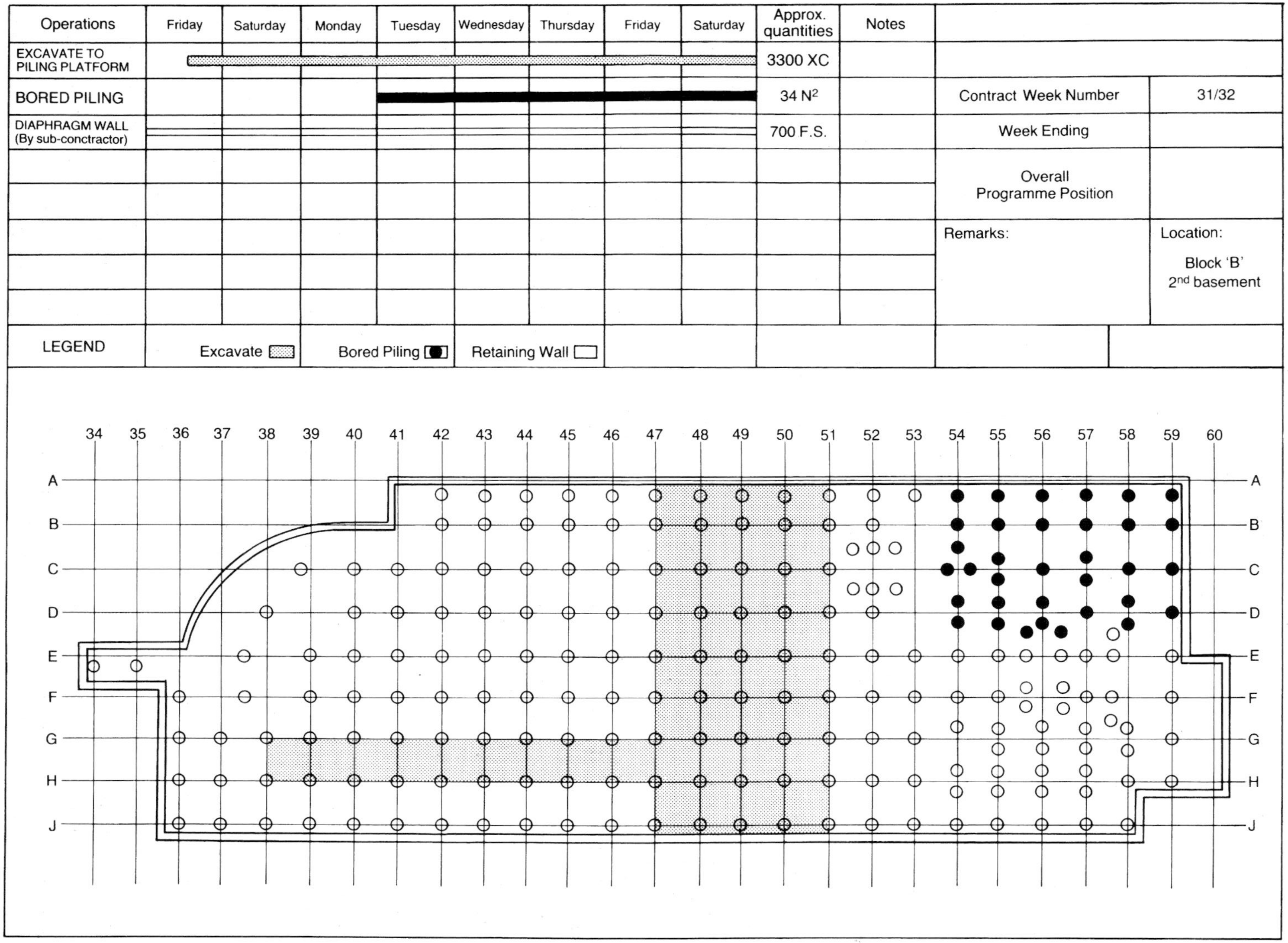

Figure 18.3 Weekly planning – method of presentation to provide communication to subordinate management.

PRELIMINARIES TARGETS and VALUES		
CONTRACT: TYPICAL	DATE OF ISSUE: AUG '80	SHEET No. 145.
DESCRIPTION	RECOVERY	TOTAL VALUE £
SUPERVISION		
PROJECT MANAGER 1 × 58 WKS		
SENIOR AGENT 1 × 58 WKS		
AGENT 1 × 58 WKS		
SENIOR ENGINEER 1 × 50 WKS		
ENGINEERS 1 × 58 WKS / 2 × 50 WKS		
PLANNING ENGINEER 1 × 58 WKS		
SERVICES ENGINEER 1 × 29 WKS		
GENERAL FOREMAN 1 × 58 WKS		
SECTION FOREMEN 2 × 58 WKS		
GROUND WORKS FOREMEN 2 × 68 WKS		
QUANTITY SURVEYORS 2 × 68 WKS		
OFFICE MANAGER 1 × 58 WKS		
STOREKEEPER 1 × 58 WKS		
TYPIST 1 × 54 WKS		
[TOTAL 1045 M/WKS]		137 651
SERVICE PAY		3 785
MECHANIC'S WAGES		812
INSURANCES		13 754
CARS		13 380
CAR ALLOWANCES		5 440
FARES		580
	[£25000 ADDED AT TENDER]	175 402
PETTY CASH [INC. PETROL]		
PETTY CASH		1 740
PETROL		13 539
		15 279
CHAIN MEN		
3 No @ £126.86 × 64.8%]a 58 WKS		14 304
ATTEND OFFICES & CLEANING		
ATTEND - 1 No @ £126.86 × 64.8%]a 58 WKS		4 768
CLEAN - 58 WKS @ £20/wk		1 160
		5 928

Table 18.2
Example format for preliminary values programme

PRELIMINARIES TARGETS and VALUES		
CONTRACT TYPICAL	DATE OF ISSUE AUG '80	SHEET No 245
DESCRIPTION	RECOVERY	TOTAL VALUE £
PLANT - NOT IN RATES		
5/3½ MIXER — 140 WKS @ £16·62/wk		2243
POWER FLOAT — 20 WKS @ £38·50/wk		770
GRINDER — 20 WKS @ £34·80/wk		696
DOUBLE-BEAM TAMPER — 10 WKS @ £36·55/wk		366
POKER VIBRATOR — 70 WKS @ £18·35/wk		1285
ROLLER - BW6S — 20 WKS @ £70·26/wk		1405
30T MOBILE CRANE — 2 x 1 DAY VISITS @ £155/day		310
15T " " — 35 x 1 DAY VISITS @ £125/day		4375
DUMPER — 20 WKS @ £87·00/wk		1740
COMPRESSOR — 38 WKS @ £111·70/wk		4245
GENERATOR — 10 WKS @ £32·00/wk		320
		17755
HOISTS		
MOONRAKER III — 2No @ £40·95)a 10WKS		819
— DRIVERS - 2 @ £116·92 x 64·8%)a 10wks		2163
		2982
UNLOAD & DISTRIBUTE		
J.G.B. 520 — 1No @ £196(wk))a 52WKS		10192
DRIVER — 1No @ £166·92 x 64·8%)a 52WKS		5625
LABOURER — 1No @ £151·47 x 64·8%)a 52WKS		5104
		20921
MAINTENANCE		
FITTER - 12WKS @ £170·22 x 64·8%		1324
CLEAR SITE		
SKIPS — 150No @ £30/each		4500
LABOURERS — ½ LAB for 58WKS @ £151·47 x 64·8%		2846
		7346
TEMPORARY LIGHTING		
PLANT, WIRING, ETC		2235
LABOUR 6WKS @ £170·22/wk x 64·8%		662
		2897

Table 18.2 continued

359

Figure 18.4 Preliminary value programme.

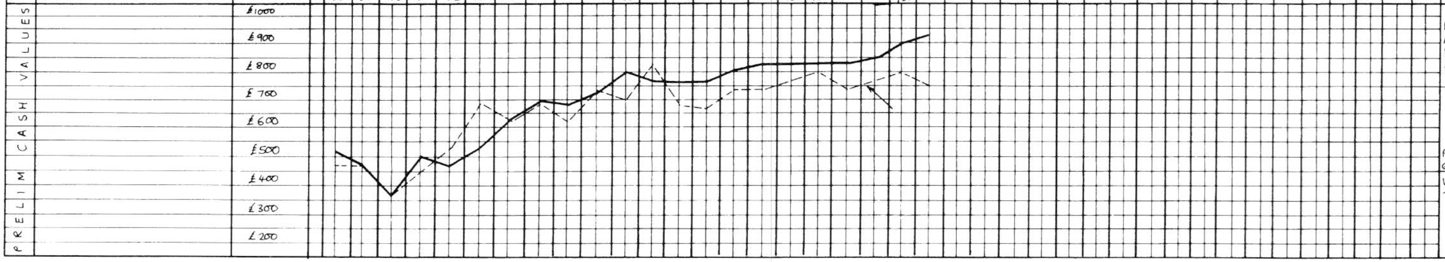

them throughout the period of the contract. Such spreading is done by the planning engineer and site manager to allocate sums of money to the periods of time when they actually occur on the operational programmes. While the example shown relates to general preliminaries, it is equally applicable to situations where plant is priced as a preliminary item when provided for general use by the main or managing contractor.

This type of chart needs the following conditions to apply:

1. An accurate programme needs to be prepared for the period of preliminaries – to include manpower requirements.
2. An accurate spread of the preliminaries to be carried out by the planning engineer and the site manager.
3. Weekly costing of the preliminaries is carried out.

As will as be seen from Figure 18.4, each item has two lines which allow programmed value against actual to be recorded. These, in turn, allow an overall weekly profit/loss to be assessed.

In association with the above, the graphical representation allows a clear financial appraisal and the questions to be asked when the actual preliminary value is running ahead of the programmed value.

Subcontractors programmes

An example of subcontractor control programmes is given in Figure 18.5. This example is for an electrical contractor. Where possible, subcontractors should be asked to include details of labour anticipated on site for the items shown on the programme.

All parties involved in the contract have commitments to carry out their work within the parameters laid down in the master programme. To make this happen, it must be insisted upon that all major subcontractors submit their detailed programmes showing completion

within the periods allowed by the main contractor to ensure the proper progress of the main contract. Key dates for starts, finishes etc. will need to be included in the subcontractor documents.

Format and content
Subcontractors programmes need to follow the presentation used by the main contract system, with matching week numbers and the same date reference. Separate information required schedules will usually be needed for the mechanical and electrical works.

Subcontractor commitments
The subcontractors programme represents a commitment to carry out the work as stated. From a site management control point of view, it is helpful if the subcontractor programme shows the labour resources required, week by week. On the other hand, if the subcontractor is unwilling to do this, the main contractor must not insist on the numbers of labour and plant to be brought on site. To do so would infer that the main contractor agreed to keep such resources employed. If in the event this could not be realized, the subcontractor could claim for standing time for his unused resources.

While the main contractor has no responsibility for the subcontractor's materials procurement programme, he needs to know that a procedure is being operated for getting his labour and materials to arrive on site when required.

In today's world, where almost all trades, both building and specialist services, are sublet, further control beyond just the subcontractors programme is highly desirable. A suitable method is to monitor the estimated cash flow month by month against monthly valuations submitted. This is best done by the use of S-curve presentation for each individual subcontractor. The general method is described later in this chapter.

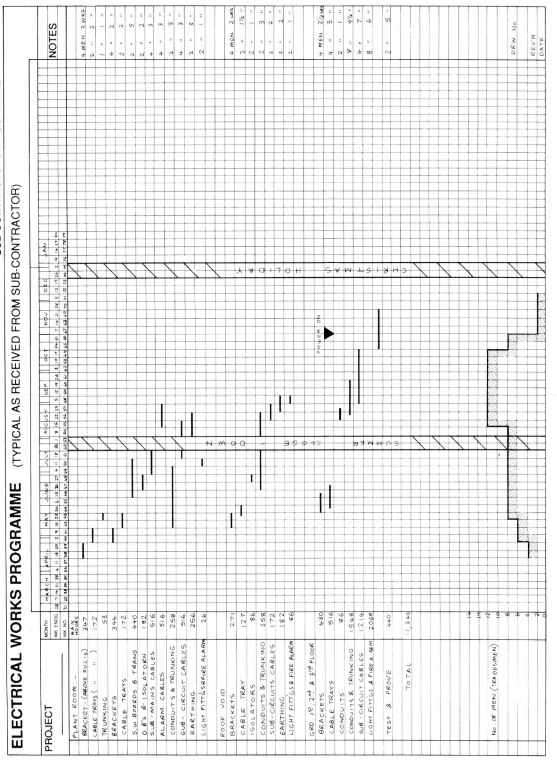

Figure 18.5 Example of subcontractor control programme.

Materials programmes

However well a contract has been planned in terms of its labour, plant and method aspects, it will be of little avail if the same degree of attention is not paid to the materials. Materials planning must, therefore, be adequately and realistically dealt with.

Materials programming normally falls into two phases on site: first, materials delivery schedules to meet the requirements of the master programme and the section programmes within it; secondly, the monitoring and control of deliveries together with ensuring that they are efficiently used.

Materials delivery schedules

These schedules will be derived from the overall construction programme and the detailed take-offs. Schedules must show the quantities involved and the delivery dates required from manufacturers and suppliers. In compiling such schedules, account must be taken of long deliveries and time taken to obtain approval from the client, where appropriate. Properly assessed, materials schedules are an invaluable aid in ensuring that materials are on site to guarantee progress and minimize the need to hold excess stocks. This latter is always undesirable as excess stocks increase the possibility of loss and damage, while taking up unnecessary site space.

Materials need to conform to a standard format, so that all levels of the site management team are familiar with the method used. A typical approach to materials scheduling is shown in Figure 18.6. The use of the legend symbols will be self explanatory, but note that a second line is used to record actual progress against planned, together with any rescheduling that may become necessary. If this is so, revisions should be annotated with a letter 'f' to denote a revised forecast to the original programme. This sample programme was originally intended for covering the materials needs for two identical blocks of flats (Blocks 15/1 and 15/2). For the purpose of this example, the materials are scheduled for block 15/1 only for reasons of clarity.

Taking the example as given, the following notes amplify the way in which the programme provides materials delivery control:

Item No. 1.	Has proceeded as planned. The schedules of delivery have been given a Ref. No. because the items required are too numerous to be shown in the Notes column.
Item No. 2.	Quotes arrived sooner than expected. Client took longer to approve. Order was placed on time. Delivery for 4th and 5th floors behind schedule.
Item No. 3.	Delivery commenced a week early and two floors delivered at a time for the convenience of the supplier.
Item No. 4.	Behind schedule. The quotations have not been approved by the client, therefore the order cannot be placed.
Item No. 5.	The materials for four floors were delivered on one load. Therefore next delivery required during W/E 12th February.
Item No. 6.	Has proceeded as planned.
Item No. 7.	In this case the delivery has been split because two different suppliers are involved.
Item No. 8.	This is a specialist item and only the overall construction time is shown. Quotations were late in being received thus delaying the order.
Item Nos. 9–15	None of the materials as yet due. Some of the ordering procedure has varied from the programme, in some cases this is because materials required at an earlier date have been placed on the same Order No.

363

GENERAL NOTES
1 Upper line shows original schedule.
2 Lower line shows actual performance up to current date and revised forecast if reqd for future dates.
3 The date of the latest re-appraisal is shown by a thick line down the schedule. Information shown to the left of this line is therefore historical, and to the right the latest revised forecasts.
4 Revised forecasts should be followed by the letter 'f'.

LEGEND
R = Request for quotations
T = Quotations received
T (boxed) = Quotations and/or drgs to client for approval
Q (boxed) = Quotations approved
O = Order placed
Δ = Delivery required
= Subcontractors construction time

MATERIAL SCHEDULE

15 STOREY BLOCKS – Example only

No	ITEM		Order No.	DATE	SUPPLIER	SCHEDULE REF NO	NOTES
1	Pre-cast conc ring beams balcony units staircases	15/1	29546	8 Nov 59	Universal Concrete Ltd	G F S/1 Int. floors S/2 Top floor S/3	See separate schedule (INTERMEDIATE FLOORS)
		15/2					
2	Facing bricks to infill panels	15/1	29530	5 Nov 59	London Brick Co		External facings: Ground floor 194 YS, Floors 1-14 88 YS; Commons: Ground floor 158 YS, Floors 1-14 88 YS; Blue facings: Ground floor 246 YS, Floors 1-14 Nil (T G F)
		15/2					
3	Window frames	15/1	29600	16 Nov 59	Evan Williams	S/4	(T G F)
		15/2					
4	Asphalt roof and balconies	15/1				Subcontract	(ROOF; BALCONIES)
		15/2					
5	Joiners 1st fix materials	15/1	29620	7 Dec 59	Builders Supplies Ltd	S/5	
		15/2					
6	Internal Breeze	15/1	29625	3 Dec 59	Bonsall	see notes	Ground floor 2.5" breeze Nil, 4" breeze Nil
		15/2					Floors 1-14 2.5" breeze 277 YS, 4" breeze 18 YS
7	Plumber and Electrician 1st fix materials	15/1	29634 29635	10 Nov 59 10 Nov 59	Plumber – Hargreaves Electrician – Hill & Sons	Subcontract	(Plumber; Electrician)
		15/2					
8	Lift installation	15/1	29567	3 Nov 59	Evans Lifts	Subcontract	
		15/2					
9	Under floor heating materials	15/1	29622	23 Nov 59	Hill & Sons	Subcontract	
		15/2					
10	Fibre glass installation	15/1	29723	29 Dec 59	Pilkington Ltd	see notes	251 YS per floor
		15/2					
11	Wall tiling and sanitary goods, hot water cylinders etc.	15/1	29631 29632	3 Dec 59	Tiling – S. Brown & Sons Sanitary goods – Gratrix Ltd	Subcontract Wimpey supply	Tiling – 12 Ys per flat WC pans } Hot and cold cylinders per floor
		15/2	29633		Cylinders – Hargreaves	Plumber fix	Lavatory Basins } 4 per floor Belfast Sinks
12	Joiner 2nd fix materials Door furniture	15/1	29620	7 Dec 59	Builders Supplies Ltd Ornimental Ironwork Ltd	S/6	
		15/2					
13	Electrical fittings	15/1	29622	23 Nov 59	Hill & Sons	Subcontract	
		15/2					
14	Thermoplastic floor tiling	15/1	29750	2 Feb 60	Marley Tiles Co Ltd	Subcontract	
		15/2					
15	Entrance doors to landings	15/1	29620	7 Dec 59	Builders Supplies Ltd	see notes	Ground floors see Arch schedule No 1036/70c. Upper floors — do —
		15/2					
16	Kitchen fitments	15/1	29648	12 Jan 60	Home Units Ltd	see notes	Per floor 8 No. Wall Units 3' 0" × 9' × 2' 0" high
		15/2					8 No. Cupboards 3' 0" × 4' 6" × 2' 6" high 8 No. Sinks — do —

Programme columns: Oct 2, 9, 16, 23, 30; Nov 6, 13, 20, 27; Dec 4, 11, 18, 24; Jan 1, 8, 15, 22, 29; Feb 5, 12, 19, 26; Mar 5, 12, 19, 26; Apr 2, 9, 16, 23, 30; May 7, 14, 21, 28; June 4, 11, 18, 25; July 2, 9, 16

GENERAL INSTRUCTIONS
1 Give as much time as possible between the various stages of placing orders etc.
2 Fill in the actual history on the bottom line in ink.
3 Fill in any necessary revised forecasts on the bottom line in pencil.
4 Shade in RED pencil the area of the schedule which has been dealt with.

Figure 18.6 Scheduling and programming materials.

Item No. 16. This shows how quotations which are sent for approval and are not accepted should be shown. In this case two revised quotes were necessary and the client took a long time in giving final approval. This has delayed the order 5 weeks. Because of this delay, the supplier is unable to give the required delivery. His latest promised delivery dates have been reforecast on the lower line.

Housing and other domestic accommodation

Housing and other domestic accommodation requires some variation of approach to that applied to industrial and commercial work. Housing, in particular, is spread over a wide area and unless careful control is exercised, management can spend a great deal of time trying to find out where the many subcontractors are working. This is especially true, today, when many housing estates are totally sub-let.

The tender stage will normally consider the anticipated sequence of work when dealing with local authority contracts. A master programme showing sequence of construction and time scale would be produced in a similar way to those already described for industrial and commercial works.

In the case of private development housing, an overall master programme is rarely appropriate. This is because, to avoid excessive monies being tied up, the contractor will usually start a part of the site to see how sales will go. If there is seen to be a good demand, further areas of the site will be released for construction. In these situations, the times for construction of individual house types and the basic cost will be the key areas for control systems.

Whether public or private housing, manage-ment control needs the ability to keep track of what is going on at all times.

Control systems

The basis of all control systems for housing is to establish from the start a list of elements in the dwelling which make up the total activity sequence from start to completion. All subcontractors should be asked to agree or request any modification they may feel is needed before work starts. Once agreed, all those involved must stick to it as all control methods will be based upon it.

Weekly planning

Housing is best controlled on a weekly basis. The format for this is shown in Figure 18.7. Not only does the programme provide for up-dating at a subcontractors' meeting every Friday, say, it is also recording past performance, relationship to programmed completions required and the coming week's performance needed to catch up with any delays that may have arisen in the previous week.

Perhaps more importantly, the approach shown indicates to all subcontractors where they should be working at a given time – indicating which block, where terraces are being built, and the dwelling in the block. Where houses are detached or semi-detached (especially in the private sector), it is more usual to use plot numbers. Once agreed at the weekly subcontractors' meeting, site management can visit where the various trades are programmed to be, without the need to search in endless dwellings to find them. If the trade in question is not where it should be appropriate action can be taken.

Note that the items shown on the weekly programme sheet should be numbered to correspond with those on the agreed work sequence list (Figure 18.7).

With this sort of detail, it may well be felt

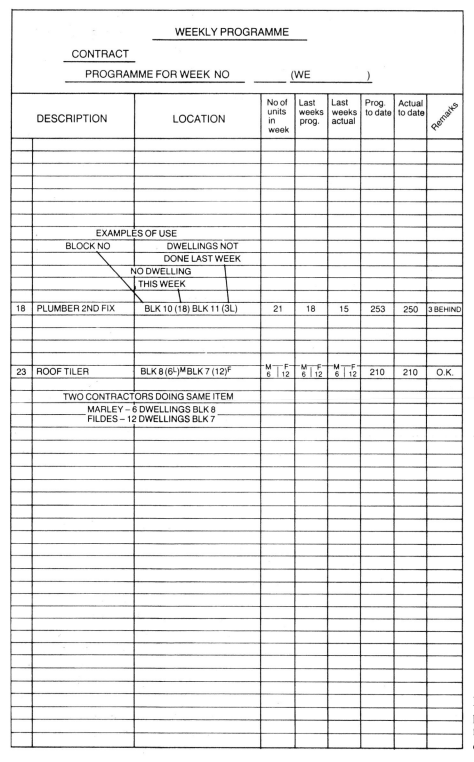

WEEKLY PROGRAMME								
CONTRACT								
PROGRAMME FOR WEEK NO _____ (WE _____)								

DESCRIPTION	LOCATION	No of units in week	Last weeks prog.	Last weeks actual	Prog. to date	Actual to date	Remarks
	EXAMPLES OF USE						
	BLOCK NO ⟶ DWELLINGS NOT DONE LAST WEEK						
	NO DWELLING THIS WEEK						
18	PLUMBER 2ND FIX	BLK 10 (18) BLK 11 (3L)	21	18	15	253	250	3 BEHIND
23	ROOF TILER	BLK 8 (6L)M BLK 7 (12)F	M—F 6 | 12	M—F 6 | 12	M—F 6 | 12	210	210	O.K.
	TWO CONTRACTORS DOING SAME ITEM						
	MARLEY – 6 DWELLINGS BLK 8						
	FILDES – 12 DWELLINGS BLK 7						

Figure 18.7   Weekly programme method for housing contracts.

that control is well exercised. Where work is so spread out, however, this is not necessarily so, especially when dwellings are in blocks. In such cases, in local authority work, the complete block has to be complete before acceptance. Even in private sale situations, a further check is desirable to ensure that all items are complete.

Progress recording

An effective form for progress recording is shown in Figure 18.8. The agreed work sequence is shown on the left hand side of the form, while the block numbers with dwelling content are shown at the top.

Each block division is divided into the number of dwellings in the block. Progress of each activity number is recorded weekly by shading in the appropriate rectangle. If a re-entrant white space occurs, it indicates that one or more items have been missed in a particular dwelling in a particular block. Thus, while the programmed number of dwellings may have been completed, certain blocks cannot be handed over.

This method of progress recording has been found to be very effective and, indeed, provides the check for assessing the next week's weekly programme.

Record drawings and financial progress recording

The final area needing attention under the heading of Contract Planning Control relates to the production of progress record drawings together with the recording of financial progress.

Record drawings

The preparation of record drawings showing when areas of work were completed is becoming increasingly more important as Quality Assurance requirements are being introduced into contract conditions.

Records of this type are best illustrated on parts of the construction drawings themselves. By reducing such drawings to A4 size, a manageable file can be built up covering particular parts of the contract. Two examples of such record drawings are shown in Figures 18.9 and 18.10. While these cover horizontal work, sections and elevations can be produced to deal with vertical progress. Any set of plans or elevations can be used in several roles. For example, the floor record drawing in Figure 18.9 can be used in the context of progress under the floor slab in relation to drainage and underground services. External works can also be dealt with in this way. Service layout drawings can have colour codes to represent the stage reached – excavation, service installation, manholes complete and trenches back-filled and so on. Site managers often show great ingenuity in devising clear and effective formats to use.

A master set of record drawings should be kept on site.

Referring to Figure 18.10, covering ground beams and pile caps, the preparation of the quantities involved also provides the planning data necessary to assess the area covered for a single pour to achieve a balanced pouring programme. Dating then takes place when the pour has been completed.

Financial progress recording S-curves

Today, when large numbers of subcontractors are involved in a contract, many site managers find their control presents considerable difficulty. Even when subcontractor programmes have been produced, the work may be so complicated in relation to specialist services that assessment of the progress can be difficult.

In addition to the subcontract programme, a second check can be achieved on the basis of comparing the contract price against monthly valuations. An assessment is made of the monthly breakdown of the contract sum in question and an estimated cash flow curve

367

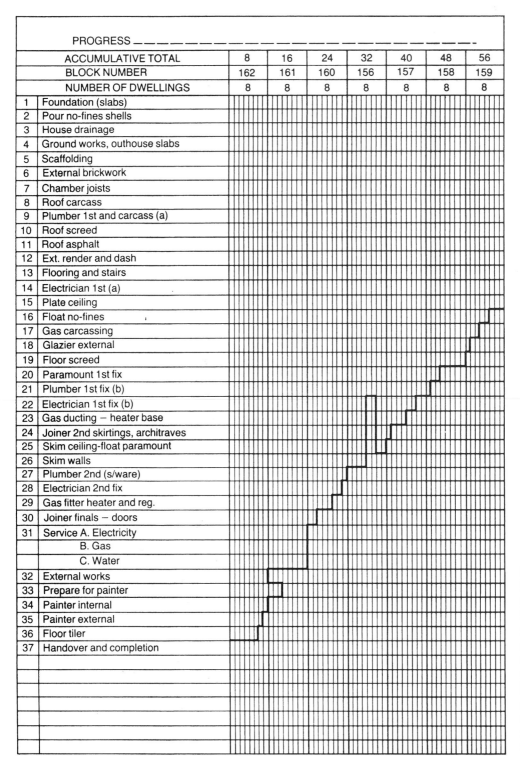

		PROGRESS —						
	ACCUMULATIVE TOTAL	8	16	24	32	40	48	56
	BLOCK NUMBER	162	161	160	156	157	158	159
	NUMBER OF DWELLINGS	8	8	8	8	8	8	8
1	Foundation (slabs)							
2	Pour no-fines shells							
3	House drainage							
4	Ground works, outhouse slabs							
5	Scaffolding							
6	External brickwork							
7	Chamber joists							
8	Roof carcass							
9	Plumber 1st and carcass (a)							
10	Roof screed							
11	Roof asphalt							
12	Ext. render and dash							
13	Flooring and stairs							
14	Electrician 1st (a)							
15	Plate ceiling							
16	Float no-fines							
17	Gas carcassing							
18	Glazier external							
19	Floor screed							
20	Paramount 1st fix							
21	Plumber 1st fix (b)							
22	Electrician 1st fix (b)							
23	Gas ducting – heater base							
24	Joiner 2nd skirtings, architraves							
25	Skim ceiling-float paramount							
26	Skim walls							
27	Plumber 2nd (s/ware)							
28	Electrician 2nd fix							
29	Gas fitter heater and reg.							
30	Joiner finals – doors							
31	Service A. Electricity							
	B. Gas							
	C. Water							
32	External works							
33	Prepare for painter							
34	Painter internal							
35	Painter external							
36	Floor tiler							
37	Handover and completion							

Figure 18.8
Progress
recording on
housing sites
designed to show
up loss of
sequence in
terraces and
blocks of
dwellings. See
also value in
refurbishment,
Chapter 2.

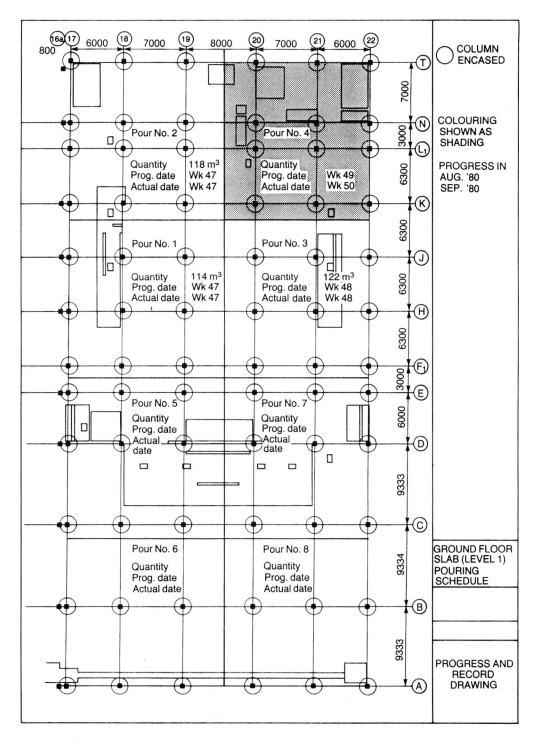

Figure 18.9   Record drawing for completion of ground floor slabs.

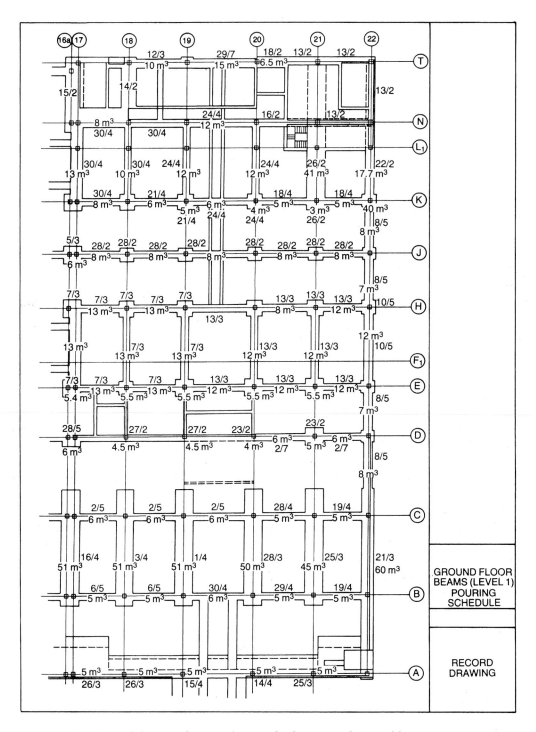

Figure 18.10 Record drawing for completion of pile caps and ground beams.

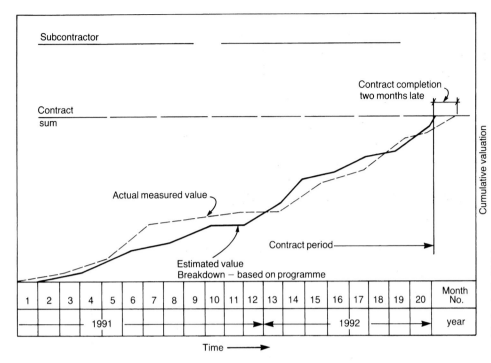

Figure 18.11
S-curve record for subcontractors' valuations against estimated performance.

drawn (Figure 18.11). This results in what is usually known as an S-curve. As monthly valuations are received they are plotted against the estimate. While descrepancies are inevitable, over valuations will show up if carried out on a regular basis and questions raised to find out why.

On complex contracts, an S-curve should be prepared for each subcontractor. Not a difficult matter when desk top computers are available.

The S-curve has already been discussed in Chapter 16, when its use was described in relation to Materials Reconciliations.

Further information

1. Many computerized programmes are on the market for dealing with programme presentation and recording progress and financial control. The latter especially in relation to subcontractors' control. Bar charts are the usually favoured presentation and can be produced in many forms and in colours for presentation.

2. Weekly planning is best done manually on the lines of the examples already illustrated. Such programmes are essentially for site management use only and should not be disclosed to the client or his representatives.

3. The opposite applies with record drawings. As these may form the basis of a claim, it is useful to have them initialled by the client's representative to provide factual evidence.

Reference

1. Chartered Institute of Building (1991). *Planning and programming in construction – a guide to good practice*. CIOB, Ascot.

Chapter nineteen

The role of planning in claims

Mention has already been made, in the Introduction to this book, that those involved in the planning of construction methods have a particular role to play in this context. How such matters fit into the construction planner role, overall, has been shown in diagrammatic form in Figure 2, in the Introduction.

In this chapter, the whole topic is examined in more detail. The Standard Form of Building Contract terms are taken as an example[1].

All forms of contract contain provision for redress if the contractor is involved in extra cost over the tender sum by virtue of alterations and delays by the architect or his representatives compared to the original basis on which the contract was priced. At the same time the main or management contractor needs to be aware that subcontractors, equally, can claim against them if alterations and delays are caused by the main or management contractor.

Although the responsibility of preparing and presenting a claim is that of the particular contractor's Quantity Surveyor, success in this field depends very much on the efforts of the whole construction team – Contracts Manager, Agent, Engineer, Planner etc.

Those involved in planning and contract control are often closer to actual site operations than the Quantity Surveyor. In consequence,

they have a very important contribution to make to any well prepared claim.

An important part of the process of monitoring the supply of information and construction progress is the identification of potential areas for claims. With the need, in present conditions, to maximize profitability, it is essential that all extras to which a company is entitled under the contract are identified and extra monies resulting recovered.

The areas that may prove to be claim material are project conditions of all kinds that differ from those set out in the Tender documents, particularly the Bill of Quantities and the Tender Programme. If such differences occur, through no fault of the contractor, a case for claiming under the terms of the contract in relation to time and money can usually be made. The type of events from which such can arise will include:

(a) Site possession problems;
(b) Variation between tender and actual construction conditions, sequence of operations etc.;
(c) Influence of Architects Instructions/Variation Orders;
(d) Provision of late information;
(e) Variations from accepted or standard practice/Code of Practice recommendations, etc.;

(f) Delays by Nominated Subcontractors;
(g) Contractor/Subcontractor relationships;
(h) Extension provisions.

Each of the above headings can now be examined in detail.

Site possession problems

If the contract conditions have laid down that the complete site will be available on a specific date, the Master Programme and Method Statement (Chapter 17) will have been based on this premise. On starting the contract, a check needs to be made to see that site possession conditions stated in the tender documents have been met.

Complete possession not only applies to ground surface conditions, but also to underground and overhead services and utilities. To determine the status of services in and around the site, it is essential that contact with the Utility Companies is made as soon as possible after the contract is awarded, and not later than the date of site possession. Sometimes 'unknown' pipes or cables can be found that require termination or diversion. Alternatively, services may have to be maintained across the site because diversions may not have been completed by the date stated in the tender documents. Terminations, diversions or maintenance of services not adequately described in contract documents should be the subject of an Architect's Instruction. Even where such work is adequately described, drawings from utility services must be issued through the Architect and work undertaken only after receiving his agreement to the course of action intended.

If buildings or obstructions should have been cleared or demolished by the date of site possession, a check is necessary that this has been done. Delays in road closures or footpaths or service diversions can be particularly disruptive to site progress. Careful records of dates of availability of areas of the site that were not part of the initial handover must be kept.

In refurbishing work, a check that the building is handed over in the condition and at the time stated in the contract documents is essential.

When assessing the effect of delays related to site occupation, it is important to remember that besides the direct delays attributable to lack of possession, indirect effects like inefficient working of plant, additional visits by subcontractors need to be identified and taken into account.

Variations between tender and construction information

The ease of assessing the changes in construction information issued to the site compared to that included in the tender scheme depends greatly on the amount of tender information available. Often tender documents state that drawings information are available at the Consultant's office for inspection, without including them in the tender package sent to the contractor. A list of completed drawings etc. which are available at the Consultant's office for inspection should be available from the Estimator. Frequent cases arise where requests for copies of such drawings are refused, even though offers are made to pay for them. Where this happens, note should be taken of content and any form of numbering that appears on the drawing. If the Architect or Engineer is willing to allow copies, they should be made and filed against any later amendments.

Such drawings/information should form part of the signed contract documentation. If they do not, then careful consideration has to be given to the degree of influence such items had on the tender thinking, and what recognition is to be given to this information when reviewing ongoing site problems and claims.

The comparison of tender drawings/contract documents, which will be available to site, with working drawings must be undertaken on a continuous basis throughout the construction period, so that ambiguities, amendments and matters requiring clarification are highlighted for further investigation by the site team.

One recurring source of variation between actual and tender conditions is related to site ground conditions. The designers normally provide either bore hole or trial pit information for the site, which should have been checked at tender stage. Problems can occur if only bore hole information is provided. This will give information suitable for piling, but not necessarily a good idea of ground conditions for a metre or so at the surface, where foundations may occur.

If ground conditions are different from those described, full records of quantities excavated etc. must be kept by the site to ensure that any claim is factual. Similarly, full records should always be maintained of the concrete placed in bored piles and all other foundation work. See also Chapter 16, dealing with the control of waste on site.

Influence of Architect's Instructions/Variation Orders

Following the receipt of Architect's Instructions/Variation Orders, site management must identify the effects that such instructions are likely to have on the progress of the works, or any section thereof, as quickly as possible. If the resulting opinion is that delays are likely to occur, in whole or in part, the Architects or Engineers should be notified in the simplest terms. In such initial advice, premature observations on the detail or extent of delay must not be made. If subcontractors are involved, they should be informed as soon as possible and asked for their view of the impact on their part of the works.

When commenting on Architect's/Engineer's Instructions and progress generally, optimistic predictions regarding progress of completion dates must be avoided. The contractor must reserve his rights to up-date the Architect/Engineer as the situation develops. Only when it is

practicable to do so should further particulars be given. A definition of the term 'Variation' is provided in JCT 1980, Clause 13.

Alterations to the Master Programme which vary the date of completion should not be made until an official extension of time has been granted by the Architect. As noted above, the Architect must be kept advised of matters which affect the Master Programme. When such matters, in a contractor's opinion, warrant a revision of programme, then a well-disciplined course of action should be agreed upon by the whole construction team, the object being to ensure that the necessary extension of time is secured. Only then can a revision to the Master Programme take place.

It should be remembered that informal or unsubstantiated alterations to the Master Programme will have far reaching contractural and financial effects upon the main contractor by way of subcontractors' and suppliers' claims.

Provision of late information

An Information Required Schedule should be supplied to the Architect displaying how a contractor needs information to be provided, to progress the works in accordance with the Master Programme and achieve the contract completion date given in the contract documents.

The Architect and through him his professional team should be reminded at regular intervals of the information which remains outstanding, and that due to be issued in the near future, say, the following two months.

Should information not be provided in accordance with the schedule, and the contractor feels that this will have a bearing on contract progress, then the Architect should be so informed by letter. It is stressed that an initial letter advising the late provision of information does not necessarily need to have contractural overtones.

Variations from standard practice

This grouping of potential circumstances for claim would arise when the actual construction techniques used on the project differed from those that could have been expected from the tender information and result in extra time and cost to execute.

An example of this type would be storey-height walls that needed to be poured in two lifts instead of one, the need arising from reinforcing details not available at the time of tender, nor covered in any description in the tender document.

In this example, the effect would be to increase both time and cost, over what would normally have been allowed in the tender method planning, for formwork, reinforcement and concrete placing. Clearly the basis for a claim.

Similar circumstances could exist, for instance, if an Architect insisted on a particular pattern of pour and stop-end joint lines, that had not been indicated at the tender stage. Such requirements usually involve additional formwork and greater effort in the concrete pouring.

Extreme tolerance requirements would also come under this general heading. For guidance in this respect as to whether the Architect's specifications are unreasonable, BS 5606, 'Accuracy in Building'[2] and BS 5964, 'Methods for setting out and measurements of buildings – Permissible measuring deviations'[3] provide the necessary chapter and verse.

Various circumstances like these, which can occur at any stage of the project, should be recorded and brought to the Quantity Surveyor's attention as the basis for a claim.

Delays by nominated subcontractors

At the time of tender, the construction planner rarely has any information about the extent of a nominated subcontractor's work. Usually, the Bill of Quantities will provide a P.C. or Provisional Sum to cover a nominated subcontractor's share of the contract. At the same time it is rare for the specialist subcontractor's name to have been established at this stage.

Specialist contractors, in building construction, are usually determined after the main contract has been let – and only then do such contractors settle down to detailed design.

The main or managing contractor therefore faces a dilemma at the tender stage. He has a bulk cost assessment (common to all tenderers), but is unable to contact a name for discussions on anticipated time for the specialist's completion period. In providing the Tender Programme (to become the submission programme), assumptions have to be made in relation to time scale for all specialist subcontractors. The tender document submitted has of necessity to point out that, while the main contractor has used his best endeavours, it will be necessary for all specialist subcontractors to conform to the Master Programme submitted with the tender, if the stated contract time is to be met. If this proves to be unattainable, the main contractor will need to receive an extension of time and reimbursement, unless a nominated subcontractor's cause of delay has been occasioned by the main contractor or his other subcontractors failing to comply with the Master Programme.

Subcontractors' claims

Subcontractors' claims really fall into two categories: claims against the main contractor for delay or failure to pass on information, or claims which pass back through the main contractor to the client due to the sort of contractual matters already discussed.

In the first case, the claim results from the main contractor's failure to provide access for particular operations at the agreed time or to supply materials or use of plant as agreed in the subcontract document. In such cases, the main contractor needs to proceed with caution. Subcontractors in general are usually more claims conscious than the main contractor. They will, perfectly correctly, endeavour to capitalize upon any deviation from an agreed programme and from what the subcontractor envisages to be their proposed method of working. While the main contractor may wish to tie subcontractors to key dates (although not in the subcontract documents), they, the subcontractors, have a broad obligation to undertake and complete the the work in conjunction with the progress of the main contract works. Before the planner gives consideration to the viability of a subcontractor's claim, the contractural validity must be established with the main contractor's Quantity Surveyor.

At this point it is important to distinguish between delays alleged to have been caused by the Architect or his professional team and those blamed on the main or other contractors working on site. Accurate recording of the project progress throughout its duration will obviously greatly aid in the checking of any subcontractor's claim. Such records are equally important in the reverse case of a main contractor wishing to pursue a claim/extension of time against a nominated subcontractor. Very detailed monitoring and documenting of the subcontractor's progress against the agreed commitment is essential if the claim is to be secured.

Extension provisions

Claims for extension of time come under the Quantity Surveyor's role. The planning engineer has a role to play in drawing to the Quantity Surveyor's attention cause for claim in any area in which he is involved.

References

1. Joint Contracts Tribunal (1980) *Standard Form of Building Contract (JCT 80)* London. 3rd Edition.
2. British Standards Institution. *Code of Practice for accuracy in building* BS 5606 : 1978 BSI London.
3. British Standards Institution. *Methods for setting out and measurement of buildings.* BS 5964 : 1980 BSI London.

Chapter twenty

Facade retention and structural alteration

The redevelopment of commercial properties in the light of present-day needs is a major aspect of construction today. One or both of two major items usually come into play: facade retention and structural alteration. The first item is often a condition of planning consent, namely maintaining the architectural appearance of the area, preserving a listed facade while alterations to the inside take place. The second item usually arises in old buildings, Victorian, Edwardian and commercial structures built in the early 1930s. In such times, the storey height was considerably more than the law requires today, even allowing for the needs of modern communications systems. In such cases, it is often possible to achieve a further storey without altering the building height, which is of considerable value in high rental areas of cities. In either or both situations, the facade will have to be retained while a new structure is being erected within the shell.

In replacing the internal structural form, in either of the situations above, the retaining of the outer shell will be paramount, both in relation to the work force and the general public passing by. Suitable temporary works specially designed for the one-off situation will be crucial. Such Temporary Works can be seen as falsework in reverse: 'Any temporary structure erected to support a permanent structure before it is made non-selfsupporting.'

Legal requirements

The major legislation applying, as far as the new works are concerned, are:

1. The Health and Safety at Work etc Act 1974, particularly in relation to safety of the Public;
2. The Management of Health and Safety at Work Regulations 1992;
3. The Construction (Design and Management) Regulations 1994;
4. The Construction (Health, Safety and Welfare) Regulations 1996;
5. The Lifting Operations and Lifting Equipment Regulations 1998, effective from 5 December 1998.

Responsibilities

As with the use of permanent works as temporary works, in Chapter 10, clear areas of responsibility need to be established with all parties involved. Since the passing of the Construction

(Design and Management) Regulations 1996, it is a legal requirement for all parties to the contract to liaise and communicate with each other, especially the professional team with the contractor in relation to any areas where design or construction have to deal with Health and Safety. Appendix A gives a breakdown of the Regulations' requirements and the records that have to be kept. Note also that the Client's representative has a significant role to play.

In particular, an accurate structural survey is an essential first step before anything else is started. This needs to have been very thorough under the auspices of the structural engineer. There must also be agreement in relation to any Party Wall award with adjoining owners. All the above needs to be clearly specified to the main contractor before he begins planning the construction sequence and the way in which necessary temporary works will be carried out.

The importance of the survey of the facade to be retained is exemplified by a contract carried out in the early days of facade retention. A row of Nash terraces in London had been bought for office development. All Nash terraces in London are protected and the planning consent required the facade to be retained to match the other terraces in the street. The method adopted was to use raking shores outside with windows and doors framed and braced in timber. Key parts of party walls were retained to support the inside of the facade, with some corner bracing. Everything went well, the new structure completed on time and the time came to confirm the tying in method of the facade to the new work. To the horror of the structural engineer and the Building Inspector, it was discovered that the lintels over the window openings were oak and riddled with dry rot. The secondary effect was the dry rot tendrils had eaten into the lime mortar of the brickwork rendering it soft and unable to hold the tie-in strength needed. The Building Inspector declared the facade a dangerous structure. The only way the work could be finished involved

taking casts of all the facade details and making glass fibre copies. While this was going on, the facade had to be cleared of dry rot, demolished and re-built to accept the glass fibre architectural features applied. The moral was clear – a detailed inspection at the start by properly qualified surveyors is an essential feature of facade retention.

Methods of facade retention

The approach to securing the building facade has to be decided in relation to available established methods and any conditions laid down by the local authority.

Five basic methods are available.

1. As shown in Figure 20.1. External raking shores in association with buttresses internally formed by leaving sections of any party walls, as shown in the figure, which abut the facade. Horizontal struts behind the facade can be used to connect through to the external shores. The disadvantage of this method is the space needed to accommodate the raking members in front of the building. Such space often cannot be made available unless partial road closure is permitted, which local authorities are apt to frown upon.

   To-day, the method is rarely used. The method shown in Figure 20.2, occupying no space in front of the building, is preferred by all where walls exist.

2. As shown in Figure 20.2. The party wall buttresses are maintained, but are strutted internally by horizontal shores giving both lateral stability and a means of securing them through window openings. Some obstruction internally to new works will arise.

   The situation in Figure 20.2 need not be as onerous as first anticipated. When converting old town houses to office accommodation, the existing window arrangements have to be maintained. With Victorian town houses, the storey heights are higher than

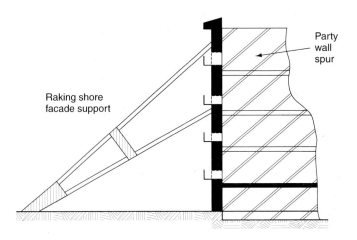

Figure 20.1 Party wall buttresses and raking shores

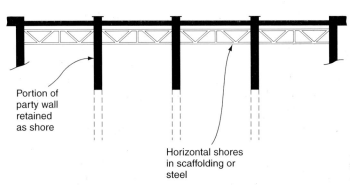

Figure 20.2 Party wall buttresses and horizontal bracing beams

required for modern offices and often allow the possibility of gaining an additional floor without altering the original building height. The immediate area behind the facade will have to maintain the original floor levels but can be accommodated by a split level design.

3. As shown in Figure 20.3. This approach provides for external support, but limits the

space needed outside the facade. It also has the advantage that no internal buttresses are needed. The facade is tied back to the external support system and unobstructed working space becomes available, an important factor where a site is a very restricted one. This situation is likely to apply when altering commercial structures. With this method,

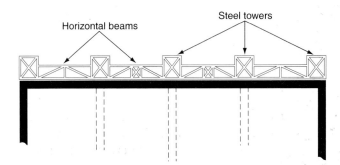

Figure 20.3 External steel towers and horizontal steel bracing beams.

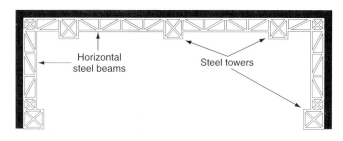

Figure 20.4   Internal steel towers and horizontal steel bracing.

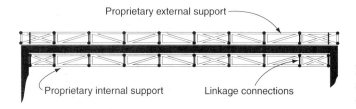

Figure 20.5   External and internal steel support systems linked through window openings

Figure 20.6 External facade support using proprietary equipment. Note the covered pedestrian way in the street as the support system covers the pavement. (Rapid Metal Developments Ltd).

the steel towers and linking horizontal beams can either be designed in structural steel specially for the one-off situation or more frequently by firms who supply form-work and falsework. Several firms have developed structural lattice systems, based away from the original steel or aluminium soldiers associated with wall formwork.

With accessories designed for the purpose, most facade retention problems can be dealt with. Design to high standards is provided with the supply of equipment on a sale or hire basis. Later figures illustrate this method. To-day, this approach is arguably the one most used, the supply firm taking responsibility for the design (see Figure 20.6).

4. As shown in Figure 20.4. It will be clear that this is simply the reverse of Figure 20.3. The same principle is used but internally instead of external.

5. As shown in Figure 20.5. A method to be avoided if at all possible. It is very expensive, and normally only used where one wall has to be supported in isolation, solely by the falsework. No other building support affords help. Figure 20.5 does not illustrate the provision needed for lateral stability against wind and unforeseen loads. If adequate foundations cannot be provided, heavy kentledge will have to be used in lieu.

Of the above options, choice will depend on any restrictions imposed by the local authority. From the contractor's point of view, Method 3 has to be the preferred option. Demolition can be completed once the falsework is in place, giving unrestricted working for the new structure.

Factors affecting choice of method

The initial action will be to discuss with the local authority their views when putting forward what you would like to do. Items for discussion would be:

1. If the facade fronts on the pavement, can the support structure bridge over the pavement (Figure 20.7)?
2. If the answer is yes, what requirements are there for protection of people walking underneath? Anything in excess of legal requirements for normal pavement gantries?
3. What standard of drape netting is required to stop things falling onto the roadway?
4. If necessary, can part of the roadway be used.

If the answers are favourable, external support is viable. If not, an internal system will be needed. In this case there are several options available:

Figure 20.7 External facade support – towers and horizontal support beams. Set back but pavement provided with covered walkway. (Mabey Ltd).

1. If the convertion is of old Victorian or prewar town houses, the facade can be dealt with by leaving a part of each party wall to form buttresses with linking horizontal beams or struts to provide lateral stability (Figure 20.8). If need be, the new structural design can be detailed to allow construction party wall to party wall in sequence (Figure 20.9). Guidance on tradition construction in this area is contained in a Construction Industry Research and Information Association Report 111[1].

2. If the existing building was offices or other commercial use, no party walls will be in place to help. In this case, an internal support system will have to be used based on vertical towers linked together by horizontal

Figure 20.8
Demolition of four
town houses leaving
stub party walls.
Flying shores in
scaffolding tie
everything together.
(G. Wimpey plc).

lattice beams providing lateral stiffness. Inevitably some restriction will occur in internal movement to the new construction (Figure 20.10).

Legal requirements in relation to adjoining owners

Rights of support

Halsbury's Laws of England defines the natural right of support 'as giving every owner of land as an incident of his ownership the right to pre-

Figure 20.9 Site as in Figure 20.8 but demolition complete. Note two 'house' bays structurally complete with foundations in hand for the remaining two 'bays'. (G. Wimpey plc).

vent such use of the neighbouring land as will withdraw the support which the neighbouring land naturally affords to his land.'

It follows that any adjacent excavations must not cause withdrawal of the support to the adjoining land or cause damage to buildings which may rest upon it.

Agreements with adjoining owners

Taking the situation in inner London, which is as demanding as any. The GLC General Powers Act 1966 deals with (amongst other things) excavations within 9 m of a highway. Elsewhere, in the UK, the same control is taken over by the Highways Acts.

Figure 20.10
Mixture in internal and external supports to large departmental store conversion. The left hand side has external support and the right hand side internal support (external support was forbidden on this face). (Trollope and Colls Ltd).

Under Part VI of the London Building Acts (Amendment) Act 1939, a party wall award has to be agreed between adjoining owners when new construction is to take place against an existing building. The award will contain details of the measures required to protect the existing building while the adjoining property is being developed, both in respect of stability and protection from the weather. Such matters are normally agreed between the architects representing both parties. At this stage, it is prudent for the contractor involved to contact the adjoining owner for permission to do a schedule of conditions within his building. This is to protect the contractor against bogus claims for damage to adjoining property. Such schedules are ideally supported by good quality commercial photographs with date information on them. Occasions arise when an adjoining owner refuses entry for such purposes. In such situations, the contractor needs to be very vigilant when working on or adjacent to a party wall.

Safety of buildings is regulated in London by the London Building Acts (Amendment) Act 1939; elsewhere in the UK by sections 76–83 of the Building Act 1984. If the new building owner wishes to place his foundations at a lower level than that of the adjoining owner's and within 3 m or below an angle of 25° and within 6 m, or in any case where a party wall is cut into, underpinned, raised etc., notices must be given under Part VI of the London Building

Acts (Amendment) Act 1939. Elsewhere common law applies, but quite often building owners and adjoining owners (or rather their representatives) agree to operate the procedures under Part VI.

Provision of support structure

Once the principle of support has been agreed by all the parties involved, consideration has to be given to letting out the work of designing and erecting the support structure. Today two main options are available: Structural Designers who have begun to specialize in facade retention and structural refurbishment, which may include a demolition option as well. Secondly, formwork and falsework companies who have developed formwork equipment which with special accessories can become a structural Meccano set for facade retention. Such firms have computer-aided design facilities and take responsibility for design, erection and tying into the facade. There still remains the option for contractors with experienced temporary works design teams to do their own thing. But the ability to hire support equipment usually will lead to the formwork companies being in the favoured position when quotations are received.

Divisions of responsibility

What requires careful examination, when quotations are studied, is to make sure all items are covered and each participant knows the extent of work for which he is responsible. This will include: facade retention contractor, Main contractor, main Structural Engineer, and the local Building Inspector. A meeting needs to be held between these parties and written agreements recorded. With the advent of the Construction (Design and Management) Regulations, the over-riding control now rests with the Planning Officer operating as the Client's representative.

A sample quotation, for a relatively small facade retention job, is shown in Example 20.1. This gives a clear division of responsibilities. On receipt of such a tender, the main contractor must check to see that no item has been unaccounted for.

Demolition

Once the facade retention system has been installed to everyone's satisfaction, demolition of the internal structure can begin. First, the facades shored will need a suitable mesh draping to prevent debris from falling into the street or on pavements to protect pedestrians and passing traffic.

Whenever structural demolitions are required, clear instructions must be provided by the structural engineer. The contractor, for his part, must make a point of checking whether or not instabilities may arise in the course of demolition, which may not be readily apparent and would need additional temporary works.

Demolition is covered by bylaws in inner London in respect of noise, dust, hours of work etc. made under the London Government Act 1963. Elsewhere, equivalent bylaws occur in the 26 different areas of England and Wales. Notice of demolition must be given to the local authority under Section 80 of the Building Act 1984. In addition, notice must also be given to the Health and Safety Executive under the Health and Safety at Work etc. Act 1974.

It will be clear that Sections 3 and 4 of the above Act which relate to safety of the public and others who may be employed, visit or require access to the site, must be carried out to the letter.

Decisions will need to be made on the methods of cutting out of structural members and the most effective means of their disposal. Noise factors will arise in closely occupied areas and the contract method will need to take

Example 20.1

Dear Sirs,

Re: Oxford Street/Wardour Street, London W.1

Please find enclosed our tender drawing No. 904/SK01 and draft programme 904/0P01, for works to be carried out on the above Contract. The tender figure comprises the following items:

1.	Provision of bases for shoring;	9,899.00
2.	Facade Retention of Nos. 111–125 Oxford Street and No. 186, Wardour Street, as B.O.Q. item 4/D/13A;	43,068.00
3.	Demolition of buildings behind facade of Nos. 111–125 Oxford Street and No. 186 Wardour Street, as B.O.Q. item 4/D/14A;	33,475.00
4.	Strip out shop fronts and windows above including bracing and boarding up – B.O.Q. items 4/D/2C–4/D/6L inclusive 4/D/8A–4/D/12F inclusive	9,705.00
5.	Erect hoarding as B.O.Q. item 4/D/1A:	13,219.50
	Total:	109,366.50

We have based our quotation on the following items to be completed by yourselves:

- facade survey;
- remove to store gable and finials;
- divert services if required;
- provide tower crane 2.5 ton hook capacity – Oxford Street;
- break-out bases on completion and reinstate if required;
- raise pavement locally at towers.

Our tender includes for:

- steelwork to be shop painted 75 microns zinc phosphate 1 coat of any paint, damage areas made good on site;
- all steelwork is to be workshop fabricated and part assembled for quick site erection;
- we undertake full design responsibility in all aspects of the shoring.

If you wish to discuss the above or have any queries regarding our tender figure, programme of works and scheme, please do not hesitate to contact us.

Yours faithfully,
for James Brailsford Associates

Figure 20.11 Site with very tight access which required support to part of the facade to get the piling rig, ready-mixed truck and backhoe in. Note party wall protection. (Westpile Ltd).

notice of the Control of Pollution Act 1974, Sections 60 and 61, which cover noise and its limitations, both in relation to the public and operatives on site. In the case of demolishing concrete structures, the use of diamond saws or thermic lancing will be the favoured methods. Both are virtually silent, but thermic lancing produces large quantities of unpleasant fumes. The establishment of generous ventilation is essential. Where steel structures are involved, oxy-acetylene cutting will be the simplest solution. While noiseless, it, too, generates fumes which require ventilation facilities and should never be carried out in confined spaces. Finally, collection and disposal of broken out material has to be organized, often in very restricted spaces.

Construction of the new works

Foundations

Once the support system is in place and the necessary demolition has been completed, work can be started on the new structural requirements. New foundations will undoubtedly be needed and may well require piling. Bearing in mind the need to avoid undue vibration in the facade support and any requirements of the party wall award, the use of continuous flight auger, or rotary auger systems will be essential. Getting the rig onto the site and muck away may be a problem. Figure 20.11 illustrates a number of points in this respect. Note the structural steel framing in the facade to allow access to the site. Figure 20.12 shows the inside of the support system: horizontal beams connected through window openings to the main support structure outside the building.

Figure 20.13 gives a vivid picture of the congestion of plant on this narrow site, with piling

389

Figure 20.12 Same site as in Figure 20.11 but showing internal strapping to tie into steel tower externally. (Westpile Ltd).

Figure 20.13 A vivid picture of plant congestion on this very narrow site. In the foreground is a newly complete base for a tower crane. Note clear view of party wall protection method. (Westpile Ltd).

Figure 20.14
Proprietary flying
shores in conjunction
with party wall stubs.
(Mabey Ltd).

rig, back hoe and in the foreground a newly completed base for a tower crane. Figure 20.14 shows flying shores used for facade retention.

Note, in all the above figures, the bituminous felt weather proofing to the adjacent party walls, held in place by battening.

It will be clear that no further work will be possible until all aspects of the foundations are completed, on a restricted site such as the one illustrated.

Superstructure

Whatever the superstructure type, erection can proceed as normal – with the tower crane as illustrated above or with local Police agreement (and Local Authority) to use mobile cranes from the street and, in in-situ concrete structures, pumping concrete from the street.

Once structurally complete, the facade can be tied back to the new structure. The architect, in conjunction with the structural engineer and local building surveyor, will be responsible for providing all necessary details.

Reference

1. Construction Industry Research and Information Association (1986) *Structural renovation of traditional buildings Report No. 111*. CIRIA, London.

Further reading

1. Rolph-Knight, L. (1984) The facade can be a nightmare. March 1984, *Civil Engineering.*
2. Highfield D. (1984) Building behind historic facades. *Building Technology and Management*. January 1984.
3. Goodchild, S.L. and Kaminski, M.P. (1989) Retention of major facades. *Structural Engineer*. Vol. 67, No. 8, 18 April 1989.

Chapter twenty-one

Refurbishment of dwellings

The refurbishment of dwellings relates to those occupied by people for domestic purposes. As such, the circumstances surrounding the work are quite different from those described in relation to facade retention and structural alteration. There is no change of use, merely the upgrading of the accommodation for the tenants. (Refurbishment is primarily for local authority property.)

This type of work is now big business for contractors as more and more sub-standard dwellings require attention to avoid the need to demolish and rebuild at greater cost than the refurbishment option.

The decanting or otherwise of tenants

Whatever the client's brief may say to the contrary, anything more than a small degree of demolition and structural alteration cannot be undertaken satisfactorily with the tenant in residence. The advantages of tenants being in residence while refurbishment takes place are:

Advantages

1. Vandalism is eliminated.
2. Theft from dwellings re-construction is reduced virtually to nil.
3. Greater economy in materials is effected because more careful repair takes place of walls, floors, skirtings etc.
4. Quality is noticeably better because of the ever-present 'Clerk of Works'.
5. Better personal relationships with the tenant are developed, leading to virtual elimination of criticism and trivial remedial work.
6. Clearing up is done by the insistence of the tenant as the work progresses and this discipline obviates the need for expensive clearing up by a separate gang later on.

The disadvantages, however, are very real if the extent of the refurbishment requires a lot of replacement and major alteration:

Disadvantages

1. Enforced conversation with tenants can lead to lost time. It has been recorded in one company that it amounts to 1 hr lost per 8 hr day.
2. Protecting and working around furniture leads to inefficiency. This could amount also to 1 hr per day.
3. Services have to be maintained. Normally this involves a separate electricity from the meter box for heat and light, a temporary cooker and toilet facilities.
4. Demolition is difficult because it cannot be tackled man fashion.
5. A proportion of dwellings will always suffer inefficiency because of social problems – sickness, invalids, young children, the elderly, night shift workers etc.
6. It is found in practice that a small but incalculable percentage of dwellings has to be decanted of necessity because of problems encountered after work commences.

More generally, if tenants have to be moved into transit dwellings, it is of the utmost importance to allow a longer possession period for the contractor, ideally an extra two weeks. This is needed to allow for notification of the move to tenants, organizing removables etc. These factors will have a substantial effect upon the number of dwellings required for occupation by the contractor, using the formula given below.

Greater likelihood exists of restrictions being imposed upon the planned output where decanting takes place than where it does not. Such restrictions invariably are due to the client experiencing difficulty in finding alternative accommodation for tenants. This is a common occurrence in refurbishing work and, whilst a clear-cut claims situation should present itself, it is prudent to have categorically told the client in the tender letter the number of houses to be decanted at one time to meet the demands of the programme.

Number of working fronts proposed

The decision to allow one set up over a long period, or two fronts over a shorter period, will depend on whether or not the client has imposed a contract period or a hand-over rate. If the contractor is required to state a contract period and the hand-over rate, the number of fronts is determined in accordance with the principles recommended below under the heading 'Continuity for Trades'.

Opinion is often divided regarding the relative merits of single or double front working. A single front producing 10 dwellings a week has the advantage of greater flexibility for labour (which is of great importance in refurbishing work) than a double front decision, producing five dwellings a week. On the other hand, the latter method induces competition between gangs.

The physical location of the dwellings

This obviously has a bearing on planned output. If houses are released in the same street, greater management control is possible with higher output. Scattered release causes less control of labour with a loss in efficiency, while materials security and control is made more difficult. To counter this, extra supervision will be required.

Unless specific information to the contrary is contained in the client's brief, or Bill of Quantities, the need for the release of dwellings in reasonable sequence is of paramount importance.

The number of dwellings available for contractor occupation at any one time

On house refurbishing, the minimum number of dwellings required from the client at any one time can be calculated:

Weekly handover rate ×
Number of weeks in cycle
= Number of dwellings required.

If the number for occupation is given in the brief, this formula can be used either in calculation of the programme period or in checking the client's contract period for feasibility. The calculation is based on a perfect situation. Additional dwellings for occupation should be allowed to account for any operational fluctuations or delays. This may well vary on different tenders, depending on different circumstances. An additional 20% is not excessive.

Continuity for trades

The most efficient approach when aiming at continuity is to determine the optimum weekly output of dwellings and plan the resources around it. When the required momentum has been achieved, improvements should be directed towards trimming labour levels, not attempting better outputs with the same labour. The temptation to push one trade ahead because scope exists will lead to lack of continuity for, of necessity, refurbishment work involves many minor activities, with large time gaps between. By way of example, if the man weeks for an average house refurbishing are calculated, they will be found to total 20 man-weeks. On this type of work, a reasonable gang size per dwelling is 8 operatives. Thus:

20 man weeks ÷ 8 operatives = 2.5 man weeks per cycle.

In practice, this is impossible and the average number of men able to work each week in a dwelling will be found to be only 2 or 3.

As dwellings are released by the client only on receipt of a similar number of completed dwellings, the importance of getting the cycle time correct at the start is obvious.

It is important, at the contract stage, not to accept the keys of a dwelling too early and commence piecemeal working because it is uneconomical and it allows the client later to claim that the cycle time commenced on the date of hand-over of the keys.

Assessing work content

Careful examination of work content should be made at tender stage, because of its bearing on the approach to the contract (Figure 21.1). For example, it is common to have to demolish and rebuild rear extensions containing bathrooms, kitchens etc. Yet this work does not normally apply to every house. The result is to have labour requirements fluctuating unacceptably. An economic solution is to have a separate gang, working in advance to complete the shells of the extensions as a continuous function.

The cycle time per block or dwelling

It is essential that this is determined with the best possible accuracy since underestimation will result in progressive losses against programme throughout the contract with resultant over-run, and overestimation will render the tender less competitive.

To stress the importance of cycle time estimation, it must be remembered that lost time is far more difficult to retrieve on refurbishment contracts than on new construction, primarily because of rigid programmes agreed with the client for occupation/decanting of the dwellings. In addition, liquidated damages are sometimes geared to the over-run on each dwelling or block of dwellings, and therefore directly related to the cycle programme.

A simple bar chart showing sequence of work in detail is a straightforward method of illustrating cycle time. While target hours from previous experience will serve as a guide, they must not be used when calculating labour content and cycle time for the reasons stated above in paragraph 'Continuity for Trades'.

Tender programme format

Having produced a detailed bar chart in arriving at the unit cycle time, it will normally be sufficient to produce only an elementary tender programme indicating the following information:

1. number of dwellings commenced;
2. cycle time;
3. number of dwellings occupied;
4. rate of handover of houses;
5. contract period.

A typical format is illustrated in Figure 21.2.

A consensus of site management experienced in this type of work is that no learning curve is needed on straightforward refurbishing due to the simplicity of the operations. A learning curve allowance will be essential in situations where there is substantial demolition, temporary support of floors, removal of structural walls and building in steel joists etc. Such work will need very careful planning in detail at the tender stage. A detailed case study of such a contract is given at the end of the chapter.

House type ____________ Address ____________________ date ________

Internal	Living Room S	Living Room F	Hall & Stairs S	Hall & Stairs F	Landing S	Landing F	Bedroom 1 S	Bedroom 1 F	Bedroom 2 S	Bedroom 2 F	Bedroom 3 S	Bedroom 3 F	Bedroom 4 S	Bedroom 4 F	Bathroom S	Bathroom F	W.C. S	W.C. F	Sitting Room S	Sitting Room F	External	S	F
Remove Tenant's Fixtures																					Replace Tiles/Slates		
Ceilings – Make good cracks m																					Point Ridge or hip		
" " " patches m^2																					Repoint brickwork		
" Renew entire ceiling m^2																					" stacks		
Walls Make good cracks m																					Replace Air Bricks		
" " " patches m^2																					Replace facing bricks		
" " " whole wall m^2																					Replace render/roughcast		
Remove Air grate & fix plaster vent																					Remake gutter joints		
Floors. Replace damaged boards																					Renew gutters		
" Punch in floor brads																					Renew downpipes		
Doors Replace																					Remove vent pipes		
" Repair																					Replace M.H. covers		
Frames Replace																					Flaunch " "		
" Repair																					Replace gulley grate		
Architraves																					Replace gulley surround		
Skirtings																					Make good cracked paths		
Handrails																					Renew path		
Stairtreads & risers																					Repair gate latch/stile/keep		
Celing trap door																					Replace gate ironmongery		
Windows – Replace beads																					Replace line posts		
" Replace sash																					Lower paving 150 below dpc		
" Replace glass																					Bulk excavation to last		
" Replace W/Board																							
" Replace ironmongery																							
Metal Window repairs																							
Woodworm treatment																							
Dry or wet rot																							
D.P.C. renewal																							
Park central heating																							
Rewiring																							
Replumbing																							

Notes

Title

Repair Schedule

Drg. No.	Rev.
Scale	Drn.
	Date

Figure 21.1 Sample repair schedule.

Preparation of tender programme

The critical factors affecting the preparation of the tender programme are:

1. the cycle time for each dwelling or block;
2. the number of dwellings available for occupation;
3. the decanting of tenants or otherwise of buildings;
4. the number of working fronts proposed;
5. the physical location of dwellings.

Once decisions have been made on the items above, the most convenient format for the Tender programme is illustrated in Figure 21.2.

Assuming the contract has been won, detailed planning can begin in earnest.

Services and the Statutory Authorities

1. Immediately a refurbishment contract is secured, a planning meeting should be held with all the statutory authorities involved. Experience suggests that assuming contact has already been made between the Client and the Authorities is often proved false.
2. Where new services are to be provided in lieu of old, the integration with other work will follow the same lines as for new work.
3. Where disconnection is required, the notice required needs to be checked. Otherwise all other activities could be delayed.
4. Reconnection will need to follow the same lines.
5. A form of systematic documentation is highly desirable to establish that disconnection has taken place, i.e. a form of permit to work. Such a procedure is essential in the case of gas and electricity supplies. NEVER ASSUME THAT DISCONNECTION HAS TAKEN PLACE.
6. In tenanted flats, it will often be the case that electricity has to be disconnected for the working period and then reconnected until the following day. A clear delegation of authority to do this should be given in writing.
7. In respect of water services internally, the planning should always allow the water services to be installed and tested before any final finishes are undertaken. If this is not done, a failure in a joint could cause a good deal of extra expense in redecoration, for example.
8. Where old services have to be removed, especially those in the ground, written assurance that they are 'dead' should be obtained from the statutory authority concerned.

Agreement of work content

The methods of agreeing work content in each dwelling are reasonably standard. They will, however, show some variation depending on whether or not Bills of Quantities were prepared by the client. Each dwelling will need the following procedure:

1. Carry out an initial survey with the Clerk of Works and record the work to be done and enter on a pre-printed survey sheet. A typical sheet is shown in Figure 21.1. This forms the basis of work to be done in that dwelling and is signed by the Clerk of Works.
2. During occupation of the dwellings, further repairs may become necessary. The agreement of the Clerk of Works is obtained and the survey sheet adjusted accordingly. This addition becomes the authorization to carry out the works.
3. After completion of the dwelling, a final joint inspection is carried out with the survey sheet as a reference. Any missing items are measured and entered. Upon agreement, the Clerk of Works completes two copies of the agreement. One copy is given to the contractor's Quantity Surveyor and the other to the Client's Quantity Surveyor.
4. The final account for each dwelling is prepared by the contractor's Quantity Surveyor.

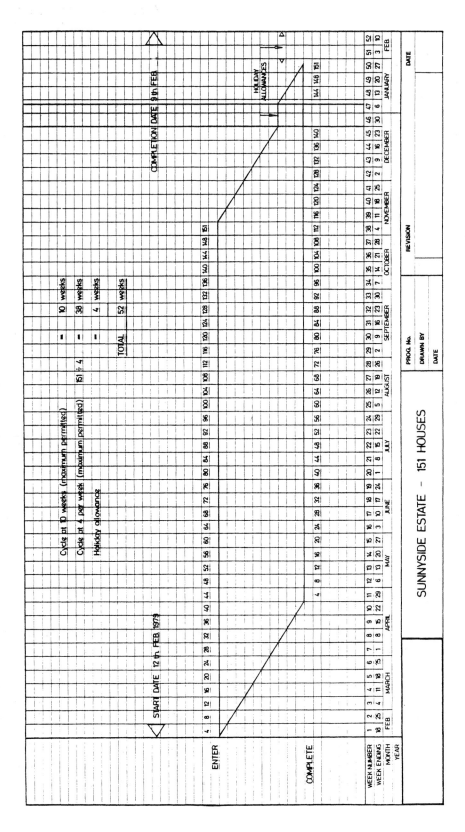

Figure 21.2   Sample tender programme for housing.

If the Bills of Quantities were prepared by the Client and their accuracy checked by the contractor's Quantity Surveyor for each dwelling type, the survey sheet need only cover items items found not to be in the Bills and provide a Schedule of Works.

Because of the variations between houses in work content which may not be apparent until work is under way, it is imperative that a *separate* file is kept for *each* dwelling in the contract, not only containing the up-to-date inspection sheet, but all correspondence and instructions relating to the particular unit. Only in this way will the financial position be safeguarded.

The practical implementation of the above procedures is critical to the progress of the contract. Delays in waiting formal instructions on repair work can disrupt a job because such repair work forms a vital part of the whole. Such repair work should be itemized as it arises and the Clerk of Works pressed for approval.

The construction programme

In producing the actual construction programme, the usual programme type as used on new dwellings will not be satisfactory. It will not be possible, for instance, to produce two programmes, one covering the contract period for submission to the client, the other covering a target period for use by the company only. Once the construction period has been decided at the contract stage, it must be agreed by the client and adhered to. Not only that, the fullest co-operation must exist between the client and contractor for the duration so that releases, occupation and hand-over procedures are effected smoothly and with least disruption to the tenant. Site supervisors agree that the greater the co-operation with the client/contractor the better the progress will be.

Times do arise when the client may request an increase in output while the contract is in progress. While this is normally undesirable because of the disruption factor, it can be effected provided the client clearly understands:

1. He must release appropriately more dwellings to the contractor.
2. He will not benefit in hand-overs for at least a period equal to the unit cycle time calculated from the first day of increased production.

The co-operation factor must involve the following organizations:

1. the Local Authority or other Client;
2. statutory Authorities;
3. subcontractors;
4. suppliers;
5. tenants.

Where tenants remain in residence, the importance of establishing good relationships cannot be over-emphasized as this has a very real bearing upon adherence to the construction programme and the overall success of the contract.

Construction programme format

With housing, a Line of Balance presentation is quite adequate. A typical programme of this type is illustrated in Figure 21.3. It has the benefit of being three-dimensional in illustrating:

1. time scale;
2. output in number of dwellings;
3. individual house addresses.

In this type of work, it is not unusual for the client to make changes in house priority. To avoid changing the construction programme in terms of addresses, which is inconvenient, it is better to leave the construction programme unchanged and use it as the Master Programme on which the contract was based. A supplementary chart can be introduced as shown in Figure 21.4. This type of presentation needs no alteration if the sequence of individual dwellings is altered, yet keeps a close view of the situation. From this, it can be gauged if completions are

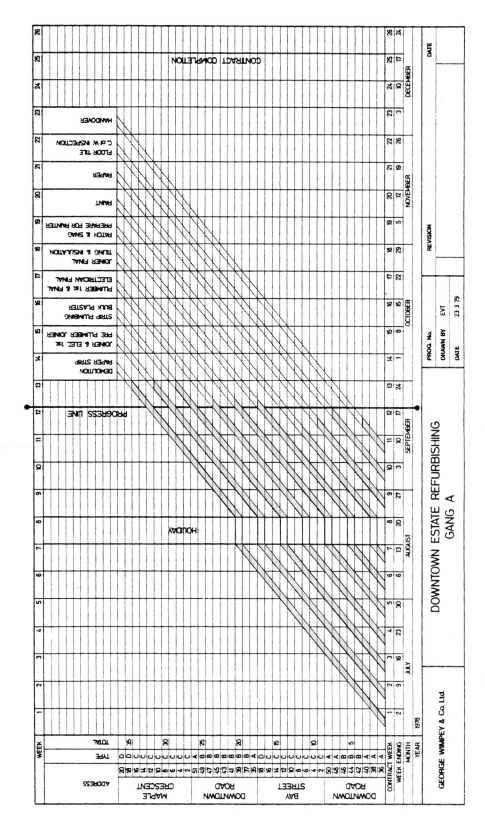

Figure 21.3 Format of construction programme.

	STAGES	ADDRESS	BEECHWOOD DRIVE								ELM DRIVE							BEECHWOOD DRIVE								ELM DRIVE							
			36	38	40	42	44	46	48	50	2	4	6	8	10	12	14	37	39	41	43	45	47	49	51	1	3	5	7	9	11	13	15
		TYPE	A	A	A	B	B	B	A	A	C	C	C	C	C	C	C																
								6		8				12			15						21		23				27				31
A	HOUSE																																
1	ARCHITECT'S INSPECTION	Client	X	X	X	X	X	X	X	X	X	X	X	X	X	X																	
2	DEMOLITION	G.W.	X	X	X	X	X	X	X	X	X	X	X	X	X																		
3	STRIP PAPER	G.W.	X	X	X	X	X	X	X	X	X	X	X	X	X																		
4	CUT OUT & FIX METER BOXES	Sub-con.	X	X	X	X	X	X	X	X	X	X	X	X																			
5	REMOVE FIREPLACES ETC.	G.W.	X	X	X	X	X	X	X	X	X	X	X	X																			
6	STRIP OUT PLUMBING	Sub-con.	X	X	X	X	X	X	X	X	X	X	X	X																			
7	STRIP OUT ELECTRICS	Sub-con.	X	X	X	X	X	X	X	X	X	X	X	X																			
8	REMOVE KITCHEN FITTINGS	G.W.	X	X	X	X	X	X	X	X	X	X	X																				
9	HACK WALL TO KITCHEN & W.C.	G.W.	X	X	X	X	X	X	X	X	X	X	X																				
10	REMOVE FLOOR AND WALL TILES	G.W.	X	X	X	X	X	X	X	X	X	X	X																				
11	BRICK UP DOOR TO KITCHEN	G.W.	X	X	X	X	X	X	X	X	X	X																					
12	REMOVE EXISTING WINDOWS	G.W.	X	X	X	X	X	X	X	X	X	X																					
13	FIRST FIX JOINER	G.W.	X	X	X	X	X	X	X	X	X																						
14	FIRST FIX PLUMBER	Sub-con.	X	X	X	X	X	X	X	X	X																						
15	FIRST FIX ELECTRICIAN	Sub-con.	X	X	X	X	X	X	X	X	X																						
16	FLUE TERMINALS	G.W.	X	X	X	X	X	X	X	X																							
17	FIBRE GLASS	G.W.	X	X	X	X	X	X	X	X																							
18	CONCRETE & ARDIT FLOOR	G.W.	X	X	X	X	X	X	X																								
19	PLASTERER	Sub-con.	X	X	X	X	X	X	X																								
20	ROOF REPAIRS	Sub-con.	X	X	X	X	X	X	X																								
21	SECOND FIX JOINER	G.W.	X	X	X	X	X	X																									
22	SECOND FIX PLUMBER	Sub-con.	X	X	X	X	X	X																									
23	SECOND FIX ELECTRICIAN	Sub-con.	X	X	X	X	X	X																									
24	WALL TILES & GLAZIER	Sub-con.	X	X	X	X	X																										
25	PAINT & PAPER	Sub-con.	X	X	X	X	X																										
26	PLUMBER SNAG & TEST	Sub-con.	X	X	X	X	X																										
27	ELECTRICIAN SNAG & TEST	Sub-con.	X	X	X	X																											
28	FLOOR TILES	Sub-con.	X	X	X	X																											
29	C. OF W. SNAG	Client	X	X	X																												
30	WASH OUT AND HAND OVER	G.W.	X	X	X																												
B	EXTERNALS																																
1	REBUILD CHIMNEY	G.W.	X	X	X	X	X	X	X	X	X	X	X	X	X																		
2	CUT OUT & RENEW D.P.C.	G.W.	X	X	X	X	X	X	X	X	X	X	X	X	X																		
3	INJECT D.P.C.	Sub-con.	X	X	X	X	X	X	X	X	X	X	X	X																			
4	REPLACE CONC. PATH	G.W.	X	X	X	X	X	X	X	X	X	X	X	X																			
5	CONNECT S.W. & F.W. MAINS	G.W.	X	X	X	X	X	X	X	X	X	X	X																				

GEORGE WIMPEY & Co. Ltd.	MIDWAY ESTATE REFURBISHING	PROG. No.	REVISION	DATE
		DRAWN BY EVT		
		DATE		

Figure 21.4 Record of Progress Chart. This shows progress in individual dwellings.

maintaining the rate desired. While presented slightly differently, this chart is in fact the same as that in Figure 18.8.

Constitution of labour gangs

The most efficient method is the formation of composite gangs. Each gang should consist of a number of operatives geared to the output required. Their individual responsibilities should be allowed to overlap from trade to trade. This approach avoids bringing in particular tradesmen for minor items of work and productivity is improved and collective bonus improved.

Such an approach will depend on the attitude of the local unions. It is desirable to proceed slowly and explore what can be achieved. Where demarcation is insisted upon, benefit will be derived from detailed planning of each operative's responsibility in the dwelling. Two approaches can be tried:

1. Allocating the maximum possible work to a trades operative during one visit. For example, one carpenter would repair floor boards, fix studding, fix skirtings, ease doors, first fix etc.
2. Allowing each trade operative to 'specialize' on the same tasks in each dwelling.

The first approach has been found to produce far better outputs and less antagonism where tenants remain in residence.

Choice of labour

Sorting out of labour should follow a different pattern to that on new building work. Operatives of high outputs are not necessarily satisfactory. The aim should be to develop gangs capable of reasonably good output coupled with good quality in awkward circumstances.

Continuity for trades

The most efficient approach when aiming for continuity is to determine the optimum weekly output of dwellings and to plan the resources around it. When this momentum has been obtained, improvements should be directed towards trimming labour levels and not attempting better outputs with the same labour.

The temptation to push one trade ahead because the scope exists will lead to lack of continuity for that trade and disruption of other trades, and this must be avoided.

Continuity for subcontractors is equally important. If not well managed, and work does not flow, they are liable to take labour off site. It then often becomes difficult to get them back again.

It is therefore necessary for the Site Agent to be the one who holds the keys of all dwellings made available and has sole power to decide when buildings are to be opened for work. Furthermore, any pressure from the Client to take on more dwellings than the optimum agreed should be resisted.

It will be found that planning the continuity for major operations, such as plumbing and heating, is relatively easy. But some of the minor operations can be very difficult if the weekly output of dewllings is very low, especially if the demarcation between trades is rigidly applied. This may mean that a certain operation may only need 75% of an operative's time. Something has to be found to occupy the remaining 25%. This is not easy to arrange if there is lack of scope. Where the choice lies with the contractor, the weekly output of dwellings must be carefully geared to providing sufficient work within the minor activities to eliminate such problems. Above all, one trade must not be allowed to push ahead, nor must individual trades be allowed to proceed out of sequence. Keeping to the planned cycle time is essential for overall control. Equally, in order to minimize disturbance to tenants remaining in occupation, trades must operate to the sequence planned.

Check-list for maintaining continuity of trades

1. *Preparatory Work*. Push paper stripping and demolition as hard as possible. No productive activity can proceed until these items are complete.
2. *Paper stripping*. To avoid damage to walls, use steam stripping. It is also much quicker. To avoid abortive work, this item should *follow* demolition items.
3. *Demolition*. Where structural alterations are required in relation to demolition, pre-planning of falsework needs is essential. The gang is to be fully briefed before commencement.
4. *Hacking of plaster*. Rapid clearing of rooms after this operation is necessary to allow following trades in as soon as possible.
5. *Scabbling of walls*. Do not use percussion tools on partition walls. There is a danger of loosening plaster on the opposite side, especially where a partition has been partly cut away for inserting new damp proof courses.
6. *Windows and glazing*. Replacement item. Time to ensure that weather proofing is in place before trades that require a weather proof dewlling.
7. *Joiner first-fix*. Typical items listed below. Those with asterisk affect the immediately following trades. Priority in areas involved.

   Stud partitions
   Cylinder bearers*
   Loft tank bearers*
   New internal door frames
   New cylinder cupboard
   Grounds for skirting boards
   Removal/repair of floor boards*
   Pipe boards*
   Loft access traps
   Cleats for radiators*

   New external doors and frames
   Plasterboard ceilings
   Drilling and notching of joists*

It is desirable to itemize in this way the various trades activities and recognize where the priorities are.

8. *Electrical work*. Where tenants remain in occupation, detailed sequences of work are necessary to ensure that the electricity supply is always reconnected at the end of each working day.
9. *Plastering*. Ideally, the plasterer should be given the best possible access. Sink units, cookers disconnected to give clear runs and avoid the need to make good in small areas. If tenants remain in occupation, the sequence needs to be arranged to cause as little disruption to them as possible.
10. *Plumbing*. As with the electrician, timing again becomes critical if the tenants remain in residence. The critical period is between dismantling the old plumbing and the installation of the new. The preferred solution is to install the new plumbing and heating alongside the old. The old is disconnected and connected to the new. This method calls for high discipline but has been found to accelerate the cycle as well as create a more agreeable situation where tenants are in residence.
11. *Service mains*. It is profitable to have a small team ahead of the main team fitting new meter boxes and other preparatory for service mains, which removes the work from the critical path.
12. *Joint plumbing/joinery gang*. In order to avoid loss of continuity by joiners work lagging behind plumbers' needs, the use of a composite gang is a solution. The gang would take the form of: 2 No. Plumbers, 1 No. Joiner and ½ labourer.
13. *Major and minor work items*. Some site managers like to separate major items from minor items. Major items are classed as

403

those which are critical for completion to allow follow-up trades access on time. Minor items do not cause major upset if delayed to a degree.

14. *Loft insulation.* Should be fitted into the cycle between electrical second fix and ceiling decoration to avoid damage to the ceiling decoration. At this point, a final inspection of the loft needs to be made to make sure that no rubbish has been left behind.

15. *Joiner second fix.* Typical items will be:

   New doors and furniture;
   Skirting boards;
   Architraves and beads;
   Cylinder cupboard and shelves;
   Bath panelling;
   Renewal of floor boards.

16. *Floor tiling.* Rooms requiring floor tiling should be decorated first to give the tiler the best possible time to complete.

17. *Conclusion.* While there will always be a degree of making good in this type of work, planning to minimize the amount of work for wet trades is of great value from a continuity point of view.

Scaffolding and falsework

The requirements in respect of scaffolding and falsework will vary considerably from contract to contract. Decisions will need to be made in the light of the particular circumstances.

In making decisions, a careful assessment of all the factors is necessary, to ensure that proper safety provisions have been provided at all stages.

The following notes provide a basic guide, but must not be regarded as covering all situations:

1. For an average volume of external work, such as the renewal of gutters, downpipes, replacement of cills, decorations etc., the use of a simple tube and fitting scaffold may well prove more economical than some form of system scaffold.

2. If the site is suitable, an alloy tower system, easily moved by two men without dismantling may be a more suitable option. Developed as a bridged or linked span application, it is possible to replace a full scaffold to an entire elevation.

3. In some parts of the country, where the use of cripple scaffolds is a traditional method, and the operations are limited to gutter replacement, roof repair etc., the method can have considerable advantage of allowing the groundworks underneath to be carried out unobstructed.

4. *In any of the approaches above, however, considerable care needs to be taken to ensure that fixing systems into the building will be adequate. Many old buildings have crumbling brickwork, which may well be incapable of providing a safe fixing.*

5. Against the above, the tying of scaffolds through windows should be avoided if there is the least likelihood of this disrupting progress internally. *It follows that an adequate survey of the structural condition is a pre-requisite to scaffolding safety.*
   For housing, where the brickwork is in poor conditions, raking supports from the ground may be the safest solution.

6. The growth in refurbishment work has resulted in the development of special scaffolds for awkward but repetitive situations. Two such examples were originally developed by Ready Scaffolding Ltd and are now much in use[1].

   One is for work on chimney repair or rebuilding. The main advantages of this equipment are:
   (a) its weight is only 32lbs, giving one man erection;
   (b) its compactness when collapsed;
   (c) its application to any roof pitch;
   (d) the elimination of damage to ridge tiles;

(e) its speed of erection.

Notwithstanding the above, where complete renewal of felt and tiles is required, work to the chimney should be carried out immediately after stripping the roof, so that a scaffold can be erected through the roof timbers.

7. With the increasing popularity of aerial work platforms in the construction industry, the possibility of their use as an alternative to scaffolding should not be neglected. Clearly access is crucial and overall cost needs evaluating against that for the normal scaffold. Experience suggests that many short term activities can be achieved cheaper with the aerial work platform.

8. The erection and dismantling of the scaffold should be itemized in the construction programme and the elevational work carefully planned so as to allow the scaffolding to stand for the minimum time. This is desirable:

   (a) from a general economic point of view;
   (b) to enable ground works to proceed with minimum delay.

9. On steeply sloping sites, the scaffold may be designed to include barrow runs to a rubbish horizontally from an upper floor or to a strategically sited rubbish chute.

10. There is evidence that painters and glaziers on external elevations prefer to work off ladders rather than scaffolding:

    (a) because lift heights are rarely suitable;
    (b) because of general ease of access.

    Given the right conditions on site, this is worth exploring because of the saving in scaffolding.

11. When working in stairwells, there are a number of system scaffolds designed with this work specially in mind. The firm already mentioned in relation to chimney work make one version.

12. With regard to situations where limited falsework may be needed, each case will need to be dealt with on its merits. There are many methods available. *What must be rec-* *ognized, however, is that any temporary support work must be designed by a competent person. What may seem to be quite straightforward is not necessarily so in practice.*

Safety check–list and notes

Great attention must be paid to safety on site, especially in the case of the refurbishing of dilapidated dwellings. Particular danger points are:

1. collapse of walls during demolition work;
2. collapse of decayed timber ground or first floor joists during removal of boards;
3. falling chimneys during removal of roof;
4. non-use of rubbish chutes for handling of roof demolition materials;
5. failure to erect temporary handrails where staircase balustrading has been removed;
6. inadequate or out-of-plumb propping to floors and roofs;
7. poor roof-edge protection;
8. failure to check that services have been disconnected;
9. failure to protect lift shafts;
10. non-use of face masks where required;
11. dust not hosed down or removed;
12. fire in roof spaces from plumbers' soldering etc.;
13. potentially poor safety supervision on top floors of blocks of flats because of the number of flights to climb;
14. on blocks of flats, temporary removal of scaffold ties by unauthorised personnel, e.g. plasterers requiring a clear internal wall;
15. landings and balconies cluttered with debris or improperly stacked with new materials;
16. floorboards not put back, especially in tenanted houses;
17. no proper trench crossings on drainage excavations;
18. materials falling down the stairwell in flats;

19. scaffolding or cripples tied into faulty brick-work;
20. lack of temporary lighting, especially in stairwells.

Particular notes

Particularly in old properties, care should be taken not to disturb existing brickwork in the removal of window frames because often there is insufficient support to the heads, and a resultant danger of collapse.

In tenanted dwellings, wet trades inevitably have to work in the vicinity of electrical sockets and switches. The electrician should therefore switch off the current and temporarily remove the fittings until the wet trades have completed. Apart from the safety standpoint, this is desirable in order to achieve a sweet plaster line throughout.

An aid to safety in a potentially dangerous area is the new stairwell scaffold staging system marketed by Ready Scaffolding Ltd.

The problem of safe access to hip roofs for repairing tiles has been overcome by the design of an oblique roof ladder. This has proved to be most successful.

It is recommended that the Regional safety officer and/or structural engineer should inspect all structurally suspect dwellings and make recommendations regarding the means of effecting the required safety. Any extra work involved should be recorded for pricing, and the authority of the Client obtained to proceed.

Case study

The following case study of a housing refurbishment contract, while more extreme than the run of the mill contract, is of considerable interest as, unusually it involved major structural alteration to 650 houses on a large council estate in South Wales.

The background

The estate in question was built in the 1930s. Typical of that period in Wales, the brick dwellings were constructed in black mortar, with building sand containing fine coal dust. After some 60 years, the external leaves of the cavity walling were showing a high degree of cracking. A detailed examination showed that years of exposure to a wet climate had caused the black mortar to corrode the wall ties across the cavities in many places and the external leaf was parting company with the inner leaf (Figure 21.5).

The local authority was faced with a dilemma. Would all the houses have to be demolished for safety reasons or would it be possible to provide a structural repair in economic terms and allow the properties to be upgraded by refurbishment? One approach had been tried out by the council. A steel frame had been inserted within the house to pick up the first floor and roof. Once in place, the external walls had been demolished and rebuilt. While the experiment was successful, it was expensive. Local contractors were then invited to study the problem and put forward ideas which would be both cheaper and practical in time/cost terms.

The solution adopted

The contracting company, for which this author was construction planning manager at the time, got down to a detailed analysis of the problem. The architect's department, the structural design organization, the Regional office and the construction planning team formed a committee to see what would give a cheaper solution, which together with refurbishing, would be quicker than demolishing and complete rebuild.

The bulk of the dwellings were in long terraces with some semi-detached dwellings scattered about the site. Structurally, floor joists, ceiling joists and roof trusses all spanned from front to back, which meant that they were car-

Figure 21.5 Terrace of houses before refurbishment.

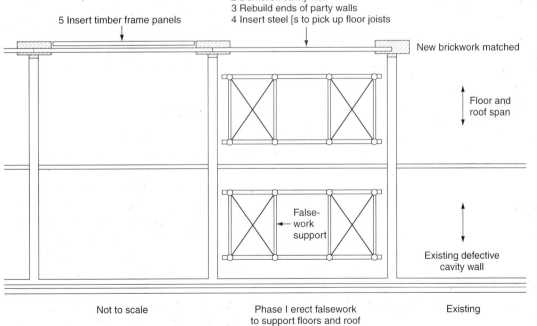

2 Demolish cavity wall
3 Rebuild ends of party walls
4 Insert steel [s to pick up floor joists

5 Insert timber frame panels

New brickwork matched

Floor and roof span

False-work support

Existing defective cavity wall

Not to scale

Phase I erect falsework to support floors and roof

Existing

Figure 21.6 Diagrammatic representation of the solution to the problem.

ried on cavity walling that had to be demolished and rebuilt. Clearly this was the key issue to be solved. The party walls were solid one brick thick, also in black mortar, but in very good condition as they had not been exposed to weather or metal wall ties.

This fact led to the idea of transferring all structural loads on to the party walls. The prob-

Figure 21.7 Falsework in place, demolition of front wall complete and first floor level steel channels in place.

Figure 21.8 Close-up showing new T brick ends to party walls with steel channels on concrete pad stones picking up floor joists.

lem was how? The final solution adopted was as follows:

1. Support first floor, ceiling and roof internally with falsework (Figure 21.6);
2. Once in place and supporting all loads,

demolish front and back walls (Figure 21.7);
3. Make good ends of party walls to a T-shape to carry pad stones at first floor and ceiling;
4. Bridge between party walls with steel channels at first floor and ceiling levels (Figure 21.8). Channels jacked to pick up floor and

Figure 21.9 Timber frame walls in position complete with window and door frames. Waterproof sheeting is being positioned.

Figure 21.10 Completed terrace showing also external ground works. Compare Figure 21.5.

Specification typical example
Modernize kitchens, including the requisite structural alterations and adaption. New kitchen wall, floor and sink units.
Make good solid ground floor areas.
Remove all fireplaces, brick up openings and ventilate flues.
Form new bathroom on first floor on 75% of dwellings, and build bathroom extension on 25%.
Install complete gas central heating system consisting of independent boiler and radiators in all rooms.
Remove old plumbing throughout and install new.
Insulate roof space.
Remove old electrical system and install new. Provide additional power points and lighting points.
Cut out and replace defective floor boards, joists etc., and generally carry out repairs incidental to the above improvement works.
Renew all DPCs by injection.
Replace all windows on all houses.
Repair internal and front and back doors.
Repair ground floor and first floor ceilings.
Internally, decorate 100% to exceptional standard. Externally, decorate 100%.
Wallpaper to exceptional standards including perfect matching of shades etc. on choice of 20 different papers.
Wall tiling to kitchen and floor tiling to solid ground floor areas.
Re-route all services.
Repair fascias, gutters and downpipes.
Make good plastering generally and completely plaster new extensions.
Extensive regrade to rear of dwellings plus associated drainage.
New garden paths, fences etc., tree planting to new communal areas.
Repair roofs including re-bedding of hip and ridge tiles 100%.

Table 21.1

Case study number:	4
Type of tender:	Competition
Location of site:	North
Number of dwellings:	187 houses
Number of types:	4

Site configuration:

    Two sites situated approximately 4 miles apart.
Dwellings made up of terraces and semi-detached.
Terrace-end extensions cut off access to major groundworks to rear, and resulted in elongated cycle-time.

Tenants:	Decanted
Construction carried out	1979/1980
Handover rate:	3 per week
Cycle time:	15 weeks

Supervision actually used:
Personnel marked thus included in rates

Agent	1
General Foremen	2
Trades Foremen	2
Quantity Surveyors	2
Trainee Quantity Surveyor	1
Admin. Asst./Storekeeper	1
Office Cleaner/Canteen Attendant	1
Cost and Bonus Surveyor	1/3

Remarks:

The above supervision was equally shared between the two sites.

Table 21.2

ceiling loads, shimmed in place onto pad stones and grouted.

5. With structural loads taken to party walls, the infill between party wall ends was filled in with full height timber frames carrying windows, doors, and external rainwater goods (Figure 21.9);

6. External cladding to timber frames was shiplap plastic at upper floor level and terrazite coated marine ply below (Figure 21.10).

Study of the illustrations makes the structural sequence clear. Once the building was closed in, refurbishment internally could get under way. Central heating, fitted bathroom, fitted kitchen, rewiring, and so on. Complete refurbishing of external works, including stores, followed.

Tenant reaction

Once tenants who had, of necessity, been decanted returned, they could not believe what they were moving into. Their original homes had been rudimentary in the extreme.

Recording of progress

For this type of work, the following system is generally agreed to be the one giving an instant appreciation of progress achieved for both site and visiting management.

The record form is shown in Figure 21.4. It will be seen that, once prepared, it is effective, even if the sequence of house completion is altered, or additional dwellings become available. Continual alteration is thus avoided.

It provides a quick visual check of the spread of operations about the site. This is an important record if the client undertook to release dwellings in a reasonable sequence.

Far better control of subcontractors is possible.

The detailed information the method provides is primarily for site management. For higher management, a simple concise weekly report will usually be adequate, containing the following information:

1. number of dwellings commenced;
2. number of dwellings completed;
3. the target figures relative to the above for the particular week.

While the chart is primarily designed for house refurbishing, it can be used for flats with obvious modifications.

With this type of work, the recording of specifications for actual contracts completed can provide useful planning data for future work. A typical tabulation is shown in Table 21.1. With this information, a record of the site details and the extent of supervision provides valuable planning and pricing data for future contract tenders. Table 21.2 shows a specimen record.

Reference

1. Ready Scaffolding Ltd, Sutton, Surrey.

Legislati affecting the planning of construction methods

Construction operations are affected by a great deal of legislation. While much of it relates to periodic inspection and testing of construction plant, those involved in any way in construction planning must be aware of the implications of any legislation which may affect decision making.

The following lists and summaries of contents provide an introduction to the legislation most likely to have a bearing on construction method planning.

Key Acts of Parliament and Regulations made under Acts of Parliament

1. Factories Act 1961;
2. Health and Safety at Work etc. Act 1974;
3. Control of Pollution Act 1974;
4. The Clean Air Act 1968 and The Clean Air (Emission of Dark Smoke) (Exemption) Regulations 1969;
5. The Lifting Operations and Lifting Equipment Regulations 1998;
6. Fire Certificates (Special Premises) Regulations 1976;
7. Shops, Offices and Railway Premises Act 1963;
8. Control of Substances Hazardous to Health Regulations 1989;
9. The Management of Health and Safety at Work Regulations 1992;
10. The Construction (Design and Management) Regulations 1994;
11. The Construction (Health, Safety and Welfare) Regulations 1996;
12. Housing Grants, Construction and Regeneration Act 1996.

In addition to the above, the following are given as examples of legislation which can seriously affect construction planning in certain situations:

13. Roads (Scotland) Act 1970 – Section 26(1);
14. Ancient Monuments and Archaeological Areas Act 1979;
15. City of London (St Paul's Cathedral Preservation) Act 1935.

It follows that before making planning decisions involving construction methods a check needs to be made of any by-laws or legislation which may specifically apply to the location involved.

The following summaries of the content of the legislation listed above are intended to give an indication of the relevance of each to construction method planning. Such summaries are not intended in any way to be a substitute for possessing and understanding the documents in question.

Health and Safety at Work etc. Act 1974

This is the head legislation for the control of the health and safety of persons at work. As legislation, it is very straightforward to read and understand.

The key sections for planning are:

Section 2 – Duties of the employer;
Section 3 – Liability to the public;
Section 4 – Liability of Main contractor to Subcontractors;
Section 6 – Liability of Suppliers of Plant and Equipment for use at work;
Section 7 – Liability of employees;
Section 8 – No one to behave recklessly etc.;
Section 21 – Improvement Notices;
Section 22 – Prohibition Notices;
Section 33 – Offences & Type of penalty for each.

Reference to certain sections of the Health & Safety at Work etc. Act in more detail were made in the Introduction, while other references arise in chapters where appropriate.

Control of Pollution Act 1974 – Sections 60 and 61

Section 60
This section deals with the control of noise on construction sites. It gives local authorities power to publish notice of requirements such as:

1. Specify plant or machinery which may or may not be used;
2. Specify hours during which the works may be carried out;
3. Specify the level of noise from the premises, or at any specified point on the premises;
4. Provide for any change of circumstances.

Section 61
Here, prior consent for work on construction sites can be applied for.

1. A person who intends to carry out works to which Section 60 applies may apply for a consent, or
2. An application for consent may be made at the same time as a request for approval under building regulations is made.

It will be clear that choice of plant can be seriously affected by this Act. Noisy plant could be banned and the method planner needs to be aware of what noise limits the local authority has approved or will accept before settling his method.

Generally
With the introduction of the Control of Substances Hazardous to Health Regulations 1989, it should be recognized that the Regulations are complementary to and interdependent with the Control of Pollution Act.

The Clean Air Act 1968 and The Clean Air (Emission of Dark Smoke) (Exemption) Regulations 1969

In relation to construction method planning, these requirements mainly concern the extent to which demolition or waste material may be burnt on site.

The Lifting Operations and Lifting Equipment Regulations 1998

These Regulations became effective from 5th December 1998 and cover linear means of raising or lowering loads or persons. That is variations of Hoists. The key sections are:

1. Citation and commencement;
2. Interpretation;
3. Application;
4. Strength and stability;
5. Lifting equipment for lifting persons;
6. Positioning and installation;
7. Marking of lifting equment;
8. Organization of lifting operations;
9. Thorough examination and inspection;
10. Reports and defects;
11. Keeping of information.

There are also a series of Amendments relating to non-building and civil engineering use.

The Construction (Lifting Operations) Regulations 1961 are revoked.

Fire Certificates (Special Premises) Regulations 1976

These Regulations cover, amongst other things, when site offices require a Fire Certificate and when they do not.

Shops, Offices and Railway Premises Act 1963

All office accommodation must comply with this Act unless the period worked there does not normally exceed 21 hours. This figure is calculated, where more than one person is employed, by taking the sum of the periods of time for which respectively those persons work there.

Amongst many other items, it lays down space requirements for individuals, minimum office temperature, safety requirements and fire precautions.

Control of Substances Hazardous to Health Regulations 1989

Such substances are those which are classified as corrosive or toxic for the purpose of Regulation 7 of the Chemicals (Hazard Information and Packaging) Regulations 1993.

The stated objectives of these Regulations are to focus attention on the health problems which people at work face when they come into contact with hazardous substances, and to bring about an improvement in occupational health. These Regulations have already been preceded by The Control of Pollution Act 1974; Notification of New Substances Regulations 1982; the Reporting of Injuries, Disease and Dangerous Occurrences Regulations 1985 and the Consumer Protection Act 1987. Unlike the construction regulations, the COSHH regulations are not specific to a particular industry. Specific interpretation is necessary for the industry in question.

As the construction industry uses many substances which come into this category employers must ensure workers are made aware of and protected from potentially harmful substances used on site.

The Management of Health and Safety at Work Regulations 1992

These regulations cover the requirements needed from all employers of labour in relation to Health and Safety on site.

1. Every employer is required to make an assessment of the risks to health and safety to which his employees may be exposed.
2. Every employer must appoint one or more competent persons to assist in taking the measures necessary to comply with relevant regulations.
3. If there are risks of serious danger in parts of the site, danger areas have to be established and employees notified of the nature of the hazard and the steps taken to protect them from such situations. Procedures to be laid down about evacuation procedures and safe places to which employees can shelter.
4. All employees, whether of the main contractor or not, to be provided with comprehensive information of all risks which may arise as the result of other contractors' work.

5. Adequate training to be given where necessary for the safe use of plant, equipment and materials.
6. Employees' duties. Every employee shall inform his employer or safety representative of any situation that would seem to represent a hazard to health and safety.

The Construction (Design and Management) Regulations 1994

The summary which follows is copied from an article in the *Chartered Builder* (*The Construction Manager*) February 1995, by permission of the Chartered Institute of Building for which the Author expresses his thanks.

The new regulations will impose statutory health and safety duties on clients, designers, planning supervisors and all contractors with the aim of reducing fatalities and accidents on site. Previously the contractor had been the only party obliged to take on the responsibility for health and safety (Figures A1.1 and A1.2).

Duties

'Clients' will now be actively involved in health and safety. They will have to appoint planning supervisors and Principal Contractors who are competent in their own disciplines and in health and safety. They will also have to make all relevant health and safety information available: they must not permit work on construction to commence until they have confirmed that the Principal Contractor has prepared an adequate health and safety plan for the work. A project file must be provided for those who will be responsible for maintenance, refurbishment and demolition once the construction work is complete. Planning Supervisors will notify the HSE when they are appointed and before construction commences. They will co-ordinate health and safety issues at the design stage, ensuring that any likely hazards to the health and safety of the construction workforce are incorporated into a health and safety plan.

They will also prepare a health and safety plan and file, and advise clients as to whether the plan is suitable to permit construction to start.

Designers, of all disciplines, will have to collaborate with each other and with the planning supervisor. They will have to take account of the effect of their design on the health and safety of the construction work-force and eliminate, reduce or control hazards by applying the principles of prevention and protection. They will also have to make health and safety information available to all parties. Principal Contractors, who may or may not be the main contractor, will have to ensure that all persons on site are informed and trained in health and safety. They will have to keep an updated health and safety plan for the construction stage, ensure that all contractors and subcontractors comply with current health and safety regulations, co-ordinate the activities of each contractor or sub-contractor and ensure that they co-operate with one another. They will also have to provide the Planning Supervisor with inputs to the health and safety file.

Exempt projects

The only projects exempt from the regulations will be:

1. Those which are not expected to employ more than four persons on site at any one time, and which are not scheduled to last longer than 30 days;
2. those for domestic clients (i.e. householders);
3. and minor works in premises which are normally inspected by a local authority.

The Construction (Health, Safety and Welfare) Regulations 1996

These regulations are designed to encompass and update in one document the following separate regulations: The Construction (General

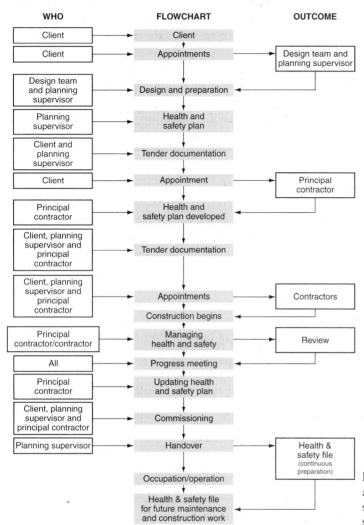

Figure A1.1 The Construction (Design and Management) Regulations have spread the responsibility for health and safety to the entire professional team.

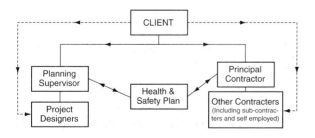

Figure A1.2 The main relationships in the CDM regulations

Provisions) Regulations 1961; the Construction (Health and Welfare) Regulations 1966 and the Construction (Working Places) Regulations, all of which are revoked.

As such, these new regulations cover: safe places of work: prevention of falls; dealing with fragile material; falling objects; stability of structures; demolition or dismantling; explosives; excavations; caissons and cofferdams; prevention of drowning; traffic routes and doors and gates; vehicle movement; prevention from the risk of fire; emergency routes and exits; emergency procedures; fire detection and fire fighting. In addition, the regulations cover requirements for welfare facilities; fresh air (related to working in confined spaces); temperature and weather protection; lighting; good order; plant and equipment and training.

These regulations also amend the Construction (Lifting Operations) Regulations as follows. After Regulation 48 add the following: *Suspended scaffolds (not power operated)* 48A section and 48B section as set out on pages 18 and 19 of this document.

They also amend the Construction (Design and Management) Regulations 1994 as follows: In regulation 2(1) of the Construction (Design and Management) Regulations 1994 the definition of 'construction work' shall be amended so that in sub-paragraph (a) of that definition, in place of the words 'regulation 7 of the Chemicals (Hazard, Information and Packaging) Regulation 1993', there shall be substituted 'regulation 5 of the Carriage of Dangerous Goods by Road and Rail (Classification, Packaging and Labelling) Regulations 1994.'

Housing Grants, Construction and Regeneration Act 1996 reprinted 1998

The Construction section of this recent Act is not directly related to the planning of construction activities. It relates to contract conditions, in particular items under the following headings:

Construction operations specifically excluded;
Transitional provisions;
Stage payments;
Set off;
Pay when paid;
Position with regard to Professional services;
Latham recommendations covered by the Act;
Resolution of all disputes to be by
Adjudication;
Dates for payment;
Suspension of work;
Interlinking main contracts with subcontracts.

Acts of Parliament of more specialized or local effect

Roads (Scotland) Act 1970 – Section 26(1)

Section 26(1) is that used in Scotland for control of tower crane overswinging the public highway. Under this section, the contractor must obtain agreement of all Public Utility companies with services in the pavement or the road, that they have no objection to loads being lifted over their plant. Once letters are received giving approval, they have to be submitted to the local authority with a statement that the crane to be used complies with a recognized code of practice for tower crane design (UK or European) and an approved stability factor.

Ancient Monuments and Archaeological Areas Act 1979

1. Covers ancient monuments protection from nearby construction operations;
2. Areas of likely archaeological interest can be designated. Such areas can then be investigated for archaeological finds before any new construction work commences. Time for investigation is laid down, in such areas delay is inevitable.

City of London (St Paul's Cathedral Preservation) Act 1935

This is an example of an Act specific to one particular building. An area around St Paul's Cathedral is designated within which the approval of the Dean and Chapter is required before any works, permanent or temporary, are carried out below ground.

Severe restrictions apply in relation to ground water flow being disturbed or altered by new foundations, basements or temporary sheet piling etc.

Further reading

Construction Industry Research and Information Association. (1998) R 171 CDM Regulations *Experiences of CDM*. CIRIA, London.

Construction Industry Research and Information Association. (1998) R 173 CDM Regulations *Practical guidance for planning supervisors*. CIRIA, London.

Construction Industry Research and Information Association. (1998) R 172 CDM Regulations *Practical guidance for clients and clients and client's agents*. CIRIA, London.

Codes of practice and British Standards

The following Code and British Standards are particularly relevant to site safety and its association with construction method planning. Failure to comply with appropriate codes and standards, while not a legal requirement, would seriously prejudice a company's case in the event of prosecution or civil actions following an accident. The author cannot guarantee that the list given is fully comprehensive as additions are always being made, or changes in numbering as revisions arise.

It should also be recognized that each code or standard will contain further lists of other codes and standards relevant to the topic, although not necessarily affecting the planning function.

CP 3010: 1972   Safe use of cranes (mobile, tower cranes and derrick cranes) but see also BS 7121: Part 1 1989; Part 2 1991; Part 4 1997 and Part 5 1997

BS 1139 Part 1 1981, Part 3, 1983   Specification for prefabricated access and working towers

BS 5228 Parts 1–3 1997; Part 4 1992   Code of practice for noise control on construction and demolition sites. Part 4 covers piling operations

BS 5328:1997   Methods of specifying concrete, including ready mixed concrete

BS 5531:1988   Code of practice for safety in erecting structural frames

BS 5744:1979   Code of practice for the safe use of cranes (overhead/underhung travelling cranes, goliath cranes, high pedestal and portal jib dockside cranes, manually operated and light cranes, container handling cranes and rail mounted low carriage cranes)

BS 5973:1993   Code of practice for access and working scaffolds and special structures in tubular steel

BS 5974: 1990   Code of practice for temporarily installed suspended scaffolds and access equipment

BS 5975:1996   Code of practice for falsework

BS 6031:1981   Code of practice for earthworks

BS 6164:1990   Code of practice for safety in tunnelling in the construction industry

BS 6187:1982   Code of practice for demolition

BS 7121   Code of practice for safe use of cranes. Part 1 1989; Part 2 1991; Part 4 1997 and Part 5 1997. Part 3 not yet issued.

	(Partly replaces BS 5744:1979 and CP 3010:1972)	BS 8093 : 1991	The use of nets on construction works
BS 7171 : 1989	Specification for mobile elevating work platforms	BS 8103 : 1995	Part 1 Code of practices for stability, site investigation, foundations and ground floor slabs for housing
BS 7212 : 1989	Code of practice for the safe use of construction hoists		
BS 8000 : 1989	Workmanship on building sites. In 15 parts. Covers all building operations	BS 8298 : 1994	Code of practice for design and installation of natural stone cladding and lining
BS 8004 : 1986	Code of practice for foundations		
BS 8081 : 1989	Code of practice for ground anchors		

Index

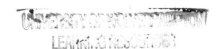